Advances in Contact Mechanics

Advances in Contact Mechanics

Editors

Haichao Liu
Haibo Zhang
Xiaoyu Ding

MDPI

Basel • Beijing • Wuhan • Barcelona • Belgrade • Novi Sad • Cluj • Manchester

Editors
Haichao Liu
State Key Laboratory of Solid Lubrication
Lanzhou Institute of Chemical Physics
Chinese Academy of Sciences
Lanzhou
China

Haibo Zhang
School of Mechanical Engineering
Beijing Institute of Technology
Beijing
China

Xiaoyu Ding
School of Mechanical Engineering
Beijing Institute of Technology
Beijing
China

Editorial Office
MDPI
St. Alban-Anlage 66
4052 Basel, Switzerland

This is a reprint of articles from the Special Issue published online in the open access journal *Lubricants* (ISSN 2075-4442) (available at: https://www.mdpi.com/journal/lubricants/special_issues/40CS2HVOH8).

For citation purposes, cite each article independently as indicated on the article page online and as indicated below:

Lastname, A.A.; Lastname, B.B. Article Title. *Journal Name* **Year**, *Volume Number*, Page Range.

ISBN 978-3-7258-1490-9 (Hbk)
ISBN 978-3-7258-1489-3 (PDF)
doi.org/10.3390/books978-3-7258-1489-3

Contents

Preface

Advancements in contact mechanics play an important role in the design of modern mechanical and bio-medical systems, enhancing their efficiency, power density, and reliability. They lay the foundation to research on the adhesion, friction, lubrication, fatigue, and wear of contacting interfaces and materials. There are numerous contact problems at different working conditions in applications such as machine elements, mechanical systems, and manufacturing processes. Each of these contact problems has specific challenges. To meet the demands, advanced analytical, numerical, and experimental methodologies have been developed. This Reprint aims to expand our understanding of the normal contact problems of layered structures and/or rough surfaces, the tribological and dynamic behavior of interacting surfaces, and the design and optimization of mechanical systems based on contact mechanics.

The Guest Editors would like to express their heartfelt gratitude to the authors, reviewers, and editorial staff at *Lubricants* for their invaluable help in publishing this Reprint.

Haichao Liu, Haibo Zhang, and Xiaoyu Ding
Editors

lubricants

Editorial

Advances in Contact Mechanics

Haichao Liu [1,*], Haibo Zhang [2,*] and Xiaoyu Ding [2]

1 State Key Laboratory of Solid Lubrication, Lanzhou Institute of Chemical Physics, Chinese Academy of Sciences, Lanzhou 730000, China
2 School of Mechanical Engineering, Beijing Institute of Technology, Beijing 100081, China
* Correspondence: liuhc@licp.cas.cn (H.L.); haibozhang@bit.edu.cn (H.Z.)

Advancements in contact mechanics play an important role in the design of modern mechanical and bio-medical systems, enhancing their efficiency, power density, and reliability. To meet the demand for these qualities, advanced analytical, numerical, and experimental methodologies have been developed. This Special Issue aims to expand our understanding of normal contact problems of layered structure and/or rough surfaces (Contributions 1–3), the tribological and dynamic behavior of interacting surfaces (Contributions 4–6), and the design and optimization of mechanical systems based on contact mechanics (Contributions 7–11).

Coated and layered materials are widely used in numerous engineering applications. The advancement of experimental, numerical, and analytical techniques for investigating the contact mechanics of these materials contributes a significant area of study. Lyashenko et al. (Contribution 1) conducted an experimental verification of the boundary element methods for adhesive contacts of a coated elastic half-space. Normal contact problems between an elastic layer of a finite thickness on a substrate and indenters of varying size/geometry were studied, and good agreement was observed when comparing the experimental results with the boundary element method (BEM). Forsbach and Willert (Contribution 2) introduced a general approximate analytical solution for the slightly non-axisymmetric normal contacts of layered and functionally graded elastic (FGE) materials. Any compact axisymmetric or nearly axisymmetric contact problem in FGE materials can be transformed into the problem of indenting the material using a rigid cylindrical flat punch, leveraging the concept of superposition. The contact problem between two interacting rough surfaces is prevalent in numerous natural and engineering phenomena. Jiang et al. (Contribution 3) proposed an efficient methodology to analyze the deformation of roughness asperities, assuming that the deformation is predominantly governed by the size-dependent plasticity. The role of strain gradient plasticity in contact is analyzed using a modified incremental contact model based on the mechanism-based gradient plasticity (MSGP) theory. The results indicate that the strain gradient plasticity does not alter the linear nature of the area-load relation but increases its slope.

Contact mechanics forms the basis of the research into friction, wear, lubrication, and even system dynamic behaviors. Although rolling friction typically has a smaller magnitude than sliding friction, it has attracted attention since the invention of wheel due to its important role in achieving high accuracy in motion and positioning through rolling systems in modern applications. Specifically, knowledge of the transient rolling resistance during the pre-rolling phase is essential. Gilavdary et al. (Contribution 4) developed a single-point contact pendulum to investigate the law of rolling resistance in cases where the dimension of the displacement zone of the rolling body is significantly smaller than that of the contact spot. Rolling resistance is shown to be determined by dissipative adhesive forces, internal friction forces, and elastic adhesion forces. Adhesive wear occurs at the contacting asperities of the mating surfaces. While models such as the Archard model can be used to calculate the wear volume in engineering applications, the wear coefficient needs to be properly calibrated in wear experiments. Furthermore, accurately predicting the

Citation: Liu, H.; Zhang, H.; Ding, X. Advances in Contact Mechanics. *Lubricants* **2024**, *12*, 179. https://doi.org/10.3390/lubricants12050179

Received: 14 May 2024
Accepted: 15 May 2024
Published: 16 May 2024

locations of crack initiation and propagation remains challenging. Li et al. (Contribution 5) present an efficient finite-element sub-model for simulating adhesive wear in elasto-plastic spherical contacts using techniques based on local mesh refinement. Both normal and tangential mechanical responses are considered. The effects of the sphere radius and normal loads on the formation and evolution of wear particles were studied, and these provide the basis for wear analysis for rough surfaces. Additionally, the dynamic behavior of a mechanical system can be strongly influenced by the contact mechanics of rough rubbing surfaces. In Contribution 6, Maaboudallah and Atalla proposed a multi-scale computational method for the prediction of friction-induced vibrations, such as those in braking systems. In contrast to smooth surface modelling, the effect of surface roughness on system dynamic instabilities can be considered in addition to other influential parameters such as the shear moduli of the pads and the coefficient of friction. It has been shown that the resulting multi-scale model incorporating surface roughness significantly improves prediction accuracy at low frequencies.

Contact mechanics serves as the foundation for the design and tribological analysis of numerous machine elements and mechanical parts, including the cylinder-liner contact in combustion engines, wheel-rail contacts in the railway industry, seals in high-speed machines, and pitch bearings in wind turbines. To predict the performance of an automotive piston ring system, Chu et al. (Contribution 7) developed a mixed lubrication model that considers elastic-plastic rough contacts and lubrication mechanics across scales. The friction at the piston ring- cylinder liner interface was modeled, analyzed, and compared with experimental results. The model, based on contact and lubrication mechanics, can be used to enhance friction-reduction technologies in the automotive industry. In the lubrication practice for pitch bearings in wind turbines, fretting corrosion is a typical mode of surface damage due to oscillating motion and dynamic high loads. Han et al. (Contribution 8) investigated the variation in film thickness in a grease lubricated contact under cyclic load-varying conditions. The decay of the lubricating film thickness was monitored using optical interferometry, and the effects of the load-varying ranges, number of cycles, grease composition, and anti-wear additives were examined. The micro-structural degradation of grease in the contact was also analyzed. In various high-speed machines, dry gas mechanical face seals are used to prevent gas leakage. The dynamic coefficients determine the vibration and stability of a rotor-seal system. Park et al. (Contribution 9) studied the dynamic coefficients of a T-grooved dry gas seal under laminar, turbulent, and slip conditions with different clearances. For a better physical understanding of the changes in adhesion coefficient in wheel-rail contacts under sanded contacts, Suhr et al. (Contribution 10) conducted a comprehensive study into the grain crushing behavior of rail sands in both dry and wet conditions. Their research focused on the initial breakage behavior, as well as the size and thickness of the formed solidified clusters and fragments (running band), providing evidence for future modelling of the sanding process and adhesion in wheel-rail contacts. In a separate study, Rong et al. (Contribution 11) investigated the mechanical properties and tribological behaviors of flash-butt welded rail joints. They found that the high proportion of ferrite generated in the weld metal resulted in increased plasticity and decreased hardness, and yielding strength. Consequently, the wear mechanism transitioned from adhesive wear and oxidation to fatigue wear with slight oxidation.

Despite over a century of research and applications in contact mechanics, marked by continuous advancements in theory, modeling, numerical methods, and related experimental techniques, new challenges in the field of mechanical-biological engineering continue to propel this area of study forward. This Special Issue is a testament to this progress. The Guest Editors would like to express their heartfelt gratitude to the authors, the reviewers, and the editorial staff at *Lubricants* for their invaluable help in publishing this Special Issue.

Conflicts of Interest: The authors declare no conflicts of interest.

List of Contributions:

1. Lyashenko, I.A.; Popov, V.L.; Borysiuk, V. Experimental Verification of the Boundary Element Method for Adhesive Contacts of a Coated Elastic Half-Space. *Lubricants* **2023**, *11*, 84. https://doi.org/10.3390/lubricants11020084
2. Forsbach, F.; Willert, E. A General Approximate Solution for the Slightly Non-Axisymmetric Normal Contact Problem of Layered and Graded Elastic Materials. *Lubricants* **2023**, *11*, 450. https://doi.org/10.3390/lubricants11100450
3. Jiang, C.; Yuan, W.; Zheng, Y.; Wang, G. Contact of Rough Surfaces: An Incremental Model Accounting for Strain Gradient Plasticity. *Lubricants* **2023**, *11*, 140. https://doi.org/10.3390/lubricants11030140
4. Gilavdary, I.; Mekid, S.; Riznookaya, N. Phenomenological Laws of Single Point Contact: Pre-Rolling Contact Resistance through Pendulum. *Lubricants* **2023**, *11*, 88. https://doi.org/10.3390/lubricants11020088
5. Li, M.; Xiang, G.; Goltsberg, R. Efficient Sub-Modeling for Adhesive Wear in Elastic–Plastic Spherical Contacts. *Lubricants* **2023**, *11*, 228. https://doi.org/10.3390/lubricants11050228
6. Maaboudallah, F.; Atalla, N. A Multi-Scale Investigation to Predict the Dynamic Instabilities Induced by Frictional Contact. *Lubricants* **2023**, *11*, 344. https://doi.org/10.3390/lubricants11080344
7. Chu, N.R.; Jackson, R.L.; Ghaednia, H.; Gangopadhyay, A. A Mixed Lubrication Model of Piston Rings on Cylinder Liner Contacts Considering Temperature-Dependent Shear Thinning and Elastic–Plastic Contact. *Lubricants* **2023**, *11*, 208. https://doi.org/10.3390/lubricants11050208
8. Han, Y.; Wang, J.; Du, H.; Li, W.; Zhao, J.; Bai, Z.; Hu, M.; Liu, H. Film Thickness Decay and Wear Behavior of Grease-Lubricated Point Contact under Cyclic Variable Loads. *Lubricants* **2024**, *12*, 42. https://doi.org/10.3390/lubricants12020042
9. Park, Y.; Hahn, M.; Jang, G. Effect of Laminar, Turbulent and Slip Conditions on the Dynamic Coefficients of a Dry Gas Seal. *Lubricants* **2023**, *11*, 98. https://doi.org/10.3390/lubricants11030098
10. Suhr, B.; Skipper, W.A.; Lewis, R.; Six, K. Sanded Wheel–Rail Contacts: Experiments on Sand Crushing Behaviour. *Lubricants* **2023**, *11*, 38. https://doi.org/10.3390/lubricants11020038
11. Rong, B.; Liu, S.; Li, Q.; Peng, J.; Shen, M. Mechanical and Tribological Behaviors of U75VG Rail Flash–Butt Welded Joint. *Lubricants* **2023**, *11*, 41. https://doi.org/10.3390/lubricants11020041

lubricants

MDPI

Article

Experimental Verification of the Boundary Element Method for Adhesive Contacts of a Coated Elastic Half-Space

Iakov A. Lyashenko [1,2,*], Valentin L. Popov [1,*] and Vadym Borysiuk [1,3]

[1] Department of System Dynamics and Friction Physics, Institute of Mechanics, Technische Universität Berlin, 10623 Berlin, Germany
[2] Department of Applied Mathematics and Complex Systems Modeling, Faculty of Electronics and Information Technology, Sumy State University, 40007 Sumy, Ukraine
[3] Department of Nanoelectronics and Surface Modification, Faculty of Electronics and Information Technology, Sumy State University, 40007 Sumy, Ukraine
* Correspondence: i.liashenko@tu-berlin.de (I.A.L.); v.popov@tu-berlin.de (V.L.P.); Tel.: +49-(0)30-314-75917 (I.A.L.)

Abstract: We consider analytical, numerical, and experimental approaches developed to describe the mechanical contact between a rigid indenter and an elastic half-space coated with an elastic layer. Numerical simulations of the indentation process were performed using the recently generalized boundary element method (BEM). Analytical approximation of the dependence of contact stiffness on the indenter diameter was used to verify the results of BEM simulations. Adhesive contacts of hard indenters of different shapes with soft rubber layers have been experimentally studied using specially designed laboratory equipment. The comparison of the results from all three implemented methods shows good agreement of the obtained data, thus supporting the generalized BEM simulation technique developed for the JKR limit of very small range of action of adhesive forces. It was shown that the half-space approximation is asymptotical at high ratios of layer thickness h to cylindrical indenter diameter D; however, it is very slowly. Thus, at the ratio h/D = 3.22, the half-space approximation leads to 20% lower contact stiffness compared with that obtained for finite thickness using both an experiment and simulation.

Keywords: indentation; elastomer; elastic layer; contact stiffness; BEM; experiment; adhesion

Citation: Lyashenko, I.A.; Popov, V.L.; Borysiuk, V. Experimental Verification of the Boundary Element Method for Adhesive Contacts of a Coated Elastic Half-Space. *Lubricants* **2023**, *11*, 84. https://doi.org/10.3390/lubricants11020084

Received: 30 January 2023
Revised: 13 February 2023
Accepted: 14 February 2023
Published: 15 February 2023

1. Introduction

Coated and layered materials are commonly used in numerous engineering applications. The main purpose of a coating technology is significant improvement of the performance of certain devices or their parts, as changing the properties of the surface can dramatically affect the behavior of the material. Coating technology is often applied to enhance the properties of the materials that are involved in mechanical contacts, to modify their tribological properties, and to improve wear resistance, adhesion, friction, etc [1]. Furthermore, coatings are also used to achieve the desired level of biocompatibility of implants [2,3], to obtain needed optical properties [4], to enable triboelectric energy harvesting [5], and for many other technologies. Therefore, development of experimental, numerical and analytical techniques for the modelling and studying of materials with a layered structure is an important topic in various fields of science and technology.

Most of the analytical theories and numerical simulation methods used for describing elastic contacts make use of "half-space approximation", wherein an object with finite sizes at certain conditions can be considered as a half-space. This approximation can be applied to bodies coated with an elastic substrate provided the characterized size, D, of the contact area is much smaller than the thickness of the coating [6]. A special case of a coated body is an elastic layer placed on the rigid substrate. If an elastic layer with thickness h is indented by a cylindrical indenter of diameter D, this layer can be considered as a half-space provided the thickness of the layer is much larger than the diameter of indenter.

Classical theories of adhesion, such as JKR [7], DMT [8] or Maugis theory [9], as well as the classic Hertz theory [10], all operate in the half-space approximation; therefore, much of our intuitive understanding of contacts is based implicitly on the results of half-space approximation. However, if the contact size is comparable with the layer thickness, the use of the half-space approximation may lead to inadequate results so that the finite size has to be taken into account. There is a number of theories considering these corrections [11–16].

In addition to purely analytical methodology, numerical simulations are also widely used for solving the contact tasks involving layers of a given thickness. Commonly used numerical methods are the finite element method (FEM) [17–19] and the Fourier-based residuals molecular dynamics (RMD) [20], among others [21]. At present, the FFT-assisted boundary element method (BEM) is considered the most powerful technique for simulation contacts. Recently, it was generalized for the case of coated elastic half-space and also takes adhesion into account [22]. In the described generalization, adhesion is considered in the "JKR-limit", meaning that the range of action of adhesive forces is much smaller than any other characteristic length of the problem (including gap and indentation depth) so that it can be considered to be zero. It is not automatically guaranteed that this condition is fulfilled in real adhesive contacts. Thus, it is important to "verify" the simulation method through comparison with experiments. This comparison is the main purpose of the present paper.

2. Materials and Methods

As mentioned in introduction, in our study we used FFT-based BEM for coated half-space [22]. This approach was recently implemented for a description of adhesive and non-adhesive contacts between rigid indenter and elastic half-space coated with a layer of different elastic properties. Within BEM, an elastic half-space with elastic modulus E_2 and Poisson ratio ν_2 coated with the elastic layer of thickness h and elastic parameters E_1 and ν_1 is considered. To solve the contact problem between coated half-space and a rigid indenter with arbitrary geometry numerically, we consider a square region on the body surface with the size $L \times L$, which has N cells in each direction, while the size of each of the N^2 square cells is $\Delta x = \Delta y = \Delta$. Pressure is assumed to be uniform in each cell. If the pressure distribution $\mathbf{p}$ is given, the displacement $\mathbf{u}$ can be calculated according to a numerical procedure [22],

$$\mathbf{u} = \mathbf{IFFT}[\mathbf{u_z} \cdot \mathbf{FFT}(\mathbf{p})], \tag{1}$$

where $\mathbf{u_z}$ is a Fourier-transformed fundamental solution (normal displacements at the contact surface), and the analytical formula for $\mathbf{u_z}$ is provided in [22]. The contact problem is solved iteratively. In each step, the displacements $\mathbf{u}$ for a given pressure distribution $\mathbf{p}$ are determined through the evaluation of Equation (1). The inverse problem of finding pressure $\mathbf{p}$ for producing given deformations $\mathbf{u}$ can be solved using the conjugate gradient method [23]. For adhesive contacts, an additional detachment criterion is needed: a surface element at the boundary of the contact area loses its contact as soon as tensile stress in this element exceeds the critical value given by

$$\sigma_c = \sqrt{\frac{E_1 \Delta\gamma}{0.473201 \cdot \Delta \cdot (1 - \nu_1^2)}}, \tag{2}$$

where $\Delta\gamma$ (J/m^2) is the specific work of adhesion between the indenter and substrate. For non-adhesive contacts, the detachment criterion is that normal pressure $p > 0$; for adhesive contact the condition $p > -\sigma_c$ must be used. Detailed information about numerical BEM procedure can be found in [22].

Experiments concerning indenter–substrate contact are a well-known challenge in many fields of tribology, contact mechanics and nanotechnology [24]. An experimental study of adhesive contacts between rigid indenters and elastic layers of different thickness was conducted on specially designed in-house laboratory equipment. Detailed description of the designed facility and examples of its performance are given in our recent paper [25]; therefore, here, we provide only brief information about the experimental setup. General view of the designed device together with an enlarged image of the indenter and sample are shown in Figure 1.

(**a**) (**b**)

Figure 1. Photo of the experimental setup: general view (**a**) and enlarged image of the indenter and the sample (**b**). Functional parts of the equipment are denoted by labels: (1) and (2)—M-403.2DG high-precision motorized linear stages (manufactured by PI); (3)—three-axis force sensor ME K3D40 with mounted indenter (4); (5)—sample being indented; (6)—tilt mechanism; (7)—digital camera Ximea 2.2MP MQ022CG-CM with FUJINON HF16SA-1, 2/3″ lens; and (8)—the 8MR190-90-4247-MEn1 motorized rotation stage.

The facility operates as a high-precision tribometer, which is capable of precise positioning of the sample in three dimensions and measuring all three components of the interaction force. All functional parts of the device that are denoted in the figure are the same for both panels: (1) and (2) are high-precision M-403.2DG motorized linear stages (manufactured by PI), which are handled by PI C-863 one-axis servo controllers; (3) is an ME K3D40 three-axis force sensor; (4) is an indenter that is mounted on the force sensor; (5) is the sample being indented placed at the 8 mm thick silicate glass plate; (6) is a tilt mechanism; (7) is a digital camera Ximea 2.2MP MQ022CG-CM with FUJINON HF16SA-1, 2/3″ lens; and (8) is the 8MR190-90-4247-MEn1 motorized rotation stage. Various modifications of the developed device had already been used to perform several studies on contact mechanics [25]. Below, we discuss experiments with indentation of hard steel indenters in soft elastic rubber sheets (elastomer) with good adhesive properties. As an elastomer, TARNAC CRG N3005 rubber sheets with linear sizes 100 mm × 100 mm × 5 mm were used (see Figure 1b, position 5). In the experiments, elastomers with thickness $h = 5, 10, 15, 20$ and 25 mm were used; to obtain the elastomer with different thickness, separate rubber sheets were stacked together. Due to the strong adhesion, these rubber sheets are firmly concatenated and did not slide over each other during the indentation. For indentation, cylinders with a flat base of diameter $D = 4, 7, 10$ and 15 mm, as well as spheres with radii $R = 30, 50$ and 100 mm, were used. All experiments were performed in laboratory under room temperature (24 °C) and relative humidity (48%).

3. Comparison of Computer Simulations and Analytical Solutions

In the case of indentation of the rigid cylindrical stamp with a flat base of radius a into an elastic half-space, dependence of the normal force F on indentation depth d is defined by a classical expression:

$$F = 2aE^*d,\ E^* = \frac{E}{1-\nu^2}, \tag{3}$$

where E and ν—elastic modulus and Poisson ratio of the elastic half-space. According to Equation (3), the contact stiffness can be expressed as:

$$K_{\text{half space}} = 2aE^*. \tag{4}$$

Within half-space approximation, it is assumed that the contact radius a is significantly smaller than the thickness of the indented elastic layer. In the case of a layer with thickness h and elastic parameters E and ν placed onto rigid substrate, the approximate contact stiffness can be estimated using the expression [11,12]:

$$\begin{aligned} K_{a\ll h} \simeq 2aE^*\Big\{1 + \tfrac{2\varepsilon a_0}{\pi}\left(1+\tfrac{2\varepsilon a_0}{\pi}\right) + \tfrac{8\varepsilon^3}{\pi}\left(\tfrac{a_0^3}{\pi^2}+\tfrac{a_1}{3}\right) + \\ + \tfrac{16\varepsilon^4 a_0}{\pi^2}\left(\tfrac{a_0^3}{\pi^2}+\tfrac{2a_1}{3}\right)\Big\}, \end{aligned} \tag{5}$$

where small parameter ε is introduced as

$$\varepsilon = \frac{a}{h} \ll 1, \tag{6}$$

while the coefficients a_i are defined as

$$a_m = \frac{(-1)^m}{2^{2m}(m!)^2}\int_0^\infty \Lambda(u)u^{2m}\mathrm{d}u, \tag{7}$$

$$\Lambda(u) = \frac{2L\mathrm{e}^{-4u} - (L^2+1+4u+4u^2)\mathrm{e}^{-2u}}{L\mathrm{e}^{-4u} - (L^2+1+4u^2)\mathrm{e}^{-2u} + L},\quad L = 4\nu - 3. \tag{8}$$

Approximation (5) is valid only in the case when the radius of contact a is smaller than the elastic layer thickness h and thus $\varepsilon < 1$ (6).

It is worth noting that in the opposite limit $\varepsilon >> 1$, analytical approximation is also possible [16,26,27]; however, this case will not be discussed here as we consider indentation into an elastomer, which is almost incompressible, and for such a material, the abovementioned analytics are in bad agreement with both simulation and experiments.

In Figure 2, solid lines show the dependence of the contact stiffness $K_{a<<h}$ (5), normalized on half-space stiffness $K_{\text{half space}}$ (4), while the thickness of elastic layer h varies from 5 to 25 mm, with increments of 5 mm. Symbols in the figure show the results of the computer simulations within the BEM method. The horizontal dashed line shows the threshold $a = h$, i.e., when the layer thickness equals the radius of the indenter. Thus, the area of the plot located above threshold line relates to the values $\varepsilon = a/h > 1$, and therefore, according to (5), (6) analytical approximation is expected not to be valid and may lead to incorrect results. However, as it can be seen from the figure, approximation (5) shows good agreement with simulations in a certain region of $\varepsilon > 1$.

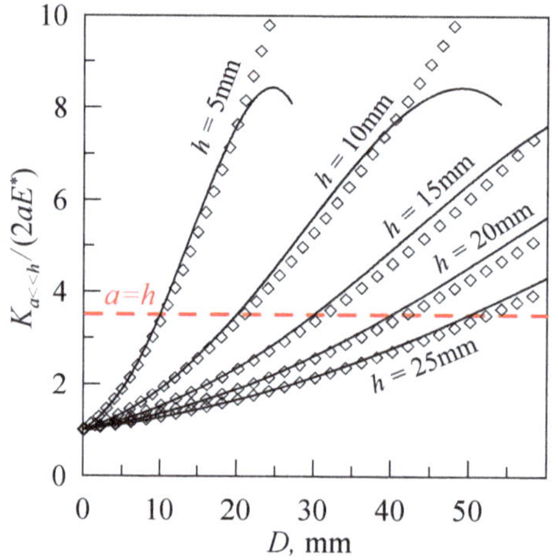

Figure 2. Solid lines—dependencies of the contact stiffness $K_{a<<h}$ (5), normalized on half-space contact stiffness $K_{\text{half space}}$ (4), on the diameter of cylindrical indenter $D = 2a$. Figure shows five curves for different magnitudes of the elastic layer thickness h varying from 5 to 25 mm with increments 5 mm. Results of computer simulations within BEM method are shown in symbols.

As follows from Figure 2, the magnitude of $K_{a<<h}/2aE^*$ decreases when the layer thickness h grows. In the limit case of infinite thickness $h \to \infty$, the solution is reduced to the half-space approximation $K_{a<<h}/2aE^* = 1$ when the contact stiffness does not depend on the indenter diameter. As it was mentioned in the Introduction, if the thickness of the indented substrate exceeds the diameter of the indenter, the substrate can be considered a half-space. However, Figure 2 shows that in real contact, the half-space approximation is valid only for very large ratios h/D. For $h = D = 5$ mm, the stiffness ratio $K_{a<<h}/2aE^* \approx 1.9$ is almost 2, meaning that the contact stiffness is almost twice the half-space approximation. Curve $h = 5$ mm reaches magnitude $K_{a<<h}/2aE^* = 1.2$ at $D \approx 1.55$ mm. Therefore, half-space approximation leads to an error of 20% when $h/D = 5/1.55 = 3.22$, i.e., when the thickness of an elastomer exceeds the diameter of an indenter by more than three times. The restrictions of the half-space approximation were discussed in detail in our recent study [28].

4. Experimental Verification of the Computer Simulations and Theoretical Model

Figure 3 shows dependencies of the normal force F_N on the indentation depth d, obtained through the indentation of the cylinders with a flat base of diameter $D = 4$, 7, 10 and 15 mm into layers of rubber TARNAC CRG N3005 of different thickness h. Each panel of the figure shows five dependencies corresponding to the different magnitudes of the layer thickness: $h = 5$, 10, 15, 20 and 25 mm. Solid lines are experimental results, while the results of the computer simulations are shown with symbols. Every dependence measured in an experiment at constant D and h consists of three curves obtained in three consecutive cycles of indentation. For all measurements, these curves visually overlap. Note that in all experiments, the velocity of indenter motion was equal to 1 μm/s in both directions. With this indenter velocity, the contact can be considered quasi-static and the viscoelasticity can be neglected [29].

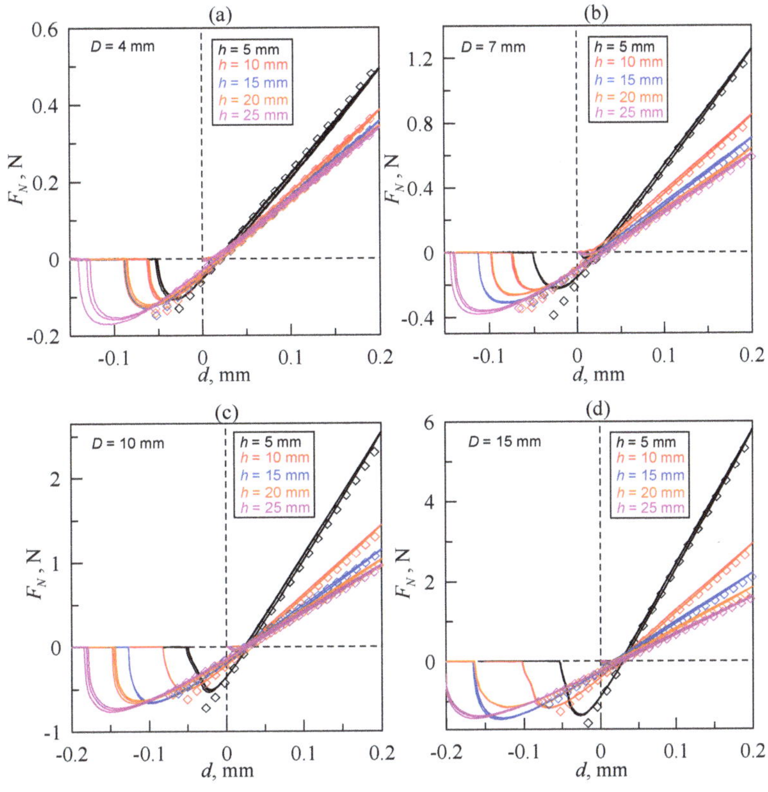

Figure 3. Dependence of the normal force F_N on indentation depth d, obtained through the indentation of cylinders with a flat base of diameter D = 4 mm (**a**), 7 mm (**b**), 10 mm (**c**) and 15 mm (**d**) into layers of rubber TARNAC CRG N3005 of different thickness h = 5, 10, 15, 20 and 25 mm.

Considered contact between rubber and steel indenter is characterized by a strong adhesion; thus, in the regions related to the detachment $d < 0$ mm, the magnitude of the normal force is negative $F < 0$ N. However, while comparing the results of experiments with theory and simulations, we will consider only magnitudes of indentation depth $d > 0$ mm where indenters with flat base exhibit exactly the same behavior in both adhesive and non-adhesive contacts. This can be explained by the fact that detachment of adhesive contact strongly depends on adhesion specific work, being a function of the surface energies of contacting bodies. Surface energy, meanwhile, is affected by the oxidation and dirt on the surface during the time of experiment. Thus, thorough cleaning of the surfaces is needed when the aim of the study is to detect the effect of the indenter radius, surface roughness, etc. Surfaces must be cleaned before each cycle of indentation strictly according to the predetermined procedure. In the presented experiments, such cleaning procedures were not performed as the adhesion phenomena were not the aim of the current study. As an example of the experiments on adhesion involving surface cleaning, we can refer to our previous work [30].

Experimental dependencies $F_N(d)$ shown in Figure 3 exhibit one distinguished feature: all obtained curves do not cross the coordinate origin, even when the zero indentation depth $d = 0$ mm is theoretically corresponding to zero normal force $F_N = 0$ N. This feature is caused by certain peculiarities of the experiment, namely the presence of the asperities of various type on the surfaces of elastomer and indenter, and the impossibility of positioning substrate and indenter exactly parallel to each other. Thus, after the appearance of the first contact point, the contact area is spreading due to the adhesion, resulting in negative normal force F_N.

In the experiment, during the indentation phase, adhesive interaction between the surfaces is weaker compared to that in the pull-off phase. This well-known fact leads to secondary adhesive hysteresis and corresponding differences in $F_N(d)$ dependencies measured during indentation and pull-off [30–32]. Such behavior can be described by introducing two different magnitudes of the adhesion specific work $\Delta\gamma$ for both phases, respectively, where $\Delta\gamma_1$ related to pull-off is significantly larger than $\Delta\gamma_0$ related to indentation. We found that the contact of the steel indenters with elastomer TARNAC CRG N3005 is characterized by the empirically estimated value of $\Delta\gamma_0$ = 0.0175 J/m^2 for indentation phase and a range of values $\Delta\gamma_1$ from about 0.3 to 1 J/m^2 for pull-off [33]. It is worth noting that larger-value $\Delta\gamma_1$ can be reached by chemical treatment of the indenter. For instance, in ref. [33], after short-time treatment of the surface of steel indenter with 40% water solution of $FeCl_3$, magnitudes of $\Delta\gamma_1$ up to 13 J/m^2 were observed. Notably, even though chemical treatment significantly increases $\Delta\gamma_1$, it has almost no effect on $\Delta\gamma_0$ (which is related to contact propagation). Dependencies plotted using symbols in Figure 3 show the results of BEM computer simulations related to the pull off of the indenter starting from maximal indentation depth d = 0.2 mm. All simulations were performed with the same values of elastic and adhesive parameters: E = 0.324 MPa, ν = 0.48, $\Delta\gamma_1$ = 0.326 J/m^2. Magnitudes of elastomer layer thickness h in simulations were chosen to be the same as in the experiments. In the simulations, an elastic layer was located at the half-space substrate with elastic modulus equal to $E_2 = 10^{100}$ Pa. Such an extremely large value ensures absolute rigidity of the substrate in simulations. At the same time, in the experiment, rubber layers were placed on the 8 mm thick silicate glass substrate with an elastic modulus exceeding that of rubber by five orders of magnitude. This glass substrate was fixed on the aluminum table as it is shown in Figure 1b.

On the other hand, within the BEM simulations, contacting surfaces are ideally flat and parallel to each other; therefore, all dependencies $F_N(d)$ obtained from simulations start from the coordinate origin, which makes it difficult to compare them to experimental data. With this purpose, theoretical curves were shifted to the right along the abscissa axis by Δd so both groups of data (theory and experiment) would overlap in the starting point $F_N(\Delta d)$ = 0. Such type of data processing is applicable in our case, as it did not change the slope of $F_N(d)$ dependencies and corresponding contact stiffness $K = \mathrm{d}F_N(d)/\mathrm{d}d$, which is the main subject of the current study. It is worth noting that another option for data correction is to shift experimental curves to the left as in ref. [29]. Therein, both experimental and theoretical dependencies cross the coordinate origin.

Another important detail of the experimental setup is the preparation of the substrates of different thickness h. To obtain the elastomer with a certain h value, separate rubber sheets, each with h = 5 mm, were stacked together. Due to the strong adhesion, these rubber sheets are firmly concatenated and did not slide over each other during the indentation. However, separate rubber sheets may have slightly different elastic properties and may thus cause an extra disagreement between experiments and simulations. Nevertheless, dependencies plotted in Figure 3 show good agreement between the experimental results and simulations.

In all four panels of Figure 3, dependencies $F_N(d)$ obtained for rubber layer with thickness h = 5 mm show a distinctly high value of contact rigidity (highest slope of the $F_N(d)$ curve), while other $F_N(d)$ dependencies are characterized by close values of related contact rigidity, especially for higher h. This situation is caused by the fact that with the growth of elastomer thickness h, the conditions of the experiment become closer to the half-space approximation limit $h \gg D$; thus, as the stiffness of the half-space $K_{\mathrm{half\ space}} = 2aE^*$ is constant (at fixed indenter radius $a = D/2$), all curves behave similarly. For instance, in the experiment with indenter of a diameter D = 4 mm (see Figure 3a), the dependencies $F_N(d)$ measured for elastomers with thickness h = 20 mm and 25 mm almost overlap as the ratio h/D equals 5 and 6.25, respectively, which practically satisfies the condition $h \gg D$. The largest difference between measured $F_N(d)$ curves was observed in the experiment with indenter of largest diameter D = 15 mm (see Figure 3d) for elastomers of different thickness h = 20 mm and 25 mm $h/D \approx$ 1.33 and 1.67, respectively.

As mentioned above (see description of Figure 2), even at magnitude h/D = 3.22, application of half-space approximation leads to a contact stiffness reduction of 20%. Thus, all dependencies shown in Figure 3d are not within the range of application of half-space approximation. For more detailed analysis, magnitudes of contact stiffness $K = \mathrm{d}F_N/\mathrm{d}d$ were estimated from the experimental dependencies $F_N(d)$ shown in Figure 3. Estimated values are plotted as symbols in Figure 4. Figure 4 also shows the contact stiffness calculated through the BEM simulations (solid lines) and theoretical approximation (5) (dashed lines).

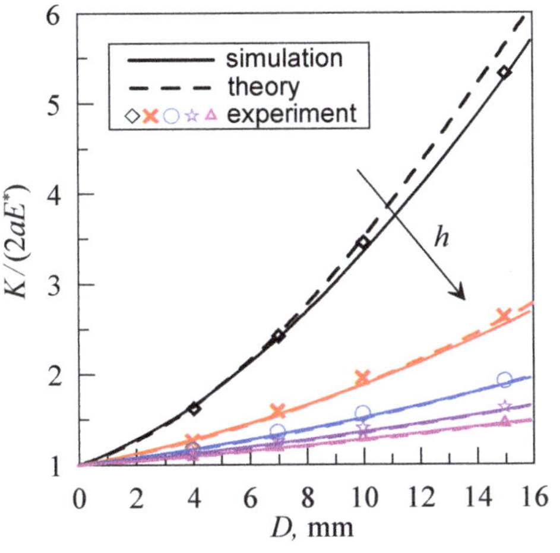

Figure 4. Dependencies of the contact stiffness normalized by the half-space stiffness $K_{half\,space}$ (4) on the diameter of cylindrical indenter D. Figure shows five curves, obtained for magnitudes of elastomer thickness h from 5 to 25 mm with increment 5 mm, arrow shows the increasing of h. Solid lines denote BEM simulations, while dashed lines and symbols denote theoretical approximation and experimental data, respectively.

Comparing dependencies obtained from the three different methods (experiment, simulations and theory), we can confirm the range determined earlier where the analytical solution can be applied, and also conclude that the experimental data are in good agreement with the computer simulations.

5. Discussion

In the closing part of our study, we discuss an additional series of experiments concerning indentation of the spherical indenters with different radii. In these experiments, steel spheres with radii R = 30, 50 and 100 mm were indented into the rubber sheets with thickness h = 5, 10, 15, 20 and 25 mm. The obtained results are shown in Figure 5. As in the previous case, solid lines represent the experimental data obtained during three cycles of indentation in every experiment (measured curves are overlap), while symbols show the results of BEM simulations.

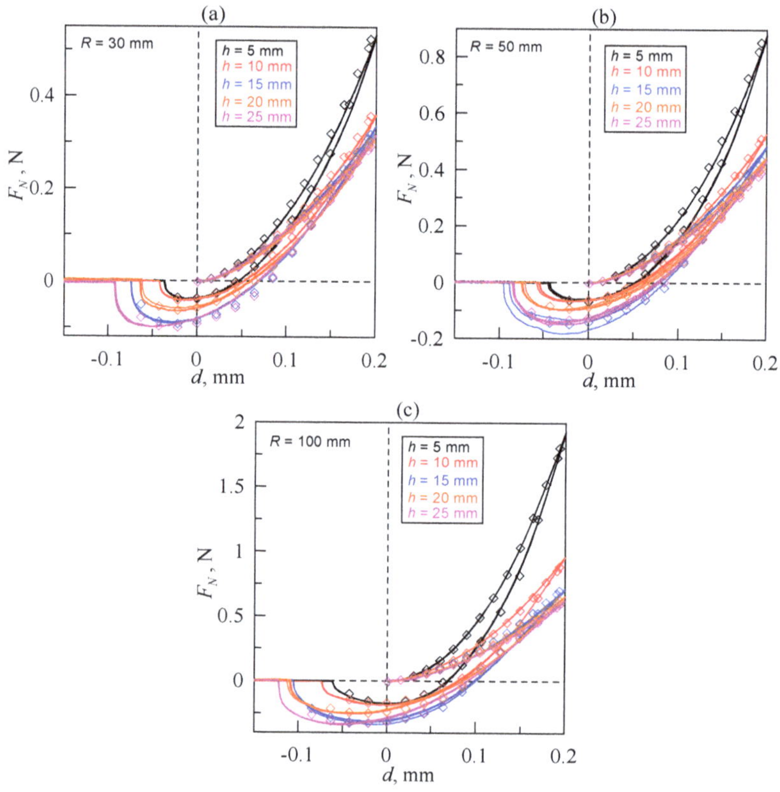

Figure 5. Dependencies of normal force F_N on indentation depth d, obtained in experiments on indentation of steel spheres with radii R = 30 mm (**a**), 50 mm (**b**) and 100 mm (**c**) into layers of rubber TARNAC CRG N3005. Each panel shows five dependencies, related to different thicknesses of the rubber substrate being indented: h = 5, 10, 15, 20 and 25 mm (shown in different colors). Experimental data are plotted in solid lines, while results of BEM simulations are shown with symbols.

The conditions of the performed experiments are the same as in the experiment, the results of which are presented in Figure 3, with the only difference being indentation with spherical indenters instead of cylindrical. The elastic parameters used for simulations are also the same: $E = 0.324$ MPa, $\nu = 0.48$. However, in the computer experiment, both indentation and pull-off were also simulated. For the indentation phase, adhesion specific work was set equal to $\Delta\gamma_0 = 0.0175$ J/m^2, while for the pull-off phase, it was chosen from the experimental data. Dependencies shown in Figure 5 were obtained with the magnitudes of $\Delta\gamma_1$ varying in the range from 0.27 J/m^2 to 0.722 J/m^2. Such a range of the $\Delta\gamma_1$ values is caused by the specific feature of the experimental procedure, where surfaces of elastomer and indenter were not cleared after every cycle of indentation, which significantly affects the $\Delta\gamma_1$.

Figure 5 shows good agreement between the simulation and the experiment, which confirms the accuracy of the performed calculations. It is important to note that elastic parameters of the elastomer were the same in each numerical experiment and were not adjusted to reproduce the experimental data. The only parameter that was adjusted in the simulation shown in Figure 5 is the specific adhesion work in the pull-off phase $\Delta\gamma_1$, which naturally changes after each cycle of the experiment.

It is worth noting that the theory of the adhesive contact of the parabolic indenter and elastic layer fixed on the elastic half-space is also presented in [11]. The developed solution therein allows for the attainment of the dependence of the normal force F on indentation depth d, similar to the data shown in Figure 5. Here, we are not comparing the results of simulations and experiments with the abovementioned work, as the accuracy of analytical approximation from [11] has already been confirmed for cylindrical stamps. We expect similar agreement of the results obtained through the experiment and theory for any other shape of the indenter, as the theoretical solution obtained in ref. [11] is based on the analytical formalism that defines elastic properties of the elastomer and is valid for any indenter shape [11,12]. Nevertheless, in our previous study [34], we compared the results of BEM simulation with the theoretical solution from ref. [11]. The comparison showed good agreement of all the data (theory, experiment and computer simulations) obtained in the experiment on indentation of the steel sphere with a radius $R = 33$ mm into the rubber layer with a thickness $h = 25$ mm.

6. Conclusions

We performed a theoretical, numerical and experimental study of the normal contact between the elastic elastomer layer of a finite thickness placed on the hard substrate and rigid indenters of different geometrical shapes (cylindrical stamps with a flat base and spheres of different radii). The main focus of the performed research was to investigate how both indenter radius and elastomer thickness affect the contact stiffness. The study shows that the magnitudes of contact stiffness measured in the series of performed experiments on indentation of the elastomer layers are in a good agreement with the ones that were calculated through the computer simulations with the boundary elements method (BEM). In addition, the results obtained from both simulation and experiments were compared with the existing analytical solution. Such comparison showed partial agreement between the theoretical and experimental data, namely when the radius of the indenter is smaller or slightly larger than the thickness of the elastomer layer. Thus, it seems that the assumptions behind the numerical procedure based on the BEM as formulated in ref. [22] are confirmed experimentally. It is worth noting that adopted BEM simulations can be preferred over all existing analytical solutions, as these are valid for any values of elastic layer thickness and any geometrical shape of an indenter.

In addition to a verification of the numerical method experimentally, the present study reveals the magnitude of error that occurs when the half-space approximation is used to describe the indentation of a plate with finite thickness. In particular, it was shown that when the ratio of substrate thickness to indenter diameter equals 3.22, the half-space approximation gives a value of contact stiffness reduced by 20%.

Furthermore, the verification of the developed BEM opens up the possibility of its application for scientific and engineering purposes. As an example of possible applications, we can refer to experiments concerning nanoindentation of materials coated with thin films. In this case, BEM can be used for additional analysis of the data, obtained from experiments performed with a relatively low ratio of substrate thickness to indenter diameter. Such a type of analysis may help to save experimental time and does not require the usage of more expensive nanosized indenters.

Author Contributions: Conceptualization, supervision, project administration, writing—review and editing, V.L.P.; methodology, hardware, software, validation, experiments, simulations, experimental data analysis, visualization, writing—original draft preparation, I.A.L.; data analysis, writing—original draft preparation, V.B. All authors have read and agreed to the published version of the manuscript.

Funding: This research was funded by the Deutsche Forschungsgemeinschaft (Project DFG PO 810-55-3). V.B. is grateful to Technische Universität Berlin for support.

Institutional Review Board Statement: Not applicable.

Informed Consent Statement: Not applicable.

Data Availability Statement: The datasets generated for this study are available on request to the corresponding authors.

Conflicts of Interest: The authors declare no conflict of interest.

References

1. Liu, Y.; Yu, S.; Shi, Q.; Ge, X.; Wang, W. Multilayer Coatings for Tribology: A Mini Review. *Nanomaterials* **2022**, *12*, 1388. [CrossRef] [PubMed]
2. Kravchenko, Y.O.; Coy, E.; Załęski, K.; Iatsunskyi, I.; Pogorielov, M.; Korniienko, V.; Pshyk, A.V.; Pogrebnjak, A.D.; Beresnev, V.M. Biocompatibility and electron microscopy studies of epitaxial nanolaminate (Al0·5Ti0.5)N/ZrN coatings deposited by Arc-PVD technique. *Ceram. Int.* **2021**, *47*, 34648–34655. [CrossRef]
3. Kravchenko, Y.O.; Garkusha, I.E.; Taran, A.V.; Coy, E.; Iatsunskyi, I.; Diedkova, K.; Roshchupkin, A.; Tymoshenko, O.; Pogorielov, M.; Misiruk, I. Development of hydrophilic NbCuSi(N) & TiAlNb(N) coatings as a new strategy for medical implants modification. *Ceram. Int.* **2022**, *49*, 4099–4108. [CrossRef]
4. Qu, Y.; Yang, Y.; Wu, J.; Zhang, Y.; Jia, L.; El Dirani, H.; Crochemore, R.; Sciancalepore, C.; Demongodin, P.; Grillet, C.; et al. Photo-Thermal Tuning of Graphene Oxide Coated Integrated Optical Waveguides. *Micromachines* **2022**, *13*, 1194. [CrossRef]
5. Šutka, A.; Lapčinskis, L.; Verners, O.; Ģērmane, L.; Smits, K.; Pludons, A.; Gaidukovs, S.; Jerāne, I.; Zubkins, M.; Pudzs, K.; et al. Bio-Inspired Macromolecular Ordering of Elastomers for Enhanced Contact Electrification and Triboelectric Energy Harvesting. *Adv. Mater. Technol.* **2022**, *7*, 2200162. [CrossRef]
6. Popov, V.L. *Contact Mechanics and Friction. Physical Principles and Applications*; Springer: Berlin/Heidelberg, Germany, 2010. [CrossRef]
7. Johnson, K.L.; Kendall, K.; Roberts, A.D. Surface energy and the contact of elastic solids. *Proc. R. Soc. Lond. A* **1971**, *324*, 301–313. [CrossRef]
8. Derjaguin, B.V.; Muller, V.M.; Toporov, Y.P. Effect of contact deformations on the adhesion of particles. *JCIS* **1975**, *53*, 314–326. [CrossRef]
9. Maugis, D. Adhesion of spheres: The JKR-DMT transition using a dugdale model. *JCIS* **1992**, *150*, 243–269. [CrossRef]
10. Hertz, H. Ueber die Berührung fester elastischer Körper. *J. Reine Angew. Math.* **1881**, *92*, 156–171. [CrossRef]
11. Argatov, I.I.; Borodich, F.M.; Popov, V.L. JKR adhesive contact for a transversely isotropic layer of finite thickness. *J. Phys. D Appl. Phys.* **2016**, *49*, 045307. [CrossRef]
12. Argatov, I.I.; Sabina, F.J. Asymptotic analysis of the substrate effect for an arbitrary indenter. *Q. J. Mech. Appl. Math.* **2013**, *66*, 75–95. [CrossRef]
13. Burmister, D.M. The General Theory of Stresses and Displacements in Layered Systems. I. *J. Appl. Phys.* **1945**, *16*, 89–94. [CrossRef]
14. Burmister, D.M. The General Theory of Stresses and Displacements in Layered Soil Systems. II. *J. Appl. Phys.* **1945**, *16*, 126–127. [CrossRef]
15. O'Sullivan, T.C.; King, R.B. Sliding Contact Stress Field Due to a Spherical Indenter on a Layered Elastic Half-Space. *ASME J. Tribol.* **1988**, *110*, 235–240. [CrossRef]
16. Yang, F. Asymptotic solution to axisymmetric indentation of a compressible elastic thin film. *Thin Solid Films* **2006**, *515*, 2274–2283. [CrossRef]
17. Wald, M.J.; Considine, J.M.; Turner, K.T. Determining the Elastic Modulus of Compliant Thin Films Supported on Substrates from Flat Punch Indentation Measurements. *Exp. Mech.* **2013**, *53*, 931–941. [CrossRef]
18. Lin, Y.-Y.; Chang, C.-F.; Lee, W.-T. Effects of thickness on the largely-deformed JKR (Johnson–Kendall–Roberts) test of soft elastic layers. *Int. J. Solids Struct.* **2008**, *45*, 2220–2232. [CrossRef]
19. Shojaei, P.; Scazzosi, R.; Trabia, M.; O'Toole, B.; Giglio, M.; Zhang, X.; Liao, Y.; Manes, A. An Approach for Material Model Identification of a Composite Coating Using Micro-Indentation and Multi-Scale Simulations. *Coatings* **2022**, *12*, 92. [CrossRef]
20. Scaraggi, M.; Comingio, D. Rough contact mechanics for viscoelastic graded materials: The role of small-scale wavelengths on rubber friction. *Int. J. Solids Struct.* **2017**, *125*, 276–296. [CrossRef]
21. Sridharan, A.; Gandhi, N.; Suresh, S. Stiffness Coefficients of Layered Soil Systems. *J. Geotech. Engrg.* **1990**, *116*, 604–624. [CrossRef]
22. Li, Q.; Pohrt, R.; Lyashenko, I.A.; Popov, V.L. Boundary element method for nonadhesive and adhesive contacts of a coated elastic half-space. *Proc. IMechE Part J J. Eng. Tribol.* **2020**, *234*, 73–83. [CrossRef]
23. Vollebregt, E.A.H. A new solver for the elastic normal contact problem using conjugate gradients, deflation, and an FFT-based preconditioner. *J. Computat. Phys.* **2014**, *257*, 333–351. [CrossRef]
24. Bec, S.; Tonck, A.; Georges, J.-M.; Georges, E.; Loubet, J.-L. Improvements in the indentation method with a surface force apparatus. *Philos. Mag.* **1996**, *74*, 1061–1072. [CrossRef]
25. Lyashenko, I.A.; Popov, V.L.; Pohrt, R.; Borysiuk, V. High-Precision Tribometer for Studies of Adhesive Contacts. *Sensors* **2023**, *23*, 456. [CrossRef]
26. Li, Q.; Popov, V.L. Adhesive contact between a rigid body of arbitrary shape and a thin elastic coating. *Acta Mech.* **2019**, *230*, 2447–2453. [CrossRef]

27. Papangelo, A. Adhesion between a power-law indenter and a thin layer coated on a rigid substrate. *Facta Universitats. Ser. Mech. Eng.* **2018**, *16*, 19–28. [CrossRef]
28. Lyashenko, I.; Borysiuk, V. The effects of half-space approximation and thickness of the contact bodies on the normal contact force: Experiment and simulations. *Procedia Struct. Integr.* **2022**, *36*, 394–400. [CrossRef]
29. Lyashenko, I.A.; Popov, V.L. Hysteresis in an adhesive contact upon a change in the indenter direction of motion: An experiment and phenomenological model. *Tech. Phys.* **2021**, *66*, 672–690. [CrossRef]
30. Lyashenko, I.A.; Pohrt, R. Adhesion Between Rigid Indenter and Soft Rubber Layer: Influence of Roughness. *Front. Mech. Eng.* **2020**, *6*, 49. [CrossRef]
31. Greenwood, J.A. Reflections on and Extensions of the Fuller and Tabor Theory of Rough Surface Adhesion. *Tribol. Lett.* **2017**, *65*, 159. [CrossRef]
32. Waters, J.F.; Guduru, P.R. Mode-mixity-dependent adhesive contact of a sphere on a plane surface. *Proc. R. Soc. Ser. A* **2010**, *466*, 1303. [CrossRef]
33. Lyashenko, I.A.; Li, Q.; Popov, V.L. Influence of chemical heterogeneity and third body on adhesive strength: Experiment and simulation. *Front. Mech. Eng. Sect. Tribol.* **2021**, *7*, 1–11. [CrossRef]
34. Lyashenko, I.A.; Popov, V.L. The effect of contact duration and indentation depth on adhesion strength: Experiment and numerical simulation. *Tech. Phys.* **2020**, *65*, 1695–1707. [CrossRef]

lubricants

Article

A General Approximate Solution for the Slightly Non-Axisymmetric Normal Contact Problem of Layered and Graded Elastic Materials

Fabian Forsbach and Emanuel Willert *

Institute of Mechanics, Technische Universität Berlin, Sekr. C8-4, Straße des 17. Juni 135, 10623 Berlin, Germany; fabian.forsbach@tu-berlin.de
* Correspondence: e.willert@tu-berlin.de

Abstract: We present a general approximate analytical solution for the normal contact of layered and functionally graded elastic materials for almost axisymmetric contact profiles. The solution only requires knowledge of the corresponding contact solution for indentation using a rigid cylindrical flat punch. It is based on the generalizations of Barber's maximum normal force principle and Fabrikant's approximation for the pressure distribution under an arbitrary flat punch in an inhomogeneous case. Executing an asymptotic procedure suggested recently for almost axisymmetric contacts of homogeneous elastic media results in a simple approximate solution to the inhomogeneous problem. The contact of elliptical paraboloids and indentation using a rigid pyramid with a square planform are considered in detail. For these problems, we compare our results to rigorous numerical solutions for a general (bonded or unbonded) single elastic layer based on the boundary element method. All comparisons show the quality and applicability of the suggested approximate solution. Based on our results, any compact axisymmetric or almost axisymmetric contact problem of layered or functionally graded elastic materials can be reduced asymptotically to the problem of indenting the material using a rigid cylindrical flat punch. The procedure can be used for different problems in tribology, e.g., within the framework of indentation testing or as a tool for the analysis of local features on a rough surface.

Keywords: normal contact problem; layered materials; functionally graded materials; almost axisymmetric contact; analytic solution; boundary element method

Citation: Forsbach, F.; Willert, E. A General Approximate Solution for the Slightly Non-Axisymmetric Normal Contact Problem of Layered and Graded Elastic Materials. *Lubricants* **2023**, *11*, 450. https://doi.org/10.3390/lubricants11100450

Received: 12 July 2023
Revised: 9 October 2023
Accepted: 15 October 2023
Published: 18 October 2023

1. Introduction

The mechanics of layered [1] and functionally graded [2] elastic materials have received a lot of interest in the framework of modeling tribological properties of, for example, articular cartilage [3], coatings [4], biomaterials [5], or soils [6]. While tribology and the materials science of layered or graded media are vast branches in science with extensive literature studies conducted—see, e.g., the review articles ([7–9]) and the references therein—we will focus on the more specific problem of the contact mechanical behavior for these materials. In that regard, one should, however, bear in mind that several tribological phenomena have their origin in the contact mechanical interaction [10].

As the material inhomogeneity severely complicates (or, to put it bluntly, apart from some special cases such as the power-law graded elastic half-space [11], inhibits) an exact closed-form solution of corresponding contact problems, analytical contact solutions of layered materials are often in asymptotic form ([12–15]). Because of the approximate character of these solutions, their predictions should be checked against rigorous numerical simulations—a step which was significantly simplified recently for the case of a coated elastic half-space, with the publication of a boundary element method (BEM) formulation of the corresponding normal [16] and tangential contact problems [17]. For the respective normal contact, the BEM results were also validated using laboratory experiments for the indentation of hard steel counter-bodies into soft elastic rubber sheets [18].

An ingenious principle to solve contact problems stems from Mossakovskii [19]. It is based on the observation that the incremental difference between two subsequent contact configurations with indentation depths d and $d + \mathrm{d}d$ can be understood as an infinitesimal indentation $\mathrm{d}d$ using a flat punch, whose planform is given by the contact region at the indentation depth d. Hence, general normal contact can be thought of as a series of incremental flat punch indentations, and therefore, the solution procedure is split into two tasks: the determination of the relation between indentation depth and contact region (which encodes the correct series of flat punches to exactly reproduce the original contact), and the solution of the corresponding flat punch problems.

For the case of axisymmetric indentation of an elastic half-space, both tasks are easily solvable, which leads to the famous solution that is often attributed to Sneddon [20], although it originated a lot earlier [21]. This (exact) axisymmetric solution was generalized some years ago for layered and functionally graded materials [22], albeit in a form where the pressure distribution under a cylindrical flat punch remains unspecified, as it cannot be calculated analytically even in the case of a single homogeneous elastic layer of arbitrary thickness. However, once this pressure distribution has been determined (numerically, e.g., based on the BEM), the corresponding (non-adhesive) normal contact problem of arbitrary axisymmetric convex indenters is solved, as well as all classes of contact problems, which can be reduced to it, e.g., the corresponding adhesive normal contact [23].

Nonetheless, let us, for a moment, turn our attention back to the indentation of a homogeneous elastic half-space. Recently, Popov [24] published an approximate analytical solution for the slightly non-axisymmetric version of this contact problem, which has proven (by comparison with rigorous numerical solutions) to give very satisfactory results. In the follow-up publication [25], the authors thoroughly tested the quality of the suggested approximate analytical solution for contact geometries that are far from axial symmetry and found that it is even well-applicable to indenters with random three-dimensional shapes, e.g., a single asperity of a rough surface.

Popov's solution rests on two fundaments: On the one hand, Barber's [26] extremal principle that the contact region at a given indentation depth maximizes a specific integral (which corresponds to the total normal force); and, on the other hand, Fabrikant's approximation [27] for the pressure distribution under a flat punch of arbitrary (compact) planform. As has been shown very recently by one of the authors of the present manuscript [28], Barber's principle applies to any elastic normal contact problem (at least, with compact contact regions), which can be thought of as a series of flat punch indentations. That is to say, it also applies to contacts of (sufficiently isotropic) layered or functionally graded materials. On the other hand, the "essence" of Fabrikant's approximation is to "scale" the axisymmetric pressure distribution under a cylindrical flat punch to the asymmetric arbitrary planform, a procedure which also can be executed very generally if the indented material is sufficiently isotropic. Hence, both fundaments of Popov's approximate solution for the homogeneous half-space can be generalized for layered or functionally graded materials (albeit in a slightly less rigorous sense, as will be discussed in the manuscript), and it is thus expected the solution itself can be generalized for the application with layered media, as well. This is the aim of the present manuscript. In other words, we extend the approximate reduction in asymmetric elastic contact problems (with compact contact domains) to the respective axisymmetric problem (and, in consequence, to the indentation problem using a rigid cylindrical flat punch, as will be demonstrated)—which was recently suggested and tested thoroughly for homogeneous and power-law graded media—to the application for arbitrary layered or functionally graded materials.

The remainder of the manuscript is organized as follows: In Section 2, we will derive the general approximate solution for the slightly non-axisymmetric normal contact of layered materials, based only on the corresponding cylindrical flat punch solution, the extremal normal force principle, and a newly suggested generalization of Fabrikant's approximation for layered media. For the convenience of the reader, the derivation of the known axisymmetric solution from the extremal principle (which is indispensable for the understanding of the asymmetric solution) is also shown. The general solution can be significantly simplified in the case of self-similar power-law indenters. In Section 3, we will compare the predictions of the approximate solution with rigorous numerical simulations based on the BEM for the case of a single elastic layer of arbitrary thickness. A short discussion of the method and the obtained results finishes the manuscript.

2. Theory

Let us consider an isotropic, layered, or functionally graded elastic medium, which is indented using a rigid counter-body without friction, as depicted in Figure 1. The layers shall have Young's moduli E_i and Poisson ratios v_i (in the case of a functionally graded material, the change in $E(z)$ and $v(z)$ is continuous rather than discrete). The rigid indenter shall have the profile $f(x,y)$ in cartesian coordinates, and we control either the indentation depth d or the total normal force F.

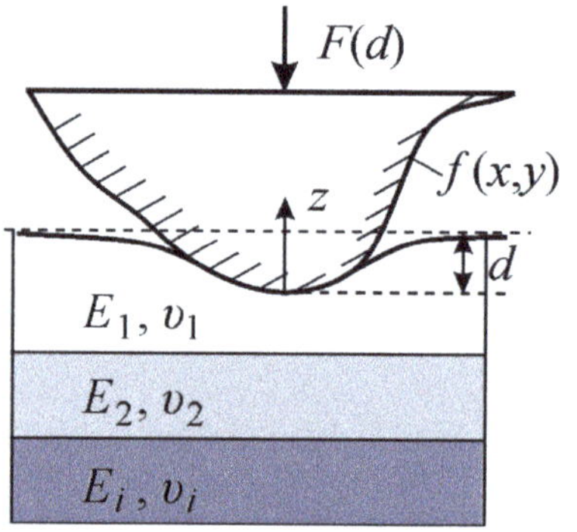

Figure 1. Indentation of a layered elastic material using a rigid counter-body. Notations are explained in the text.

2.1. *Solution for a Cylindrical Flat Punch*

For dimensional reasons, we can always write the pressure distribution resulting from the unit indentation of the elastic material using a rigid cylindrical flat punch with radius a in non-dimensional form as

$$p^*(r) = \frac{E^*}{\pi a}\overline{p}\left(\frac{r}{a};\beta\right), \quad r < a, \quad \beta = \frac{a}{l}, \quad E^* = \frac{E_1}{1-v_1^2}, \tag{1}$$

with the radial coordinate r and some length scale l intrinsic to the material, e.g., some layer thickness or grading depth. Depending on the exact form of the vertical inhomogeneity, there might be several more dimensionless quantities involved in the determination of p^*, but for the contact problem, these are just constant parameters.

The non-dimensional pressure distribution in Equation (1) must be calculated somehow, e.g., based on the boundary element method (BEM) or with finite elements (FE). However, we will not bother with the details of that calculation and postulate that this non-dimensional distribution is known and tabulated with sufficient precision.

We can also define the universal axisymmetric contact stiffness in non-dimensional form as

$$K = 2\pi \int_0^a p^*(r)\, r\mathrm{d}r = 2E^* a \int_0^1 \overline{p}(\rho;\beta)\,\rho\mathrm{d}\rho = 2E^* a\,\overline{K}(\beta). \tag{2}$$

This contact stiffness is universal because the incremental difference between two axisymmetric contact configurations is equivalent to an incremental indentation using a rigid flat punch, as has been discussed in the introductory section.

As an illustrative example, in the case of a homogeneous elastic half-space, we have

$$\overline{p}(\rho) = \left(1-\rho^2\right)^{-1/2}, \ \overline{K} = 1. \quad (3)$$

2.2. Solution for an Axisymmetric Punch

According to Betti's reciprocal theorem, the total normal force resulting from the indentation using a rigid, smooth axisymmetric indenter at an indentation depth d equals ([29]; [30], p. 52)

$$F = 2\pi \int_0^a [d - f(r)]\, p^*(r) r \mathrm{d}r. \quad (4)$$

Inserting the pressure distribution (1), we obtain

$$F = 2E^* \overline{K}(\beta)[da - G(a;\beta)], \quad (5)$$

with the transformed profile

$$G(a;\overline{a}) = T\{f(r)\}(a;\beta) = \frac{1}{a\overline{K}(\beta)} \int_0^a f(r) \overline{p}\left(\frac{r}{a};\beta\right) r \mathrm{d}r, \ \ g(a;\beta) = \frac{\partial G(a;\beta)}{\partial a}. \quad (6)$$

As can be seen, $T\{\}$ denotes a linear operator, specifically the integral transform of the profile, using the flat punch pressure as the integral kernel. Note that $\beta = a/l$, i.e., G and g, can be written as explicit functions of a, but for reasons that will become clearer later, we will treat a and β as separate (albeit not independent) variables.

As was shown recently by one of the authors [28], due to the universality of contact stiffness, Barber's maximum normal force principle—the correct contact domain maximizes the normal force for any given indentation depth—applies to any frictionless normal contact problem with a compact contact domain, that can be thought of as series of incremental flat punch indentations. Hence, the contact radius a can be determined by maximizing the expression (5) with respect to a, and therefore [22],

$$d(a) = \left[\frac{\mathrm{d}}{\mathrm{d}a}\left(a\overline{K}(\beta)\right)\right]^{-1} \frac{\mathrm{d}}{\mathrm{d}a}\left[G(a;\beta)\overline{K}(\beta)\right]. \quad (7)$$

Note that these are total derivatives, i.e., the dependencies on β must be accounted for.

Up until here, all results have been exact. Let us now turn our attention to the approximate contact solution for an arbitrary convex punch whose shape slightly deviates from rotational symmetry.

2.3. Approximate Solution for Slightly Non-Axisymmetric Profiles

In the case of a general convex punch, the normal force integral resulting from the reciprocity theorem reads ([30], p. 52)

$$F = \iint_{\Omega} [d - f(x,y)]\, p^*(x,y) \mathrm{d}x\mathrm{d}y, \quad (8)$$

with the contact domain Ω and the corresponding pressure distribution p^* for the unit indentation of the inhomogeneous elastic material using a flat punch with that planform. As before, the correct contact domain maximizes the force (8) at any given indentation depth d. However, to constructively apply that principle, we require an expression for the general flat punch pressure distribution p^*. For that purpose, let the contour of the domain

Ω be given in polar coordinates by the closed curve $r = a(\varphi)$, and let us assume that the pressure distribution p^* can be approximated well enough by the expression

$$p^*(r,\varphi) \approx \frac{E^*}{\pi a_0}\overline{p}\left(\frac{r}{a(\varphi)};\beta_0\right),\quad r < a(\varphi),\quad \beta_0 = \frac{a_0}{l},\quad a_0 = \frac{1}{2\pi}\int_0^{2\pi} a(\varphi)\,\mathrm{d}\varphi, \tag{9}$$

with the known non-dimensional pressure distribution under a cylindrical flat punch.

Equation (9) describes a "Fabrikant-type" approximation. In other words, the pressure distribution under an arbitrary flat punch is assumed to be given approximately by the distribution for a cylindrical punch, scaled to the asymmetric shape of the punch.

However, the important difference to the homogeneous half-space (and, thus, Fabrikant's original approximation [27]) is the size parameter β, which does not exist for the homogeneous half-space because the latter has no intrinsic length scale, and therefore the contact problem does not exhibit size effects. One may argue that the size parameter also varies over the contour, and therefore, instead of β_0, one should use $\beta(\varphi)$ as a variable size parameter for the pressure distribution. This, however, would severely complicate the following considerations, and we therefore decided to use the simplified version given in Equation (9). As the final approximate results will be compared with rigorous numerical simulations, anyways, one can dispense with a little mathematical rigor for the sake of the applicability of the analytical results.

Inserting the approximation (9) into the force integral (8), we obtain

$$F \approx \frac{E^*\overline{K}(\beta_0)}{\pi a_0}\int_0^{2\pi} da(\varphi)^2 - a(\varphi)G(a(\varphi),\varphi;\beta_0)\mathrm{d}\varphi, \tag{10}$$

with the transformed profile

$$G(a(\varphi),\varphi;\beta_0) = T\{f(r,\varphi)\}(a(\varphi);\beta_0),\quad g(a(\varphi),\varphi;\beta_0) = \frac{\partial G}{\partial a(\varphi)}, \tag{11}$$

as in Equation (6). The following procedure for the approximate determination of the contact contour $a(\varphi)$ operates in analogy to the homogeneous case (see [24] for details) and shall, therefore, only be given briefly.

The maximum principle requires maximizing the force (10) with respect to the contact contour. Let us first maximize the force for a fixed value of a_0 by introducing the Lagrange functional

$$L = F - \lambda\left(\frac{1}{2\pi}\int_0^{2\pi} a(\varphi)\,\mathrm{d}\varphi - a_0\right), \tag{12}$$

with the Lagrange multiplier λ. The necessary condition for an unconditional extremum of the Lagrange functional leads to the algebraic equation

$$0 = 2da(\varphi) - G(a(\varphi),\varphi;\beta_0) - a(\varphi)g(a(\varphi),\varphi;\beta_0) - \frac{\lambda a_0}{2E^*\overline{K}(\beta_0)}. \tag{13}$$

We can separate the slightly non-symmetric profile into an axisymmetric component and a small non-symmetric deviation,

$$f(r,\varphi) = f_0(r) + \delta f(r,\varphi). \tag{14}$$

Accordingly, upon expanding the transformed profile, one obtains

$$\begin{aligned} G(a(\varphi),\varphi;\beta_0) &= G_0(a(\varphi);\beta_0) + \delta G(a(\varphi),\varphi;\beta_0), \\ G_0(a(\varphi);\beta_0) &= T\{f_0(r)\}(a(\varphi);\beta_0), \quad g_0(a(\varphi);\beta_0) = \tfrac{\partial G_0}{\partial a(\varphi)}, \\ \delta G(a(\varphi),\varphi;\beta_0) &= T\{\delta f(r,\varphi)\}(a(\varphi);\beta_0), \quad \delta g(a(\varphi),\varphi;\beta_0) = \tfrac{\partial(\delta G)}{\partial a(\varphi)}. \end{aligned} \tag{15}$$

Moreover, the contact area will also be almost axisymmetric,

$$a(\varphi) = a_0 + \delta a(\varphi). \tag{16}$$

Equation (13) has exactly the same structure as for the homogeneous half-space. Therefore, we can immediately write, in perfect analogy to the homogeneous solution,

$$\delta a(\varphi) \approx \frac{\delta G(a_0,\varphi;\beta_0) + a_0 \delta g(a_0,\varphi;\beta_0)}{2d - 2g_0(a_0;\beta_0) - a_0 g_0{}'(a_0;\beta_0)}, \tag{17}$$

where the prime denotes the first derivative with respect to the first functional argument.

Also, in analogy to the solution for the homogeneous half-space, the normal force (10) simplifies to

$$F \approx \frac{E^*\overline{K}(\beta_0)}{\pi a_0} \int_0^{2\pi} d a_0^2 - a_0 G_0(a_0;\beta_0) \mathrm{d}\varphi = 2E^*\overline{K}(\beta_0)[d a_0 - G_0(a_0;\beta_0)], \tag{18}$$

which is the same as the axisymmetric result (5).

Hence, maximizing the force (18) with respect to a_0 will result in the axisymmetric relation (7) between the indentation depth and the effective contact radius a_0, i.e.,

$$d(a_0) = \left[\frac{\mathrm{d}}{\mathrm{d}a_0}\left(a_0 \overline{K}(\beta_0)\right)\right]^{-1} \frac{\mathrm{d}}{\mathrm{d}a_0}\left[G_0(a_0;\beta_0)\overline{K}(\beta_0)\right]. \tag{19}$$

Equations (17)–(19) give a complete, analytic, general, approximate contact solution for the slightly non-symmetric inhomogeneous case, based only on the pressure distribution under a cylindrical flat punch (which gives the integral kernel for the profile transformation $f \to G$).

Note that also the asymmetric pressure distribution p can be calculated by superimposing the flat punch pressure distributions (9) over the whole indentation procedure, i.e., from $d = 0$ until $d = d_{\max}$ [24].

2.4. Power-Law Indenters

The general asymmetric solution shown above can be simplified considerably if the indenter profile can be written in the self-similar power-law form

$$f(r,\varphi) = r^n \psi(\varphi), \quad f_0(r) = r^n \langle\psi\rangle, \quad \delta f(r,\varphi) = r^n[\psi(\varphi) - \langle\psi\rangle], \quad n > 0, \tag{20}$$

where the brackets denote averaging over the polar angle. It is

$$T\{r^n\}(a;\beta) = \frac{a^{n+1}}{\overline{K}(\beta)} \frac{\kappa_n(\beta)}{n+1}, \quad \kappa_n(\beta) = (n+1)\int_0^1 \overline{p}(\rho;\beta)\,\rho^{n+1}\mathrm{d}\rho, \tag{21}$$

and therefore, according to Equation (19), the relation between indentation depth and the effective contact radius is given by

$$d(a_0) = \left[\overline{K}(\beta_0) + \beta_0 \overline{K}'(\beta_0)\right]^{-1} \left[\kappa_n(\beta_0) + \frac{\beta_0}{n+1}\kappa_n{}'(\beta_0)\right] \langle\psi\rangle a_0^n, \tag{22}$$

and the asymmetric solution follows from Equation (17) as

$$\begin{aligned} \delta a(\varphi) &= a_0 \frac{\langle\psi\rangle - \psi(\varphi)}{\langle\psi\rangle} \frac{n+2}{n+1} \frac{1}{n+2-2d^*(\beta_0)}, \\ d^*(\beta_0) &= \frac{\overline{K}(\beta_0)}{\overline{K}(\beta_0)+\beta_0 \overline{K}'(\beta_0)} \left(1 + \frac{\beta_0}{n+1} \frac{\kappa_n{}'(\beta_0)}{\kappa_n(\beta_0)}\right). \end{aligned} \tag{23}$$

In the homogeneous case, we have $d^* = 1$, and therefore, of course, the respective known results [25] are recovered.

3. Case Studies

In this section, we will compare our approximate solution to rigorous numerical calculations for the indentation of a single elastic layer with thickness h, resting on a rigid foundation, based on the boundary element method [16]. The corresponding numerical solution for the pressure distribution under a rigid cylindrical flat punch—which is an indispensable prerequisite to the approximate solution—has been provided in the supplementary material of [23].

One may distinguish different boundary conditions between the layer and the rigid foundation: an "unbonded" layer rests on the foundation without (significant) friction, e.g., in the case of delaminated layers, while a "bonded" layer is fixed rigidly to the substrate. Note that for the unbonded layer, only compressive normal stresses can be transmitted to the substrate; in other words, in some configurations, loading of the layer surface could result in loss of contact between the layer and the foundation [31], which corresponds to a receding contact. We will neglect this effect in our analysis; Greenwood & Barber [32], in the framework of indentation of an elastic layer with an infinite-length cylinder, argue that the corresponding tensile stresses (between the layer and its base) necessary to establish contact are too small to significantly alter the contact configuration.

3.1. Contact of an Elliptical Paraboloid with a Single Elastic Layer

First, let us consider the contact with the elliptical quadratic indenter profile

$$f(r, \varphi) = r^2 \left(A \cos^2 \varphi + B \sin^2 \varphi\right). \tag{24}$$

Without loss of generality, we put $A < B$ and define the eccentricity of horizontal indenter cross-sections, e.g.,

$$e_g = \sqrt{1 - \frac{A}{B}}. \tag{25}$$

In general, the contact domain does not have to be—except for special cases such as the homogeneous half-space—perfectly elliptical. However, we can introduce the "half-axes" of the contact domain as $b_1 = a(\varphi = 0)$ and $b_2 = a(\varphi = \pi/2)$, and hence the contact eccentricity e as

$$e = \sqrt{1 - \frac{b_2^2}{b_1^2}}. \tag{26}$$

In Figure 2, the BEM-based numerical results are shown for the difference between the contact eccentricity and the indenter eccentricity as a function of the indenter eccentricity and the logarithmic normalized layer thickness for the indentation of an unbonded elastic layer. The scatter in the contour line with the level 0.002 is due to the finite grid length of the boundary elements. Figure 3 gives the corresponding results based on the general approximate analytical solution developed in the previous section. The agreement between the approximate solution and the rigorous numerical calculations is quite good. Note that the contour line diagrams show a difference between the contact area eccentricity and the indenter eccentricity; i.e., as the contact eccentricity is of the order of 0.1, the error of the contact eccentricity in the approximate solution compared with the rigorous BEM results is less than 10% (and in most parameter ranges significantly less than 10%).

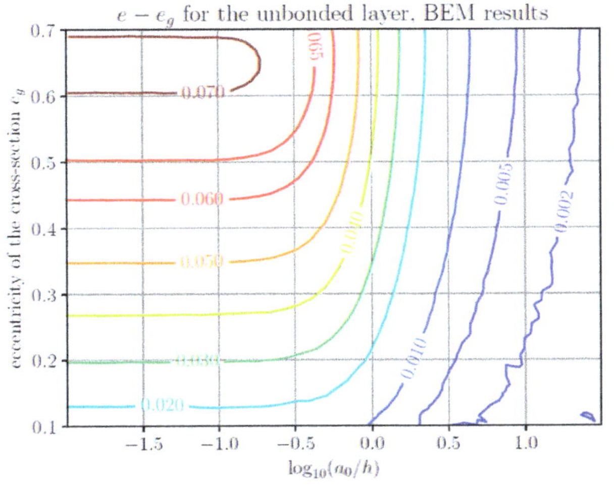

Figure 2. Contour line diagram of the difference between the contact eccentricity e and the indenter eccentricity e_g as a function of the indenter eccentricity and the logarithmic normalized layer thickness, for the contact of an elliptical paraboloid with a single unbonded elastic layer. Numerical results based on the boundary element method (BEM).

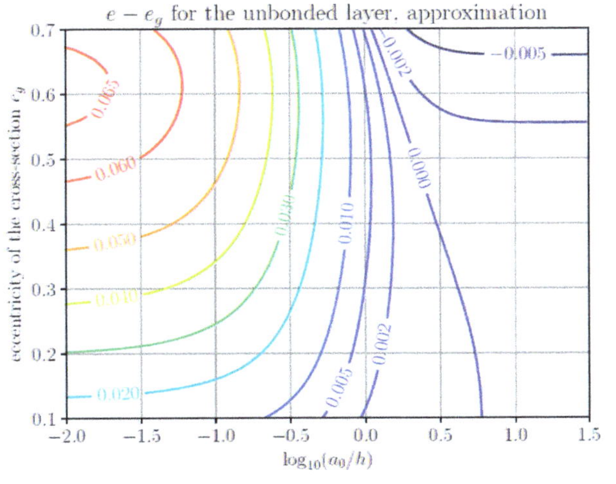

Figure 3. Contour line diagram of the difference between the contact eccentricity e and the indenter eccentricity e_g as a function of the indenter eccentricity and the logarithmic normalized layer thickness, for the contact of an elliptical paraboloid with a single unbonded elastic layer. Approximate analytical solution.

However, for the unbonded layer, the contact eccentricity will never be smaller than the indenter eccentricity. Therefore, slightly negative values for the difference, as in the upper right corner of Figure 3, are unphysical. Also, the half-space solution (at the left edge of both diagrams), according to the numerical simulation, seems to be valid a lot "longer" (until $h \approx 10a_0$) than estimated by the approximate solution.

Figure 4 (BEM) and Figure 5 (approximate analytical solution) show the results that are analogous to Figures 2 and 3 but for a bonded incompressible layer. The quality of the approximate procedure seems to be a bit worse, which is due to the fact that the material behavior is actually more complicated because of the incompressibility (as the material has to "flow somewhere" if it is pushed by the indenter) than captured by the approximate solution. However, one has to keep in mind that in the diagrams, differences between contact and indenter eccentricities are shown; accordingly, the relative error of the absolute value of contact eccentricity in the approximate solution is still less than 10%.

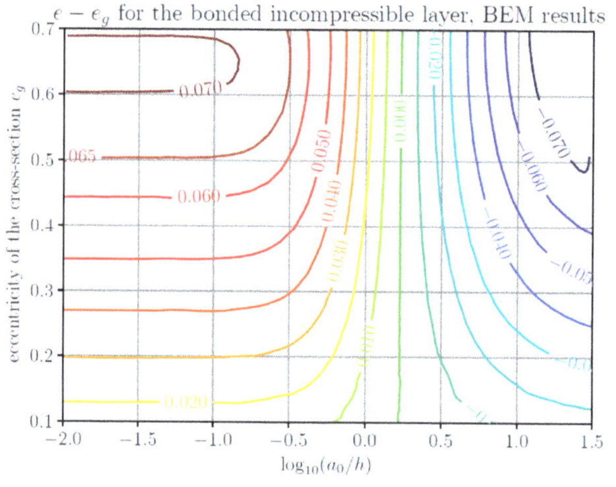

Figure 4. Contour line diagram of the difference between the contact eccentricity e and the indenter eccentricity e_g, as a function of the indenter eccentricity and the logarithmic normalized layer thickness, for the contact of an elliptical paraboloid with a single bonded incompressible elastic layer. Numerical results based on the boundary element method (BEM).

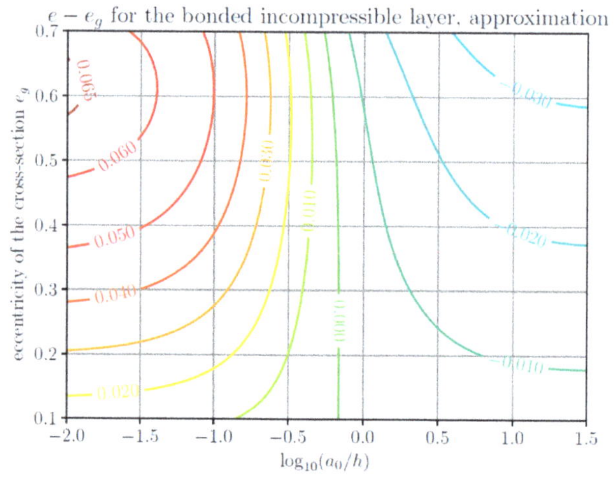

Figure 5. Contour line diagram of the difference between the contact eccentricity e and the indenter eccentricity e_g, as a function of the indenter eccentricity and the logarithmic normalized layer thickness, for the contact of an elliptical paraboloid with a single bonded incompressible elastic layer. Approximate analytical solution.

Let us now turn our attention to the relations between the macroscopic contact quantities, i.e., indentation depth d, normal force F, and average contact radius a_0. In Figure 6, the numerical and approximate analytical results are shown for the normalized average contact radius a_0/h as a function of the normalized indentation depth for different values of layer thickness in the case of an unbonded layer. Figure 7 presents the corresponding solutions for the total normal force normalized for the maximum value in the case of a homogeneous half-space. The agreement between the approximate analytical procedure and the (rather time-consuming) rigorous numerical calculations is very good.

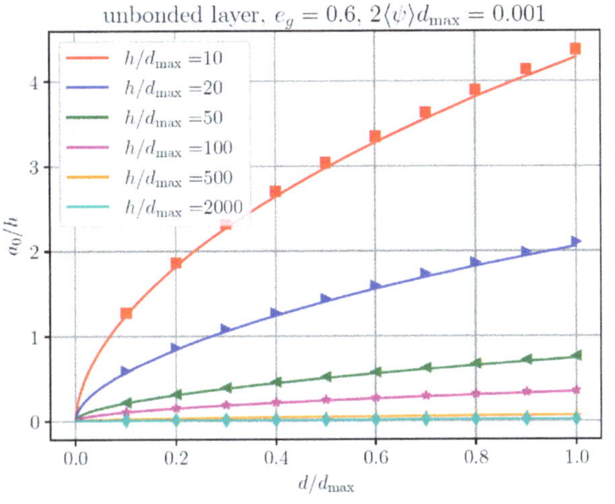

Figure 6. Normalized average contact radius a_0/h as a function of the normalized indentation depth for different values of layer thickness for the contact of an elliptical paraboloid with a single unbonded elastic layer. Lines: approximate analytical solution. Markers: Numerical results based on the boundary element method (BEM).

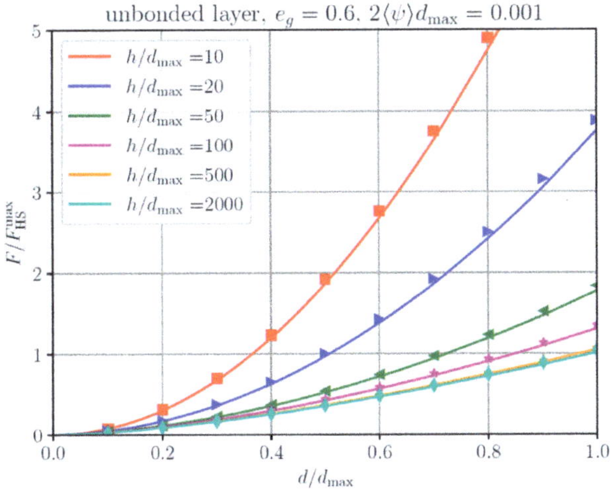

Figure 7. Total normal force, normalized for the maximum value for the homogeneous half-space, as a function of the normalized indentation depth, for different values of layer thickness, for the contact of an elliptical paraboloid with a single unbonded elastic layer. Lines: approximate analytical solution. Markers: Numerical results based on the boundary element method (BEM).

In Figures 8 and 9, the results analogous to Figures 6 and 7 are shown, but for a bonded incompressible layer. Once again, the agreement between the approximate and numerical solutions is very good.

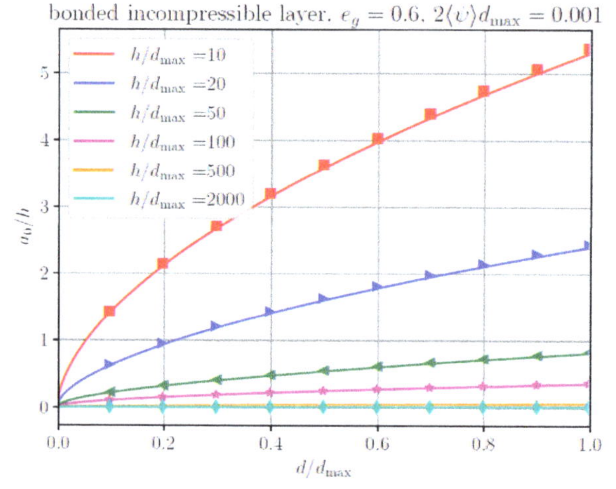

Figure 8. Normalized average contact radius a_0/h as a function of the normalized indentation depth for different values of layer thickness for the contact of an elliptical paraboloid with a single bonded incompressible elastic layer. Lines: approximate analytical solution. Markers: Numerical results based on the boundary element method (BEM).

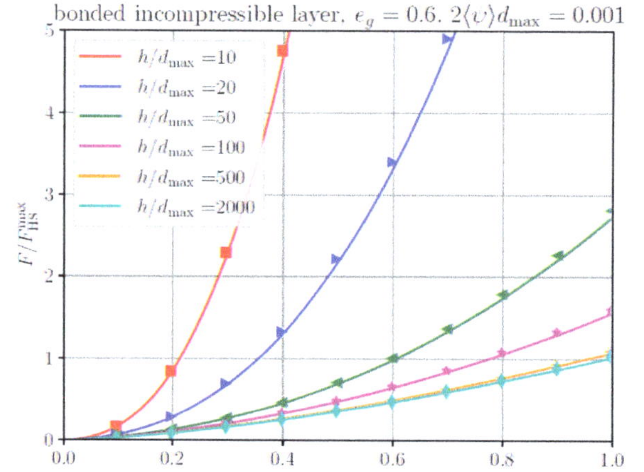

Figure 9. Total normal force, normalized for the maximum value for the homogeneous half-space, as a function of the normalized indentation depth, for different values of layer thickness, for the contact of an elliptical paraboloid with a bonded incompressible layer. Lines: approximate analytical solution. Markers: Numerical results based on the boundary element method (BEM).

3.2. Indentation of a Single Elastic Layer Using a Rigid Pyramid with Square Planform

As a second example, let us consider the indentation of a single elastic layer using a shallow, rigid pyramid with a square planform. The small inclination angle of the pyramid shall be α. Figure 10 gives comparisons between the approximate solution and BEM-based numerical results for the contact boundary in normalized variables in the case of an unbonded layer (left) and a bonded incompressible layer (right). The prediction of the contact domain by the analytic procedure is very good, except for the sharp corners of the indenter. This is expected because the "Fabrikant-type" approximation (9) can only capture edge-singularities of the pressure distribution (for flat punches that are smooth along the contact boundary); as the stress singularity at the corner is more severe (i.e., of higher order), the contact area is not estimated correctly along the corner of the indenter cross-sections. Interestingly, that effect is smaller for the bonded incompressible layer due to the special (smoother) edge and corner behaviors of the pressure distribution for that material class.

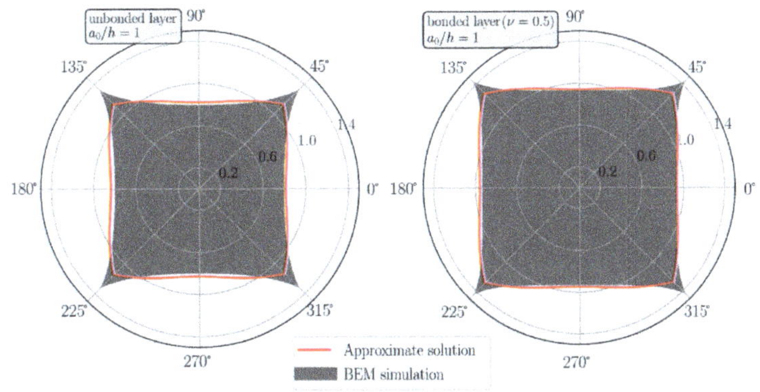

Figure 10. Normalized contact boundary $a(\varphi)\tan\alpha/d$ for the indentation of a single elastic layer of thickness $h = a_0$ using a shallow, rigid pyramid with square planform. Red line: approximate analytic solution. Grey: contact domain in the BEM simulation. (**Left**) unbonded layer. (**Right**) bonded incompressible layer.

Turning to the relations between macroscopic quantities, in Figure 11, the approximate and numerical results are shown for the normalized average contact radius as a function of the normalized indentation depth for different values of layer thickness. As before, the agreement between analytical and numerical calculations is very good.

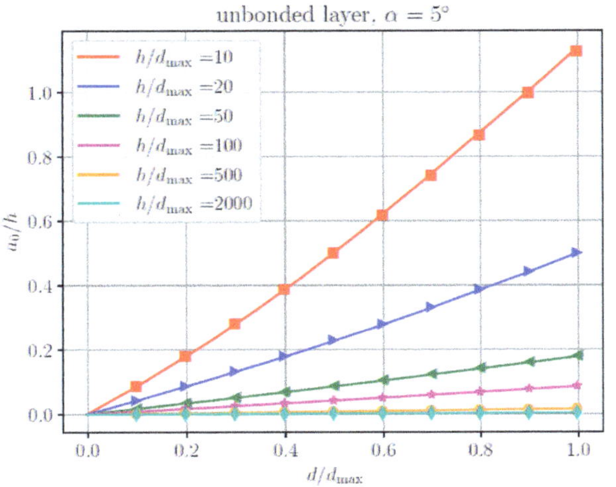

Figure 11. Normalized average contact radius a_0/h as a function of the normalized indentation depth for different values of layer thickness for the indentation of an unbonded elastic layer using a rigid pyramid with square planform and inclination angle α. Lines: approximate analytical solution. Markers: Numerical results based on the boundary element method (BEM).

Figure 12 presents the corresponding solutions for the total normal force, normalized for the maximum value in the case of a homogeneous half-space. Again, the agreement between the suggested approximate analytical solution and the numerical results is almost perfect.

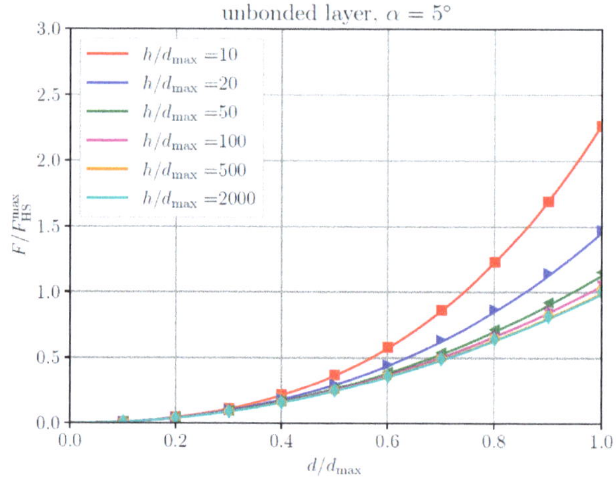

Figure 12. Total normal force, normalized for the maximum value for the homogeneous half-space, as a function of the normalized indentation depth, for different values of layer thickness, for the indentation of an unbonded elastic layer using a rigid pyramid with square planform and inclination angle α. Lines: approximate analytical solution. Markers: Numerical results based on the boundary element method (BEM).

For reasons of space, we will omit showing the analogous results for the bonded incompressible layer. However, in that case, the agreement between the analytical and numerical calculations is almost perfect.

Finally, it should be noted that, in the case of a thin elastic layer, the material asymptotically behaves similarly to a power-law graded elastic half-space, with the exponent k of the power-law grading given by $k = 1$ for an unbonded or bonded compressible thin layer, and by $k = 3$ for a bonded incompressible thin layer. Hence, the corresponding asymptotic contact solutions can be obtained from the general asymptotic contact solution for the indentation of a power-law graded elastic half-space using a slightly non-axisymmetric indenter, which has been published very recently by one of the authors [28].

4. Discussion

We have presented an analytical approximate procedure for the solution of the general (non-symmetric) frictionless normal contact problem (with a compact contact domain) of arbitrary layered functionally graded elastic materials—based only on the solution for the pressure distribution under a rigid cylindrical flat punch indenting the inhomogeneous elastic medium of interest. Thus, the flat punch superposition idea—one of the most powerful tools in analytic contact mechanics—has been extended explicitly to non-homogeneous materials and non-axisymmetric profile geometries. It should be noted that the application of the superposition idea is, generally, not straightforwardly possible for the contact of two elastic bodies if the materials involved exhibit different functional forms of inhomogeneity.

The pressure distribution under a cylindrical flat punch needs to be obtained in the beginning, either using a numerical simulation or from the literature. If one is only interested in the solution of one specific contact problem (and the respective flat solution is not immediately available), it might be debatable whether it is not easier to directly and rigorously solve that problem instead of solving the flat punch problem and constructing an approximate solution for the problem of interest from the flat punch solution. On the other hand, if many contact problems shall be solved for one specific material (e.g., in the context of parameter studies or profile optimization) or the flat punch solution is known, it will speed up the analysis by several orders of magnitude, to only work in the "superposition framework". In that regard, considering the obvious importance of the flat punch indentation problem, it might be a useful task for future work to create an openly accessible library of flat punch solutions for different types of inhomogeneous materials.

The approximate contact solution presented here is applicable to different tasks in tribology and engineering. It can be used for the fast analysis of macroscopic contacts with complex (or even random) shapes or as a tribological tool for the interaction description of single micro-contacts ("asperities") in the contact of rough surfaces—with several implications regarding friction and wear [33]. Moreover, the analytic approximate procedure can be used to solve other classes of contact problems, which can be reduced to the elastic normal contact problem, e.g., the viscoelastic contact problem—via the elastic-viscoelastic correspondence principle [34].

Applying the suggested analytic solution, one, however, should be aware of its approximate character. While the macroscopic contact relations, i.e., for the force-indentation curve and, thus, the contact stiffness, seem to be captured extremely well by the analytic solution—at least, for the case studies considered in the present manuscript—local quantities, such as the precise shape of the contact domain, or the contact stress distribution, might differ (slightly) from a rigorous numerical solution of the same problem.

Author Contributions: Conceptualization, E.W.; formal analysis, E.W.; numerical analysis, F.F.; writing—original draft preparation, E.W.; writing—review and editing, E.W. and F.F.; visualization, F.F. All authors have read and agreed to the published version of the manuscript.

Funding: This research was partially funded by the German Research Foundation under project number PO 810/66-1.

Data Availability Statement: Not applicable.

Acknowledgments: The authors are grateful to Valentin L. Popov for valuable discussions on the topic.

Conflicts of Interest: The authors declare no conflict of interest.

Nomenclature

Latin Symbols

a	contact radius; contour of the contact domain in polar coordinates
b_1, b_2	half-axes of the contact domain
A, B	profile constants
d	indentation depth
d^*	short-cut variable
$d_{\max}$	maximum indentation depth
e	contact eccentricity
e_g	eccentricity of horizontal indenter cross-sections
E	Young's modulus
E^*	effective Young's modulus
f	profile function
F	normal force
g	derivative of the profile transform G
G	transformed profile
h	layer thickness
k	exponent of the power-law of the elastic grading
K	contact stiffness
n	exponent of the profile power-law
p^*	pressure distribution under a cylindrical flat punch with unit indentation depth
r	polar radius
x, y, z	Cartesian coordinates

Greek Symbols

β	non-dimensional layer thickness parameter
ν	Poisson ratio
κ_n	stretch factor, corresponding to the exponent n
φ	polar angle
ψ	angular function

The index "0" corresponds to the axisymmetric part of a non-axisymmetric variable. A bar over a variable denotes a non-dimensional version of the variable. Brackets denote averaging over the polar angle. A "δ" denotes the deviation of a non-axisymmetric variable from the axisymmetric part.

References

1. King, R.B. Elastic analysis of some punch problems for a layered medium. *Int. J. Solids Struct.* **1987**, *23*, 1657–1664. [CrossRef]
2. Suresh, S. Graded materials for resistance to contact deformation and damage. *Science* **2001**, *292*, 2447–2451. [CrossRef] [PubMed]
3. Argatov, I.I. A general solution of the axisymmetric contact problem for biphasic cartilage layers. *Mech. Res. Commun.* **2011**, *38*, 29–33. [CrossRef]
4. Miyoshi, K.; Pohlchuck, B.; Street, K.W.; Zabinski, J.S.; Sanders, J.H.; Voevodin, A.A.; Wu, R.L.C. Sliding wear and fretting wear of diamondlike carbon-based, functionally graded nanocomposite coatings. *Wear* **1999**, *225–229*, 65–73. [CrossRef]
5. Liu, Z.; Meyers, M.A.; Zhang, Z.; Ritchie, R.O. Functional gradients and heterogeneities in biological materials: Design principles, functions, and bioinspired applications. *Prog. Mater. Sci.* **2017**, *88*, 467–498. [CrossRef]
6. Selvadurai, A.P. The Analytical Method in Geomechanics. *Appl. Mech. Rev.* **2017**, *60*, 87–106. [CrossRef]
7. Sathish, M.; Radhika, N.; Saleh, B. A critical review on functionally graded coatings: Methods, properties, and challenges. *Compos. B Eng.* **2021**, *225*, 109278. [CrossRef]
8. Khadem, M.; Penkov, O.V.; Yang, H.-K.; Kim, D.-E. Tribology of multilayer coatings for wear reduction: A review. *Friction* **2017**, *5*, 248–262. [CrossRef]
9. Zhang, S.; Ma, T.; Erdemir, A.; Li, Q. Tribology of two-dimensional materials: From mechanisms to modulating strategies. *Mater. Today* **2019**, *26*, 67–86. [CrossRef]
10. Goryacheva, I.G.; Paggi, M.; Popov, V.L. Editorial: Contact Mechanics Perspective of Tribology. *Front. Mech. Eng.* **2021**, *7*, 649792. [CrossRef]
11. Rostovtsev, N.A. On the theory of elasticity of a nonhomogeneous medium. *PMM J. Appl. Math. Mech.* **1964**, *28*, 745–757. [CrossRef]
12. Jaffar, M.J. Asymptotic behaviour of thin elastic layers bonded and unbonded to a rigid foundation. *Int. J. Mech. Sci.* **1989**, *31*, 229–235. [CrossRef]
13. Barber, J.R. Contact problems for the thin elastic layer. *Int. J. Mech. Sci.* **1990**, *32*, 129–132. [CrossRef]
14. Borodich, F.M.; Galanov, B.A.; Perepelkin, N.V.; Prikazchikov, D.A. Adhesive contact problems for a thin elastic layer: Asymptotic analysis and the JKR theory. *Math. Mech. Solids* **2019**, *24*, 1405–1424. [CrossRef]
15. Wu, J.; Ru, C.Q. An analytical solution to the adhesive cylindrical indentation of a compressible elastic thin layer. *J. Adhes.* **2021**, *97*, 1291–1309. [CrossRef]
16. Li, Q.; Pohrt, R.; Lyashenko, I.A.; Popov, V.L. Boundary element method for nonadhesive and adhesive contacts of a coated elastic half-space. *Proc. Inst. Mech. Eng. J. J. Eng. Tribol.* **2020**, *234*, 73–83. [CrossRef]
17. Burger, H.; Forsbach, F.; Popov, V.L. Boundary Element Method for Tangential Contact of a Coated Elastic Half-Space. *Machines* **2023**, *11*, 694. [CrossRef]
18. Lyashenko, I.A.; Popov, V.L.; Borysiuk, V. Experimental Verification of the Boundary Element Method for Adhesive Contacts of a Coated Elastic Half-Space. *Lubricants* **2023**, *11*, 84. [CrossRef]
19. Mossakovskii, V.I. Compression of Elastic Bodies under Conditions of Adhesion (Axisymmetric Case). *PMM J. Appl. Math. Mech.* **1963**, *27*, 630–643. [CrossRef]
20. Sneddon, I.N. The relation between load and penetration in the axisymmetric Boussinesq problem for a punch of arbitrary profile. *Int. J. Eng. Sci.* **1965**, *3*, 47–57. [CrossRef]
21. Popova, E.; Popov, V.L. Ludwig Föppl and Gerhard Schubert: Unknown classics of contact mechanics. *ZAMM Z. Angew. Math. Mech.* **2020**, *100*, e202000203. [CrossRef]
22. Argatov, I.I.; Heß, M.; Popov, V.L. The extension of the method of dimensionality reduction to layered elastic media. *ZAMM Z. Angew. Math. Mech.* **2017**, *98*, 622–634. [CrossRef]
23. Forsbach, F. A Simple Semi-Analytical Method for Solving Axisymmetric Contact Problems Involving Bonded and Unbonded Layers of Arbitrary Thickness. *Machines* **2023**, *11*, 474. [CrossRef]
24. Popov, V.L. An Approximate Solution for the Contact Problem of Profiles Slightly Deviating from Axial Symmetry. *Symmetry* **2022**, *14*, 390. [CrossRef]
25. Popov, V.L.; Li, Q.; Willert, E. Approximate Contact Solutions for Non-Axisymmetric Homogeneous and Power-Law Graded Elastic Bodies: A Practical Tool for Design Engineers and Tribologists. *Friction* **2023**, 1–16. [CrossRef]

26. Barber, J.R. Determining the contact area in elastic-indentation problems. *J. Strain Anal. Eng. Des.* **1974**, *9*, 230–232. [CrossRef]
27. Fabrikant, V.I. Flat punch of arbitrary shape on an elastic half-space. *Int. J. Eng. Sci.* **1986**, *24*, 1731–1740. [CrossRef]
28. Willert, E. On Boussinesq's Problem for a Power-Law Graded Elastic Half-Space on Elliptical and General Contact Domains. *Materials* **2023**, *16*, 4364. [CrossRef]
29. Shield, R.T. Load-Displacement Relations for Elastic Bodies. *Z. Angew. Math. Phys. ZAMP* **1967**, *18*, 682–693. [CrossRef]
30. Barber, J.R. *Contact Mechanics*; Springer International Publishing AG: Cham, Switzerland, 2018; 585p. [CrossRef]
31. Filon, L.N.G. On an approximate solution for the bending of a beam of rectangular cross-section under any system of load, with special reference to points of concentrated or discontinuous loading. *Proc. R. Soc. Lond.* **1902**, *70*, 491–496. [CrossRef]
32. Greenwood, J.A.; Barber, J.R. Indentation of an elastic layer by a rigid cylinder. *Int. J. Solids Struct.* **2012**, *49*, 2962–2977. [CrossRef]
33. Milanese, E.; Brink, T.; Aghababaei, R.; Molinari, J.-F. Emergence of self-affine surfaces during adhesive wear. *Nat. Commun.* **2019**, *10*, 1116. [CrossRef] [PubMed]
34. Khazanovich, L. The elastic-viscoelastic correspondence principle for non-homogeneous materials with time translation non-invariant properties. *Int. J. Solids Struct.* **2008**, *45*, 4739–4747. [CrossRef]

lubricants

Article

Contact of Rough Surfaces: An Incremental Model Accounting for Strain Gradient Plasticity

Chunyun Jiang, Weike Yuan *, Yanbin Zheng and Gangfeng Wang *

Department of Engineering Mechanics, SVL and MMML, Xi'an Jiaotong University, Xi'an 710049, China
* Correspondence: yuan_wk@xjtu.edu.cn (W.Y.); wanggf@mail.xjtu.edu.cn (G.W.)

Abstract: In the contact of rough surfaces, most contact patches are at the scale of micrometers, and thus, their contact deformation can be dominated by the size-dependent plasticity. In this paper, we propose a new strategy to analyze the role of strain gradient plasticity in the contact response between a realistic rough surface and a rigid plane, which modifies the incremental contact model based on the mechanism-based gradient plasticity (MSGP) theory. For several different rough surfaces with their topography measured experimentally, the relations between applied load and real contact area are derived in a simple but effective way. It is found that strain gradient plasticity significantly increases the level of mean contact pressure. The hardening effect caused by strain gradient plasticity weakens somewhat as the contact area increases. Compared with previous methods, the present model might be more efficient and of wider application.

Keywords: rough surface; contact mechanics; strain gradient plasticity; contact area

Citation: Jiang, C.; Yuan, W.; Zheng, Y.; Wang, G. Contact of Rough Surfaces: An Incremental Model Accounting for Strain Gradient Plasticity. *Lubricants* **2023**, *11*, 140. https://doi.org/10.3390/lubricants11030140

Received: 19 January 2023
Revised: 24 February 2023
Accepted: 14 March 2023
Published: 15 March 2023

1. Introduction

The contact mechanics of rough solids play a fundamental role in many physical phenomena and engineering applications. Due to the inevitable roughness, it is now widely accepted that the real contact area between contacting bodies that is intimately related to friction, wear, sealing, and lubrication is generally a small fraction of the apparent one. However, to give an accurate prediction of the real contact area for a realistic rough surface is still challenging since the surface roughness is of great randomness and irregularity, and the contacting asperities usually involve complex deformation mechanisms.

Over the past few decades, theoretical investigation on the contact of rough surfaces has experienced a flourishing development [1–3]. The statistical multi-asperity contact models that originated from the pioneering work by Greenwood and Williamson (GW) [4] take a great proportion in this field, which ideally simplify the asperities on the rough surface by smooth spheres or paraboloids with randomly distributed heights and sizes. In the elementary multi-asperity models, the contacting asperities were assumed to deform elastically without interaction and obey the classical Hertz theory [4–6]. Such assumptions were later relaxed in the improved multi-asperity models. For example, Chang et al. [7] and, later, Kogut and Etsion [8] modified the GW model by considering the elastic-plastic deformation of asperities. The non-negligible interaction and coalescence between adjacent asperities were taken into account for large contact area fractions [9–11]. To obtain the relationship between real contact area and normal load in dry and lubricated contacts, the GW model was also extended to the contact of rough surfaces in the presence of foreign particles [12,13]. Apart from the multi-asperity models, fractal contact models were established that were based on the nature of self-affine rough surfaces [14,15]. With the concept of magnification and the power spectral density of roughness, Persson [16,17] proposed a scaling contact theory of rough surfaces. Using the profilometric model to calculate the contact area, Wang et al. [18] developed an incremental equivalent model to analyze the contact of elastic rough surfaces, which was later extended to the elastic-plastic cases [19,20]. Moreover, Hyun et al. [21] applied a finite element method to address the elastic contact of self-affine fractal surfaces. Boundary element-based approaches were also widely employed to calculate the contact responses between a rigid rough surface and an elastic substrate [22]. In the contact models mentioned above, the asperity deformation at all length scales was described by the classical elastic (plastic) theories, and the real contact area was found to be generally proportional to the applied normal load at small loads. Recently, it was suggested that accounting for some new scale-dependent mechanisms, such as strain gradient [23,24], adhesion [25], and surface effect [26], could be indispensable when analyzing the asperity deformation of rough surfaces, since the size of most asperities is on the order of microns or even down to sub-micrometers.

Size-dependent plasticity has been repeatedly demonstrated in the experiments of torsion [27], bending [28], and indentation [29–31] for crystalline materials at scales below tens of micrometers. Particularly, the indentation hardness was found to be larger if measured at a smaller indentation size [29–31]. Such an indentation size effect cannot be interpreted in the framework of classical plastic theory where the flow stress is determined only by the strain. Fleck et al. [27] pointed out that the size effect should be attributed to the geometrically necessary dislocations that result from the large strain gradient in a small material volume. Therefore, besides the homogeneous strain, the flow stress is also closely related to the strain gradient [27,32]. Based on the Taylor dislocation hardening model, Gao et al. [33] proposed the mechanism-based strain gradient plasticity theory (MSGP), which links the density of geometrically necessary dislocations to the effective strain gradient. With this MSGP theory, Huang et al. [34] performed a finite element analysis of the microindentation test, which successfully reproduced the experimental observed linear dependence of the square of indentation hardness on the inverse of indention depth. The MSGP theory was later modified into a low-order version without introducing high-order stress [35]. To study the effect of size-dependent plasticity on the contact performance of rough surfaces, Song et al. [23] and You et al. [24] built three-dimensional finite element models with this low-order theory of MSGP. Numerical results were presented to illustrate the significant role of strain gradient plasticity. Compared with the classical J_2 isotropic plasticity, strain gradient plasticity leads to a higher slope of the linear relationship between normal load and real contact area.

Considering the surface roughness and the micro-scale effect of plasticity, it is a cumbersome task to derive the real contact area between rough solids by implementing the finite element simulations through the user-defined subroutines of commercial finite element software. Alternatively, a simple but effective theoretical method would be more attractive in the parametric analysis of rough surface contact. Up to now, few works have been directed to achieve this goal. Recently, Song et al. [36] succeeded in extending the original GW model to incorporate size-dependent plasticity and asperity interaction. However, the underlying assumption in the GW-based model that surface asperities are modeled by smooth spheres with their heights following Gaussian distribution confines its applicability.

In the present paper, we employ the incremental contact model proposed by Wang et al. [18] to study the contact between a rigid plane and a deformable rough surface considering strain gradient plasticity. It is assumed that the contact of rough surfaces is equivalent to the accumulation of identical circular contacts with radii estimated from the total contact area and the number of contact patches, which are directly obtained from the truncation sections of the rough surface at different heights. Different from the GW-based models, the incremental contact model does not require the ideal rough surface description using smooth spherical or paraboloidal asperities and is not limited to the isotropic Gaussian surfaces and, thus, is of wider application range. Here, we consider realistic rough surfaces of copper, which were generated by rubbing with sandpaper and measured with a white light interferometer. The plasticity of the rough surfaces is handled based on the MSGP theory. The interaction between surfaces is assumed frictionless and non-adhesive.

In Section 2, we describe the fundamental principle of the incremental contact model of rough surfaces. In Section 3, the explicit expression of the circular flat contact stiffness is derived based on the finite element analysis using the MSGP theory, which is the key in the incremental contact model accounting for strain gradient plasticity. In Section 4, the effect of strain gradient plasticity on the relation between normal load and real contact area is presented. The main conclusions are summarized in Section 5.

2. The Incremental Contact Model

For the elastic contact between a rough surface and a rigid plane, Wang et al. [18] first developed an incremental equivalent circular model to establish the relationship between normal load and real contact area, provided the surface topography and material properties are known. This model was validated through a comparison of the predicted results with the corresponding finite element simulations [18–20]. In this work, we further extend this incremental model to the contact of rough surfaces by taking into account the size-dependent plasticity.

Figure 1a schematically depicts the contact problem that a deformable rough surface is compressed by a rigid plane under a normal load F. According to the incremental contact model [18], the resulting contact area can be equivalently represented by the geometrically truncated area of the original rough surface at the separation z. As shown in Figure 1b, the contact region $A_c(z)$ consists of a series of separated contact patches. The number of contact patches is denoted by $N(z)$. Then, these irregular contact patches are equivalently simplified by identical circular contact patches (Figure 1c) with the radius determined by

$$r(z) = \left[\frac{A_c(z)}{\pi N(z)}\right]^{1/2} \tag{1}$$

For a decrement of surface separation dz, the corresponding increment of normal load dF can be obtained by the current contact stiffness of the rough surface. Neglecting the interaction between neighboring contact patches, the current contact stiffness can be expressed by [18]

$$\frac{\mathrm{d}F}{\mathrm{d}z} = N(z)k(r) \tag{2}$$

where $k(r)$ is the contact stiffness of an individual circular patch with a radius of r.

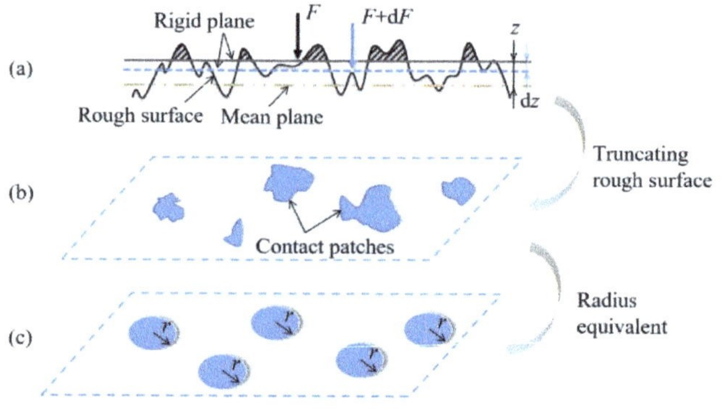

Figure 1. Schematic diagram of the equivalent circular contact model. (**a**) Contact of a rough surface with a rigid plane, (**b**) separated contact patches, (**c**) equivalent circular contact patches.

For the linear elastic material with elastic modulus E and Poisson's ratio ν, $k(r)$ is given by $2E^*r$ [37], where $E^* = E / (1 - \nu^2)$ is the combined elastic modulus. For the elastic-plastic materials, $k(r)$ relies not only on the radius of the contact patch r but also on the mean contact pressure $F/A_c(z)$ and the plastic material parameters [19,20]. In this work, the plastic contact deformation is considered based on the MSGP theory, and thus, $k(r)$ also depends on the intrinsic material length for the strain gradient plasticity. Explicit expression of $k(r)$ is presented in Section 3.

By using the fourth-order Runge–Kutta method, the differential equation Equation (2) can be numerically solved with the initial condition of $F(z \to \infty) = 0$. As a result, the normal load F can be obtained as a function of separation z. Meanwhile, the real contact area A_c is given by the truncated area at separation z. Based on $A_c(z)$ and $F(z)$, the load-area relation for the contact of rough surfaces can be established.

It should be pointed out that the contact area $A_c(z)$ and the number of contact patches $N(z)$ at different separation z are the requisites in this incremental contact model. For realistic rough surfaces, it is almost impossible to derive general analytical expressions of $A_c(z)$ and $N(z)$. As an option, they can be calculated by using a numerical technique as the surface topography is measured. More details about the computations of $A_c(z)$ and $N(z)$ can be referred to in the previous works [18,19].

3. Circular Flat Contact with Strain Gradient Plasticity

To obtain the normal contact stiffness of each circular contact patch, finite element simulations are implemented for the axisymmetric contact between an elastic-plastic substrate and a micro-sized circular flat punch. In the finite element formulation based on the principle of virtual work, the basic equations from the theory of MSGP are employed [33,34].

3.1. Material Constitutive Model of MSGP

Macroscopically, most ductile crystalline materials follow the typical strain hardening model with linear elasticity and power-law hardening plasticity. During plastic deformation, the flow stress σ_{flow} can be expressed as a function of the strain by

$$\sigma_{\text{flow}} = \sigma_y \left(\frac{\varepsilon}{\sigma_y / E} \right)^n \tag{3}$$

where σ_y is the initial yield stress in uniaxial tension, ε is the effective strain, and n is the plastic work hardening exponent.

From the microscopic perspective, the plastic hardening phenomenon of crystalline material is generally caused by the movement and stacking of statistically stored dislocations (SSD) and geometrically necessary dislocations (GND) [32,33]. Based on the Taylor dislocation model that is assumed valid in the theory of MSGP [32,33], the shear flow stress τ_{flow} can be written in terms of dislocation density as

$$\tau_{\text{flow}} = \alpha\mu b\sqrt{\rho_T} = \alpha\mu b\sqrt{\rho_S + \rho_G} \tag{4}$$

where α is an empirical coefficient, μ is the shear modulus, b is the magnitude of the Burgers vector, and ρ_T is the total dislocation density equaling the summation of SSD density ρ_S and GND density ρ_G.

With the shear flow stress, the tensile flow stress can be given by

$$\sigma_{\text{flow}} = M\tau_{\text{flow}} = M\alpha\mu b\sqrt{\rho_S + \rho_G} \tag{5}$$

where M is the Taylor factor.

In the framework of isotropic plasticity, the density of GND ρ_G is related to the effective plastic strain gradient η by [38,39]

$$\rho_G = \frac{\lambda\eta}{b} \tag{6}$$

where λ is the Nye factor [40].

In addition, the density of SSD ρ_S can be determined by combining Equations (3) and (5) in the absence of the strain gradient (η = 0), that is

$$\rho_S = \left(\frac{\sigma_{\text{ref}}}{M\alpha\mu b}\right)^2 \varepsilon^{2n} \tag{7}$$

where σ_{ref} is the reference stress in uniaxial tension given by

$$\sigma_{\text{ref}} = \frac{\sigma_y}{(\sigma_y/E)^n} \tag{8}$$

Substituting Equations (6) and (7) into Equation (5) gives the flow stress as

$$\sigma_{\text{flow}} = \sigma_{\text{ref}}\sqrt{\varepsilon^{2n} + l\eta} \tag{9}$$

It can be observed that the flow stress consists of two hardening contributions: the strain hardening ε^{2n} and the strain gradient plasticity hardening $l\eta$. Here, l is the intrinsic material length of the strain gradient plasticity given by

$$l = M^2\alpha^2 b\lambda\left(\frac{\mu}{\sigma_{\text{ref}}}\right)^2 \tag{10}$$

For the face-centered-cubic (fcc) polycrystalline materials, the Taylor factor M is equal to 3.06, and the Nye factor λ is on the order of 2 [34]. Therefore, the intrinsic material length can be approximately written as

$$l = 18\alpha^2 b\left(\frac{\mu}{\sigma_{\text{ref}}}\right)^2 \tag{11}$$

which is typically on the order of micrometers.

According to Equation (9), it can be understood that a higher strain gradient results in a higher flow stress under the same amount of strain, corresponding to the experimental observation that the sample has a higher indentation hardness measured at a smaller depth [29–31]. With the microscopic plasticity law accounting for strain gradient, the constitutive equations in the deformation theory of MSGP are given as [33,34]

$$\begin{aligned} \sigma_{ij} &= K\varepsilon_{kk}\delta_{ij} + \tfrac{2\sigma_{\text{flow}}}{3\varepsilon}\varepsilon'_{ij} \\ \tau_{ijk} &= l_\varepsilon{}^2\left[\frac{K\eta^{\text{H}}_{ijk}}{6} + \tfrac{\sigma_{\text{flow}}}{\varepsilon}\left(\Lambda_{ijk} - \Pi_{ijk}\right) + \tfrac{\sigma_{\text{ref}}{}^2}{\sigma_{\text{flow}}} n\varepsilon^{2n-1}\Pi_{ijk}\right] \end{aligned} \tag{12}$$

where σ_{ij} is the Cauchy stress tensor, ε_{ij} is the strain tensor, K is the elastic bulk modulus, ε'_{ij} represents the deviatoric strain defined by $\varepsilon'_{ij} = \varepsilon_{ij} - \varepsilon_{kk}\delta_{ij}/3$ (thus, the effective strain can be determined by $\varepsilon = (2\varepsilon'_{ij}\varepsilon'_{ij}/3)^{1/2}$), the flow stress σ_{flow} is given by Equation (9), τ_{ijk} is the high order stress tensor, η_{ijk} is the strain gradient tensor, η^{H}_{ijk} represents the volumetric part of the strain gradient tensor defined by $\eta^{\text{H}}_{ijk} = (\eta_{jpp}\delta_{ik} + \eta_{ipp}\delta_{jk})/4$, the effective strain gradient η is related to the deviatoric strain gradient η'_{ijk} (defined by $\eta'_{ijk} = \eta_{ijk} - \eta^{\text{H}}_{ijk}$) by $\eta = (\eta'_{ijk}\eta'_{ijk})^{1/2}/2)$, l_ε is the mesoscale cell size given by $l_\varepsilon = 5b\mu/\sigma_y$, and Λ_{ijk} and Π_{ijk} are given by

$$\begin{aligned} \Lambda_{ijk} &= \tfrac{1}{72}\left(2\eta'_{ijk} + \eta'_{kji} + \eta'_{kij} + \tfrac{1}{2}\eta_{kpp}\delta_{ij} + \tfrac{1}{3}\eta^{\text{H}}_{ijk}\right) \\ \Pi_{ijk} &= \tfrac{1}{54\varepsilon^2}\left[\varepsilon'_{mm}\left(\varepsilon'_{ik}\eta'_{jmn} + \varepsilon'_{jk}\eta'_{imn}\right) + \tfrac{1}{4}\eta_{qpp}\left(\varepsilon'_{ik}\varepsilon'_{jq} + \varepsilon'_{jk}\varepsilon'_{iq}\right)\right] \end{aligned} \tag{13}$$

respectively.

3.2. Finite Element Analysis

In the contact simulations accounting for strain gradient plasticity, the isoparametric element developed by Wei and Hutchinson [41] is adopted. It was shown that this type of element worked well in the finite element analysis of microindentation experiments [34]. For the considered contact problem, choosing this type of element should bring about reasonable results.

Through the user-defined element subroutine in the commercial finite element software ABAQUS, the nine-node axisymmetric quadrilateral isoparametric elements are defined to discretize the substrate. As shown in Figure 2, the mesh near the contact region is highly refined, whereas relatively coarse mesh is used for the part far from the indenter. The size of the smallest element is approximately 0.05 μm, which is much smaller than the global size of the substrate. The accuracy of the simulation results has been ensured through mesh convergence tests, which found that a mesh with approximately 33,000 elements is sufficient for our simulations. The flat punch with a radius of r is set as the analytic rigid. To reduce the stress concentration at the edge of the punch, a tiny fillet with a radius of $0.01r$ is introduced, which has a negligible effect on the overall contact response. Note that the contact interface is assumed as non-adhesive and frictionless. The bottom of the substrate is constrained in the direction of the z-axis, while the radial displacement is restricted at the axis of symmetry.

The substrate is modelled by a homogeneous and isotropic solid of polycrystalline copper. With the standard uniaxial tensile tests on a servo-hydraulic testing machine (MTS-858/2.5T, MTS), the material parameters of the copper were measured in [42]: elastic modulus E = 105.6 GPa, Poisson's ratio ν = 0.34, yield stress σ_y = 159.6 MPa, and hardening exponent n = 0.13. For the copper, the magnitude of Burgers vector is b = 0.255 nm [34]. By varying the empirical coefficient α in the Taylor model, a series of values of l can be assumed by Equation (10).

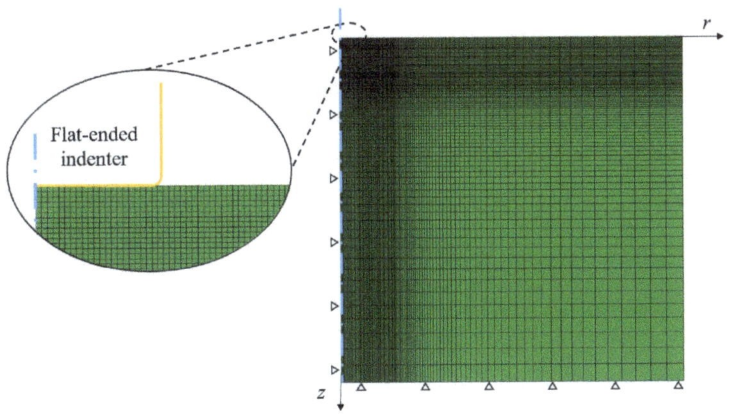

Figure 2. Finite element model of the circular flat contact accounting for strain gradient plasticity.

3.3. Explicit Expression of Contact Stiffness

Based on the dimensional analysis and the results in [19], the relationship between contact load P and indentation depth δ can be described by

$$\frac{P}{\pi r^2 \sigma_y} = f\left(\frac{2E^*\delta}{\pi \sigma_y r}, \frac{l}{r}\right) \tag{14}$$

where the ratio l/r determines the effect of strain gradient plasticity.

With the finite element simulations, the contact load can be obtained as a function of indentation depth for the circular flat contacts with different l and r. Figure 3 displays the variation of the normalized load $P/(\pi r^2\sigma_y)$ with respect to the normalized depth $2E^*\delta/(\pi\sigma_y r)$ for $l/r = 0$, 1.07, 9.65, 34.3, and 68.6. In the elastic regime of $P/(\pi r^2) < \sigma_y$, the load increases linearly with depth following the classical elastic contact solution, i.e., $P = 2E^* r\delta$ [37]. With an accumulation of plastic deformation, the load-depth relation gradually diverges from the linearity and essentially becomes dependent on the strain gradient plasticity parameter l/r. When the contact radius is much larger than the intrinsic material length, i.e., $l/r \to 0$, our results agree well with the prediction of Ding et al. [20] using J_2 plasticity. For the contact with a larger l/r, a higher load is required to generate the same depth, which implies a stronger hardening effect due to strain gradient plasticity. It should be pointed out that the finite simulations were performed for a large number of contact cases. For the sake of clarity, only several results are displayed.

Based on the curve fitting of the numerical results, it was found that the dimensionless function in Equation (14) can be expressed in the form of

$$f\left(\frac{2E^*\delta}{\pi\sigma_y r}, \frac{l}{r}\right) = \frac{2E^*\delta}{\pi\sigma_y r}\left[1 + a_l\left(\frac{2E^*\delta}{\pi\sigma_y r}\right)^{b_l}\right]^{c_l} \quad (15)$$

where a_l, b_l, and c_l are fitting coefficients that depend only on the ratio of the intrinsic material length l to the radius of contact r.

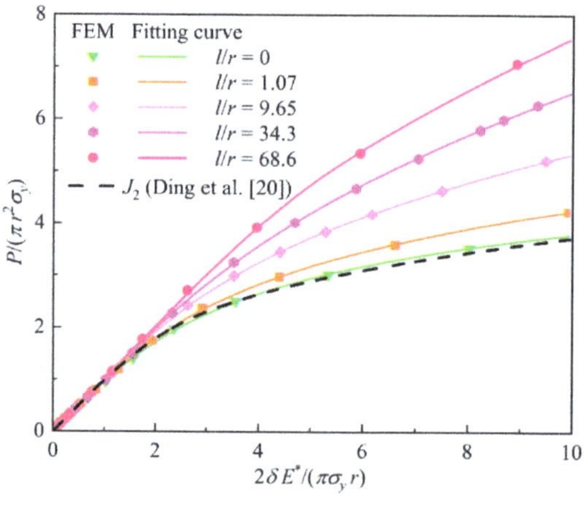

Figure 3. The dependence of the normalized load $P/(\pi r^2\sigma_y)$ on the normalized depth $2E^*\delta/(\pi\sigma_y r)$ for $l/r = 0$, 1.07, 9.65, 34.3, and 68.6.

For the circular contact between a rigid flat punch and a substrate, the contact stiffness $k(r)$ is defined by the derivative of contact load P with respect to indentation depth δ. In view of Equation (14), the contact stiffness $k(r)$ normalized by $2E^*r$ can be expressed in terms of the strain gradient plasticity parameter l/r and the mean contact pressure $P/(\pi r^2)$ normalized by the yield stress σ_y,

$$\frac{k(r)}{2E^*r} = g\left(\frac{l}{r}, \frac{P}{\pi r^2\sigma_y}\right) \quad (16)$$

For a given strain gradient plasticity parameter l/r, the variation of the normalized contact stiffness with respect to the normalized mean contact pressure can be determined from Equations (14) and (15). To avoid implicit expression, the dimensionless function in Equation (16) is expressed in an explicit form as

$$g\left(\frac{l}{r}, \frac{P}{\pi r^2 \sigma_y}\right) = \left[1 + m_l \left(\frac{P}{\pi r^2 \sigma_y}\right)^{2.2}\right]^{n_l} \tag{17}$$

where m_l and n_l are fitting coefficients.

For a series of l/r, the corresponding fitting coefficients of m_l and n_l can be obtained by curve fitting, and it is found that the dependences of m_l and n_l on the ratio l/r ranging from 0 to 2 can be further fitted by

$$\begin{aligned} m_l &= \frac{0.25(l/r)^2 + 0.022(l/r) + 0.011}{(l/r)^3 + 0.27(l/r)^2 + 0.52(l/r) + 0.22} \\ n_l &= -\frac{6.32(l/r)^2 + 2.69(l/r) + 1.17}{(l/r)^3 + 8.06(l/r)^2 + 0.72(l/r) + 0.34} \end{aligned} \tag{18}$$

For the contact of a rough surface with a rigid plane, we can derive its relationship between real contact area and normal load based on the incremental contact model as described in Section 2, where strain gradient plasticity is accounted for with the contact stiffness $k(r)$ given by Equation (16).

4. Results and Discussion

Figure 4 shows the topographies of three different realistic copper rough surfaces (S1, S2, and S3), which were generated by rubbing with sandpaper and measured with a white light interferometer over a nominal area of approximately 1 mm^2. The dependences of contact area $A_c(z)$ and contact patch number $N(z)$ on the separation z were given in [42]. The mean radii of the contact patches of these surfaces are calculated by Equation (1) with $A_c(z)$ and $N(z)$. As shown in Figure 5, the mean radii of the contact patches are on the order of micrometers for small contact areas. Thus, it is expected that the effect of strain gradient plasticity should be considerable in the contact of these rough surfaces.

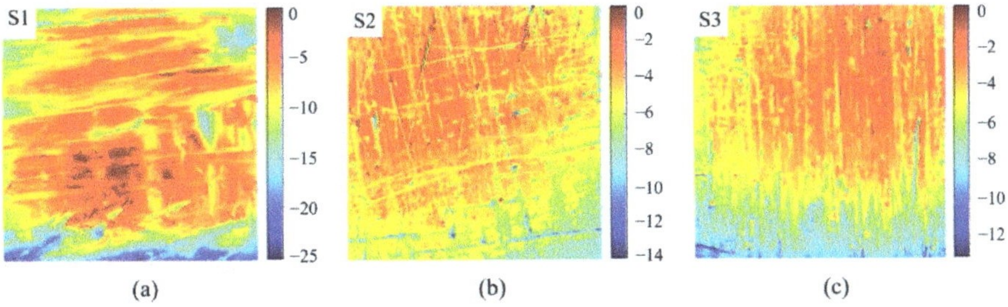

Figure 4. The topographies of rough surfaces. (**a**) Surface S1, (**b**) surface S2, (**c**) surface S3. The color scale carries unit μm.

Figures 6–8 show the area-load relation for surfaces S1, S2, and S3, respectively, with different intrinsic material length l. For the cases with small intrinsic material length, our results approach the prediction based on the classical J_2 plasticity [20]. When the strain gradient plasticity is taken into account, it is found that the area-load relation is still as close to linear at small loads as that obtained based on the classical plastic theory. However, the slope that represents the mean pressure over the real contact area increases significantly with the intrinsic material length. Such a hardening trend is qualitatively consistent with the finite element results of Song et al. [23]. It is worth mentioning that Yuan et al. [26] analyzed the contact of rough surfaces with surface effect and also found the proportionality between load and area was raised, which is a mechanism at nanoscale that is different from the present strain gradient plasticity at the scale of micrometers.

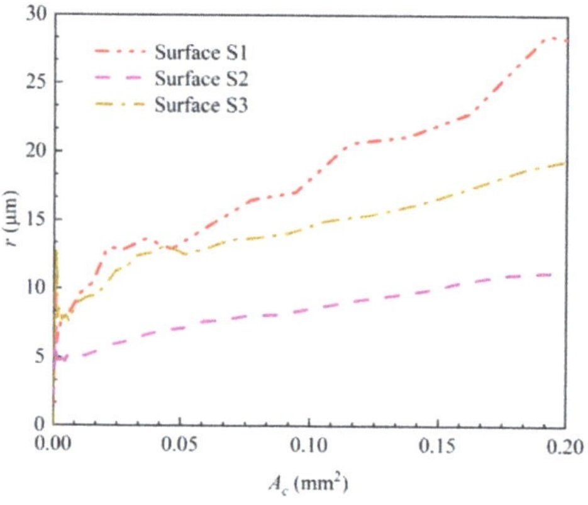

Figure 5. Variation of the mean radius of the contact patches with respect to the contact area for surfaces S1, S2, and S3.

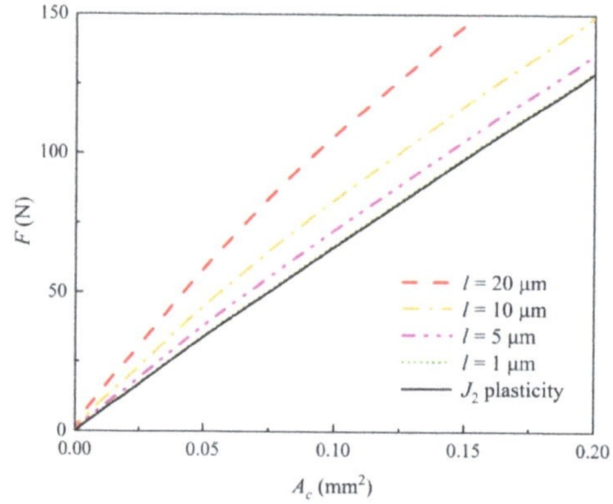

Figure 6. The dependence of the normal load on the contact area for surface S1.

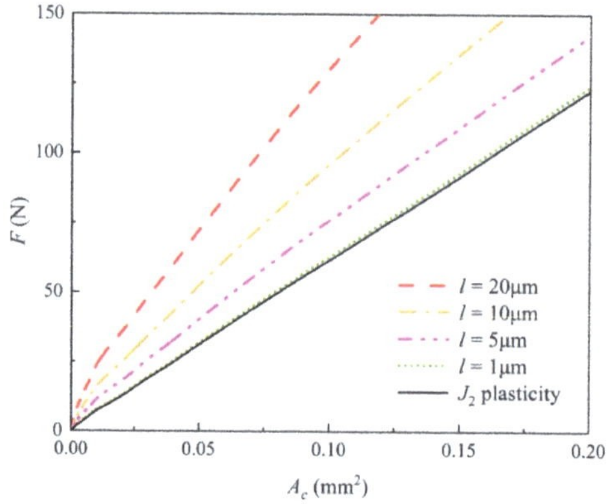

Figure 7. The dependence of the normal load on the contact area for surface S2.

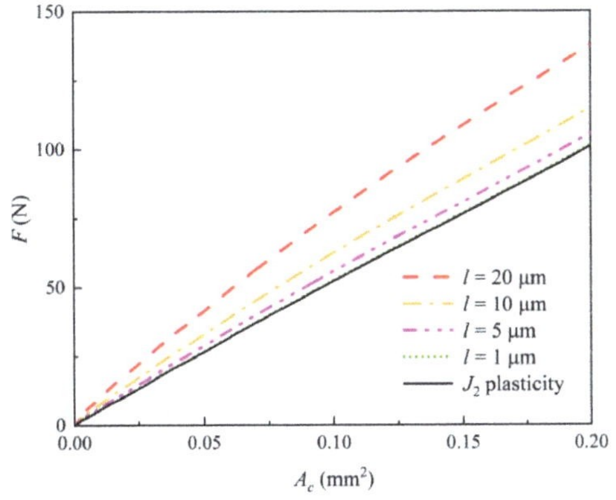

Figure 8. The dependence of the normal load on the contact area for surface S3.

To characterize the hardening degree of the rough surface contact considering strain gradient plasticity, we introduce a dimensionless factor as

$$\theta = \frac{F_{\text{MSGP}} - F_{J2}}{F_{J2}} \times 100\% \tag{19}$$

where F_{MSGP} and F_{J2} represent the normal loads obtained based on the MSGP theory and the classical theory of J_2 plasticity, respectively.

Figures 9–11 display the variation of the hardening factor θ with respect to the contact area A_c for surfaces S1, S2, and S3, respectively. Overall, the hardening factor gets smaller with increasing contact area, which is basically caused by the expanding of the contact patches. For the considered rough surfaces, it can be observed from Figure 5 that the mean radius of the contact patches for the contact area of 0.2 mm^2 (approximately 20% of the nominal area) is approximately triple that for the contact area of 0.01 mm^2 (approximately 1% of the nominal area). Correspondingly, the hardening factor for l = 10 μm is found to decrease by approximately one third as the contact area increases from 1% to 20% of the nominal area. The hardening effect of strain gradient plasticity in the contact of rough surfaces is particularly sensitive and prominent for initial contact where the size of the contact patches is relatively small.

It should be pointed out that the contact analysis of rough surfaces accounting for strain gradient plasticity was addressed earlier by Song et al. [23,36] with both direct finite element simulations and a modified multi-asperity GW model. Compared with their approaches, our model is more straightforward and convenient. With the present model, the real contact area generated by a given normal load can be handily predicted once the surface topography and the material parameters are measured. This advantage could be meaningful in the design and mechanical fabrication of solid surfaces as well as in the study of tribological problems, such as friction, sealing, wear, and lubrication.

The present contact model does not take into account both the elastic and plastic interactions between contact patches. For large contact area fractions, such interactions would be non-negligible as the stresses between the contact patches could be significantly high, and the present model should be further modified. In addition, the material of the rough surface, for simplicity, is assumed isotropic and homogeneous without any defects. The effects of local inhomogeneity and defects in the contacting asperities on the contact response is beyond the scope of the present work and deserves future investigations.

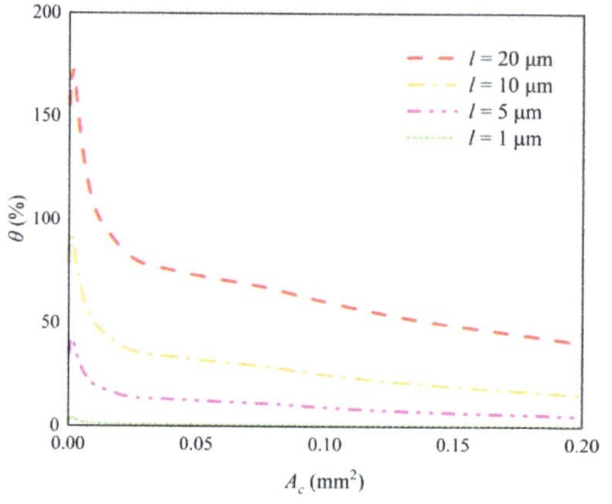

Figure 9. The dependence of the hardening factor on the real contact area for surface S1.

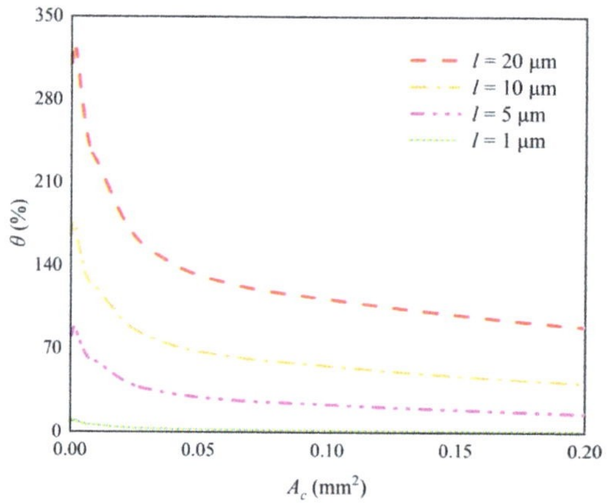

Figure 10. The dependence of the hardening factor on the real contact area for surface S2.

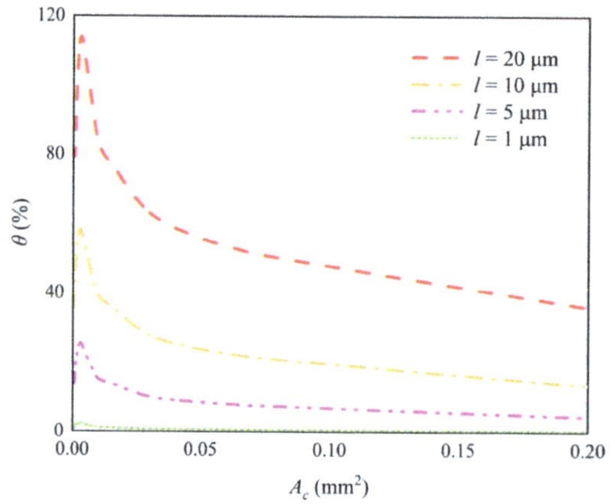

Figure 11. The dependence of the hardening factor on the real contact area for surface S3.

5. Concluding Remarks

The incremental contact model for rough surfaces is extended to account for strain gradient plasticity based on the MSGP theory. Through finite element analysis, the contact stiffness of a single circular flat punch is presented in an explicit form, which is higher as the ratio of intrinsic material length to contact radius increases. For three different realistic rough surfaces, the area-load relations are derived by employing the modified incremental contact model. Strain gradient plasticity does not change the linear nature of the area-load relation but increases the slope, which is the mean contact pressure. This trend is qualitatively consistent with the results in the literature. In addition, for the considered rough surfaces, the hardening degree of strain gradient plasticity could decrease by approximately one third as the contact area increases from approximately 1% to 20% of the nominal area. This work provides a simple but efficient approach to predict the real contact area under a given load for a rough surface considering strain gradient plasticity. On the basis of the present model, more advanced analysis can be conducted by considering other factors, such as asperity interactions and material inhomogeneity.

Author Contributions: Conceptualization, C.J. and G.W.; methodology, C.J., W.Y. and G.W.; software, C.J. and Y.Z.; validation, C.J., W.Y. and G.W.; formal analysis, C.J. and W.Y.; investigation, C.J. and W.Y.; resources, G.W.; data curation, C.J.; writing—original draft preparation, C.J. and W.Y.; writing—review and editing, W.Y. and G.W.; visualization, C.J. and W.Y.; supervision, G.W. and W.Y.; project administration, G.W.; funding acquisition, G.W. All authors have read and agreed to the published version of the manuscript.

Funding: This research was funded by the National Natural Science Foundation of China (grant no. 11525209).

Data Availability Statement: The data presented in this study are available on request from the corresponding author.

Conflicts of Interest: The authors declare no conflict of interest.

Nomenclature

List of abbreviations:

GND	Geometrically necessary dislocations
MSGP	Mechanism-based strain gradient plasticity
SSD	Statistically stored dislocation

List of symbols:

A_c	Contact area between a rough surface and a rigid plane
b	Magnitude of the Burgers vector
E	Elastic modulus
E^*	Combined elastic modulus
F	Normal load applied on the rigid plane
F_{MSGP}	Normal load based on the MSGP theory
F_{J2}	Normal load based on the classical J_2 plastic theory
K	Elastic bulk modulus
k	Contact stiffness of circular patch
l	Intrinsic material length of strain gradient plasticity
M	Taylor factor
N	Number of contact patches
n	Plastic work hardening exponent
P	Normal load applied on the flat punch
r	Radius of contact patch
z	Surface separation
α	Empirical coefficient
δ	Indentation depth
ε	Effective strain

η	Effective strain gradient
θ	Hardening degree factor of the rough surface contact
λ	Nye factor
μ	Shear modulus
ν	Poisson's ratio
ρ_G	GND density
ρ_S	SSD density
ρ_T	Total dislocation density
σ_{flow}	Flow stress
σ_{ref}	Reference stress in uniaxial tension
σ_y	Initial yield stress in uniaxial tension
τ_{flow}	Shear flow stress

References

1. Taylor, R.I. Rough surface contact modelling-A Review. *Lubricants* **2022**, *10*, 98. [CrossRef]
2. Vakis, A.I.; Yastrebov, V.A.; Scheibert, J.; Nicola, L.; Dini, D.; Minfray, C.; Almqvist, A.; Paggi, M.; Lee, S.; Limbert, G.; et al. Modeling and simulation in tribology across scales: An overview. *Tribol. Int.* **2018**, *125*, 169–199. [CrossRef]
3. Müser, M.H.; Dapp, W.B.; Bugnicourt, R.; Sainsot, P.; Lesaffre, N.; Lubrecht, T.A.; Persson, B.N.J.; Harris, K.; Bennett, A.; Schulze, K.; et al. Meeting the contact-mechanics challenge. *Tribol. Lett.* **2017**, *65*, 118. [CrossRef]
4. Greenwood, J.A.; Williamson, J.B.P. Contact of nominally flat surfaces. *Proc. R. Soc. A* **1966**, *295*, 300–319.
5. Bush, A.W.; Gibson, R.D.; Thomas, T.R. The elastic contact of a rough surface. *Wear* **1975**, *35*, 87–111. [CrossRef]
6. Greenwood, J.A. A simplified elliptic model of rough surface contact. *Wear* **2006**, *261*, 191–200. [CrossRef]
7. Chang, W.R.; Etsion, I.; Bogy, D.B. An elastic-plastic model for the contact of rough surfaces. *ASME J. Tribol.* **1987**, *110*, 50–56. [CrossRef]
8. Kogut, L.; Etsion, I. A finite element based elastic-plastic model for the contact of rough surfaces. *Tribol. Trans.* **2003**, *46*, 383–390. [CrossRef]
9. Afferrante, L.; Carbone, G.; Demelio, G. Interacting and coalescing Hertzian asperities: A new multiasperity contact model. *Wear* **2012**, *278*, 28–33. [CrossRef]
10. Vakis, A.I. Asperity interaction and substrate deformation in statistical summation models of contact between rough surfaces. *ASME J. Appl. Mech.* **2014**, *81*, 041012. [CrossRef]
11. Wang, G.F.; Long, J.M.; Feng, X.Q. A self-consistent model for the elastic contact of rough surfaces. *Acta Mech.* **2015**, *226*, 285–293. [CrossRef]
12. Chern, S.Y.; Chen, Y.Y.; Liu, W.L.; Horng, J.H. Contact characteristics at interface in three-body contact conditions with rough surfaces and foreign particles. *Lubricants* **2022**, *10*, 164. [CrossRef]
13. Horng, J.H.; Yu, C.C.; Chen, Y.Y. Tribological characteristics and load-sharing of point-contact interface in three-body mixed lubrication. *ASME J. Tribol.* **2021**, *144*, 052201. [CrossRef]
14. Majumdar, A.; Bhushan, B. Fractal model of elastic-plastic contact between rough surfaces. *ASME J. Tribol.* **1991**, *113*, 1–11. [CrossRef]
15. Yan, W.; Komvopoulos, K. Contact analysis of elastic-plastic fractal surfaces. *J. Appl. Phys.* **1998**, *84*, 3617–3624. [CrossRef]
16. Persson, B.N.J. Theory of rubber friction and contact mechanics. *J. Chem. Phys.* **2001**, *115*, 3840–3861. [CrossRef]
17. Persson, B.N.J. Elastoplastic contact between randomly rough surfaces. *Phys. Rev. Lett.* **2001**, *87*, 116101. [CrossRef]
18. Wang, G.F.; Liang, X.M.; Yan, D. An incremental equivalent circular contact model for rough surfaces. *ASME J. Tribol.* **2021**, *143*, 081503. [CrossRef]
19. Liang, X.M.; Ding, Y.; Yan, D.; Yuan, W.K.; Wang, G.F. Elastic-perfectly plastic contact of rough surfaces: An incremental equivalent circular model. *ASME J. Tribol.* **2021**, *144*, 051501. [CrossRef]
20. Ding, Y.; Liang, X.; Wang, G. An incremental contact model for rough surfaces of strain-hardening solids. *Int. J. Appl. Mech.* **2022**, *13*, 2250088. [CrossRef]
21. Hyun, S.; Pei, L.; Molinari, J.F.; Robbins, M.O. Finite-element analysis of contact between elastic self-affine surfaces. *Phys. Rev. E* **2004**, *70*, 026117. [CrossRef] [PubMed]
22. Campañá, C.; Müser, M.H.; Robbins, M.O. Elastic contact between self-affine surfaces: Comparison of numerical stress and contact correlation functions with analytic predictions. *J. Phys. Condens. Matter.* **2008**, *20*, 354013. [CrossRef]
23. Song, H.; Van der Giessen, E.; Liu, X. Strain gradient plasticity analysis of elasto-plastic contact between rough surfaces. *J. Mech. Phys. Solids* **2016**, *96*, 18–28. [CrossRef]
24. You, S.; Tang, J.; Wen, Y. Three-dimensional elastoplastic contact analysis of rough surface considering a micro-scale effect. *ASME J. Tribol.* **2022**, *144*, 011503. [CrossRef]
25. Pastewka, L.; Robbins, M.O. Contact between rough surfaces and a criterion for macroscopic adhesion. *Proc. Natl. Acad. Sci. USA* **2014**, *111*, 3298–3303. [CrossRef]
26. Yuan, W.; Long, J.; Ding, Y.; Wang, G. Statistical contact model of rough surfaces: The role of surface tension. *Int. J. Solids Struct.* **2018**, *138*, 217–223. [CrossRef]

27. Fleck, N.A.; Muller, G.M.; Ashby, M.F.; Hutchinson, J.W. Strain gradient plasticity theory and experiment. *Acta Metall. Mater.* **1994**, *42*, 475–487. [CrossRef]
28. Stolken, J.S.; Evans, A.G. A micro bend test method for measuring the plasticity length scale. *Acta Mater.* **1998**, *46*, 5109–5115. [CrossRef]
29. Ma, Q.; Clarke, D.R. Size dependent hardness of silver single crystal. *J. Mater. Res.* **1995**, *10*, 853–863. [CrossRef]
30. Poole, W.J.; Ashby, M.F.; Fleck, N.A. Micro-hardness of annealed and work-hardened copper polycrystals. *Scr. Mater.* **1996**, *34*, 559–564. [CrossRef]
31. McElhaney, K.W.; Vlassak, J.J.; Nix, W.D. Determination of indenter tip geometry and indentation contact area for depth-sensing indentation experiments. *J. Mater. Res.* **1998**, *13*, 1300–1306. [CrossRef]
32. Nix, W.D.; Gao, H. Indentation size effects in crystalline materials: A law for strain gradient plasticity. *J. Mech. Phys. Solids* **1998**, *46*, 411–425. [CrossRef]
33. Gao, H.; Huang, Y.; Nix, W.D.; Hutchinson, J.W. Mechanism-based strain gradient crystal plasticity—I. Theory. *J. Mech. Phys. Solids* **1999**, *47*, 1239–1263. [CrossRef]
34. Huang, Y.; Xue, Z.; Gao, H.; Nix, W.D.; Xia, Z.C. A study of microindentation hardness tests by mechanism-based strain gradient plasticity. *J. Mater. Res.* **2000**, *15*, 1786–1796. [CrossRef]
35. Huang, Y.; Qu, S.; Hwang, K.C.; Li, M.; Gao, H. A conventional theory of mechanism-based strain gradient plasticity. *Int. J. Plast.* **2004**, *20*, 753–782. [CrossRef]
36. Song, H.; Vakis, A.I.; Liu, X.; Van der Giessen, E. Statistical model of rough surface contact accounting for size-dependent plasticity and asperity interaction. *J. Mech. Phys. Solids* **2017**, *106*, 1–14. [CrossRef]
37. Sneddon, I.N. The relation between load and penetration in the axisymmetric Boussinesq problem for a punch of arbitrary profile. *Int. J. Eng. Sci.* **1965**, *3*, 47–57. [CrossRef]
38. Nye, J.F. Some geometrical relations in dislocated crystals. *Acta Metall.* **1953**, *1*, 153–162. [CrossRef]
39. Ashby, M.F. The deformation of plastically non-homogeneous materials. *Phil. Mag.* **1970**, *21*, 399–424. [CrossRef]
40. Arsenlis, A.; Parks, D.M. Crystallographic aspects of geometrically-necessary and statistically-stored dislocation density. *Acta Mater.* **1999**, *47*, 1597–1611. [CrossRef]
41. Wei, Y.; Hutchinson, J.W. Steady-state crack growth and work of fracture for solids characterized by strain gradient plasticity. *J. Mech. Phys. Solids* **1997**, *45*, 1253–1273. [CrossRef]
42. Liang, X.M.; Jiang, C.Y.; Wang, M.R.; Dai, W.L.; Wang, G.F. Experimental study on the load-area relation of rough surfaces and comparison with theoretical model. *Eur. J. Mech. Solid.* **2023**, *99*, 104934. [CrossRef]

lubricants

MDPI

Article

Phenomenological Laws of Single Point Contact: Pre-Rolling Contact Resistance through Pendulum

Igor. Gilavdary [1], Samir Mekid [2,3,*] and Natalia. Riznookaya [1]

[1] Faculty of Instrument-Making, Belarusian National University of Technology, 220013 Minsk, Belarus
[2] Mechanical Engineering Department, King Fahd University of Petroleum and Minerals, Dhahran 31261, Saudi Arabia
[3] Interdisciplinary Research Center for Intelligent Manufacturing and Robotics, Dhahran 31261, Saudi Arabia
* Correspondence: smekid@kfupm.edu.sa

Abstract: The development results of a single-point contact system set up as a pendulum to study the laws of rolling resistance to contacting bodies at a distance significantly reduced compared to the elastic contact spot size. The designed device uses a physical pendulum sustained by only one ball on a flat polished surface. The problem of stability of the pendulum swing plane is solved. A phenomenological theory of rolling resistance is described. The surface tension of solids on the contact zone, parameters of the frequency-independent internal friction and the pressure of the adhesion forces are found.

Keywords: single point contact; pre-rolling; contact friction; positioning; pre-rolling resistance

Citation: Gilavdary, I.; Mekid, S.; Riznookaya, N. Phenomenological Laws of Single Point Contact: Pre-Rolling Contact Resistance through Pendulum. *Lubricants* **2023**, *11*, 88. https://doi.org/10.3390/lubricants11020088

Received: 11 January 2023
Revised: 9 February 2023
Accepted: 13 February 2023
Published: 17 February 2023

1. Introduction

Rolling friction models have attracted the attention of engineers and researchers since the wheel invention. As a common rule, rolling friction is mostly studied under the condition of constant rolling speed. Many publications have emerged in this area allowing basic laws derived to describe the reliance between the rolling resistance moment and the magnitude of the load, the driving conditions and various other influencing factors. Interest of this topic is confirmed by an almost unlimited number of publications that is still growing, just to cite the recently appeared ones [1–4].

These fundamental problems are solved for the sake of understanding and immediately for specific advanced applications revolving around pre-rolling and rolling. In achieving high accuracy of rolling motion control and ultra-precision positioning through rolling systems, knowledge of the laws of rolling resistance at the pre-rolling in forward motion and when reversing the motion is required. This also led to the need to study the laws of rolling friction at low loads, low speeds, and small displacements of the rolling body. It was found that under these conditions, the main mechanisms of rolling friction are associated with the elasticity of the contacting bodies, relative slip and internal hysteretic friction, as well as adhesion [5,6]. "Pre-rolling" is a special name given to displacements in a friction pair, in which rolling friction has a nonstationary character. Here, the dependence laws of the rolling resistance moments on displacements are of a specific nonlinear character, which is still unknown [3,7–10].

The development of micro- and nanotechnology has led to the need to study the adhesion forces, as the reason for the reciprocal sticking of individual MEMS elements that disturb their work [11]. The sticking factor due to adhesion forces should also be considered when using micro- and nano-manipulators in the technique [12,13].

Despite the practical importance of the above problems, "empirical procedures for measuring and representing the laws of friction" [4] with high sensitivity and accuracy in the nano- and micro range of values have not yet been developed.

This paper presents the development results of the empirical procedure for measuring and representing the rolling friction law, in a particular case when the displacement of the rolling body is significantly less than the radius of the contact spot. We will call this rolling region of deep pre-rolling (DPR). This works aims to describe the design of the single-contact ball pendulum device developed and experienced by the authors. In this device, the physical pendulum with one ball rests on the flat surface of the test specimen allowing free swings with a stable swing plane. In addition, the description of a special technique for measuring rolling resistance forces with high sensitivity and accuracy in DPR mode is given.

2. Review of Pendulum Devices for Studying the Surface of Solids

2.1. Pendulum Devices with One Supporting Ball

The devices that use the free swing method of a physical pendulum rest upon a flat surface through a rolling body (balls, rollers, edge of a prism (or knife)) have long been used to measure the hardness and strength of materials and products [14–25]. The hardness and strength of the sample pads in these devices is being associated with a decrease in the pendulum swing amplitude as a function of time.

Mendeleev D.I., apparently, was the first who conducted a thorough study (1895–1898) of the influence on the error of weighing the nature of the damping of the swing of the pendulum, which rested on the platform with a prism edge. He made, in particular, the following conclusion; the time of an individual oscillation decreases with decreasing amplitude of the oscillation itself much faster than, according to well-known formulas, the dependence of the oscillations period of a mathematical pendulum on amplitude [14].

From the point of view of the possibility of calculating the bodies' deformations during their elastic contact, the contact of the ball and the flat surface is the simplest. This part of the elasticity theory is the most developed [6]. Several researchers have made attempts to build a pendulum device on one ball. However, when working with such a (single-contact) pendulum, the researchers faced the problem of instability of its swing plane, since the pendulum on one ball has three rotational degrees of freedom, which leads to instability of the swing plane and significantly complicates the research.

In 1923, Herbert, E developed a pendulum device to measure the hardness of metals (Figure 1) [15,21]. In this device, the pendulum rested on the test sample with a single steel or ruby ball of 1 mm in diameter inserted into the tip. The total weight of the pendulum was 4 kg, its center of gravity should have been located above the fulcrum, but below the center of the ball by 0.1 mm [17]. After installing the pendulum on the test sample surface, the pendulum was deflected to an initial position of several degrees, and the time of the first swing was measured, which was multiplied by 10, and this parameter was used as an estimate of the test material hardness [18].

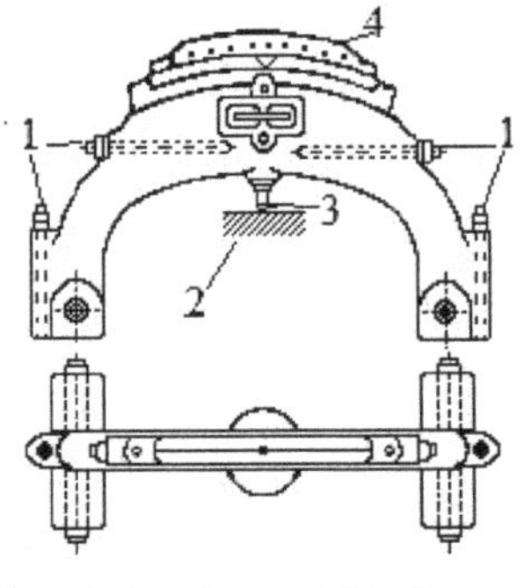

1—a balancing weights, 2—a test sample, 3—a tip with ball, 4—a scale

Figure 1. The Herbert Pendulum [15].

There is a problem with the stability of the swing plane in Herbert's pendulum. Therefore, in the instructions for the use of Herbert's pendulum, it was recommended to deflect the pendulum from the equilibrium position "using a feather" [16]. Subsequently, this device was modified by other researchers so that the pendulum had less weight and rested not on one ball, but on one roller (Figure 2). The roller had a diameter of 2 mm and a length of 12 mm as stated in [19]. This solved the problem of stability of the swing plane and allowed the use of the device to assess the hardness of brittle, viscoelastic, and biological materials [19–21]. In [22], hardness measurements of Herbert pendulums using various weights are described.

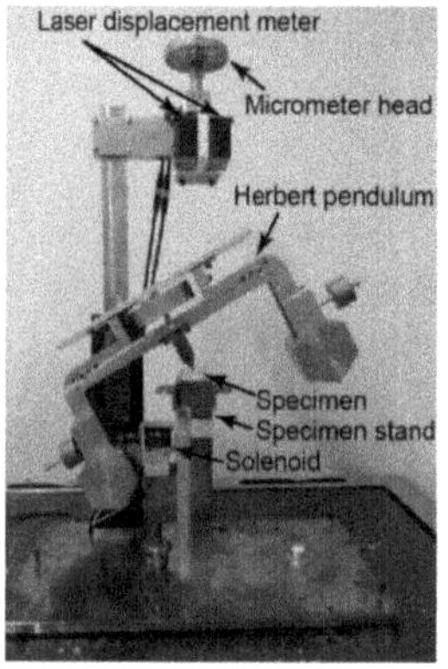

Figure 2. The Herbert pendulum device with one roller support [19].

Kuznetsov V.D. (1929) developed a pendulum device in which the pendulum was supported by one or two sharpened needles, or, two balls with a diameter of 0.5 mm (Figure 3), and the pendulums themselves had different shapes and weights. Sharp needles were used to study the strength of crystals whose surface has been damaged by these needles during the swing of the pendulum, but the balls were used to measure the hardness of materials [23].

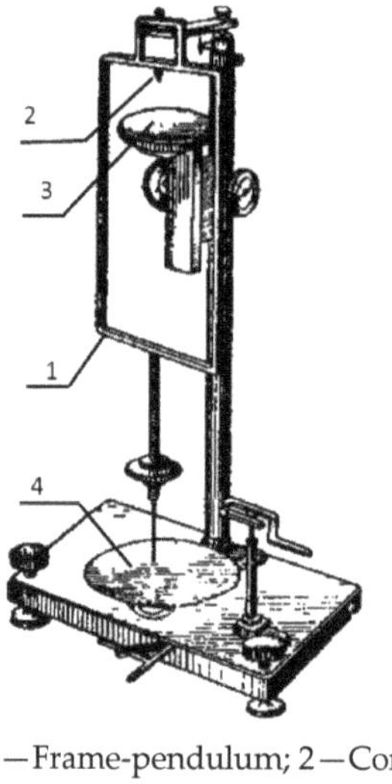

1—Frame-pendulum; 2—Conical tip of hardened steel (angle at the apex 90°); 3—Sample; 4—Scale (a concave disk with the drawn circles of various radii)

Figure 3. The Kuznetsov pendulum device with one support [17].

Due to the instability of the swing plane of the pendulum in this device (Figure 3), the amplitude of the pendulum deviation was measured here using a system of dials. The use of two pendulum support points in the Kuznetsov device made it possible to stabilize the swing plane of the pendulum [24].

Kuznetsov's two-contact pendulum [25] is widely used in modern commercially produced pendulum devices by Koenig and Persos (Figures 4 and 5), which formed the basis of standards measuring for the plastics and coatings hardness [26]. The surface hardness here is estimated as the ratio of the pendulum swings number on the test surface and on a calibrated glass plate.

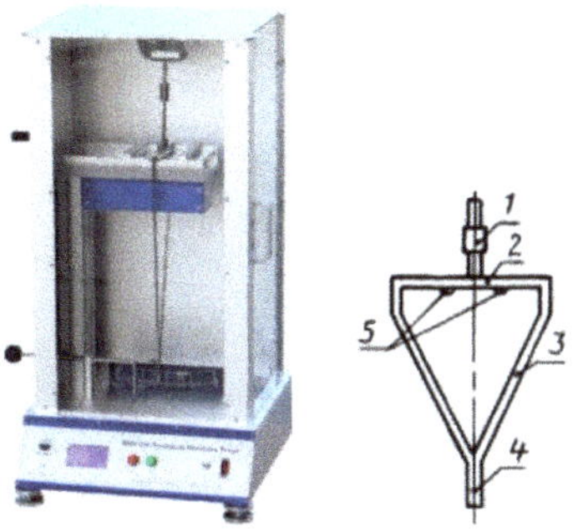

1—Counterweight for regulating the natural frequency; 2—Cross-members; 3—Frame; 4—arrow; 5—Support balls

Figure 4. Koenig device and pendulum [25,26].

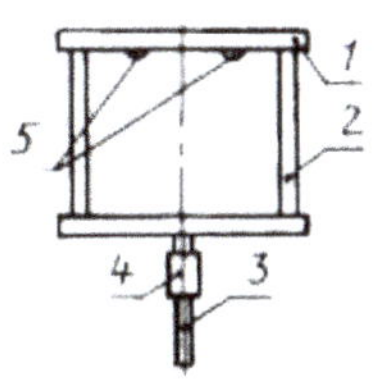

1—cross member; 2—frame; 3—arrow; 4—counterweight to control the natural frequency; 5—support balls

Figure 5. Persosa pendulum [25].

The problem with these previous one ball system is that they are too heavy with balancing issues, furthermore, no theory has been published publicly to describe the phenomena.

2.2. Pendulum Devices with Two Supporting Balls

G.A. Tomlinson was apparently, the first to use a physical pendulum with two supporting balls for systematic studies of the molecular nature of rolling friction at low swing amplitudes. In his experiments carried out in 1929, a pendulum in the form of a disk was mounted on a cylindrical axis and rested on two rolling bodies (two cylinders, or two half-cylinders, or two balls). One version of the experiment to measure the rolling friction of a cylinder along a plane is shown in Figure 6. The amplitude of the disk oscillation was approximately in the range of 221 to 6 arc minutes. The rolling friction coefficient was found by calculating the damping of the swing amplitude of the disk under the assumption that the damping decrement remains constant [27]. In his article, Tomlinson, among other things, concluded that the rolling friction coefficient is practically independent of speed.

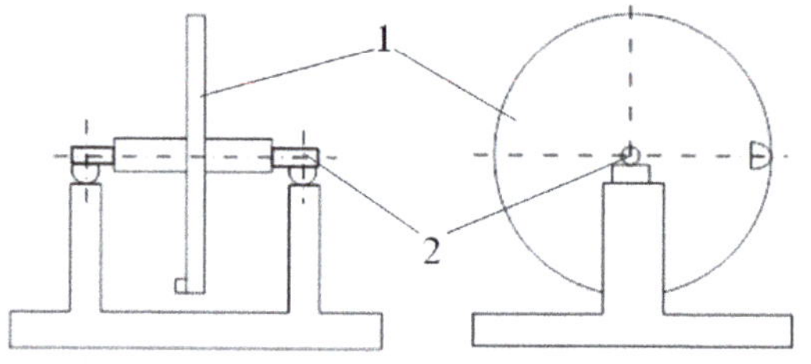

1—a disk weighing 600 g; 2—a cylindrical axis

Figure 6. Tomlinson's device [27].

Previously, the authors of this work developed a two-contact pendulum device in which a physical pendulum made free damping swings with an amplitude substantially smaller than the angle of elastic contact, resting on two identical balls on a test flat surface (Figure 7) [28]. The pendulum mass center was on ball's contact spot [29]. It was suggested that under loads referred to the field of elastic deformation and low rolling speeds, the balls movement was similar to pure rolling. In this case, the main mechanism of rolling resistance is the adhesion forces and the forces of frequency-independent internal friction arising from the deformation of the contacting bodies. A phenomenological theory of rolling resistance was constructed in the DPR mode, and it was shown that with this device it is possible to study the laws of rolling resistance on the nano—and microscale with high sensitivity and accuracy.

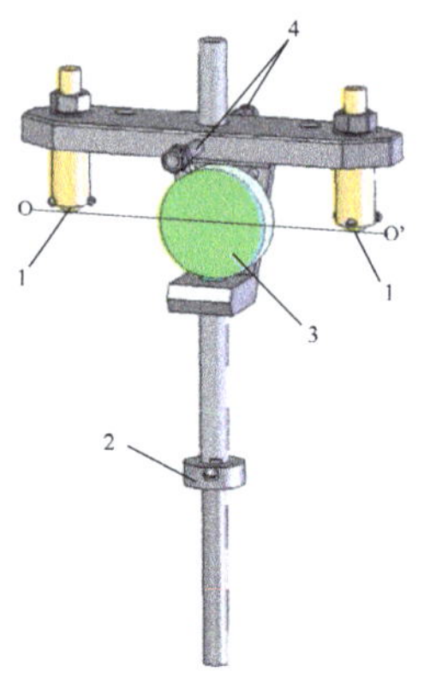

1—support balls, 2, 4—balancing weights, 3—a mirror

Figure 7. Model of the two-contact pendulum [28].

In this device, the initial pendulum swing amplitude did not exceed 300 arcsec, and the final pendulum swing amplitude was 2 arcsec. We used polycrystalline sapphire balls with a diameter of 10 mm, and the pendulum weight was about 1.2 kg. Depending on the material of the supporting surface (tempered steel, hard glass, electrotechnical silicon), the angle of elastic contact was in the range of 37 to 47 arcmin. During the experiments, the amplitude and time of each swing were measured allowing the construction of the amplitude vs. time and period vs. the number of swings or time. The constructed mathematical models made it possible to approximate with high accuracy these experimental dependences with analytical dependences as presented below.

However, a two-contact pendulum has an obvious drawback: it can only be used if there are two identical test samples, placed under each ball, or one sample with a relatively large uniform surface. This significantly reduces the range of materials tested. In addition, there is a problem of identity, both of the balls themselves and of the conditions for their fixation were not always exactly similar.

3. The Phenomenological Theory of Rolling Resistance in DPR Mode

This theory is built on the assumption that the adhesion forces can be represented as some force bonds (springs) that connect the contacting bodies. When the pendulum swings, one part of these forces breaks, and in this case, a certain fraction of the pendulum energy is spent on the work of tearing off the surface of the ball from the surface under consideration. Also, part of the pendulum energy must work against the forces of internal friction during the deformation of the contacting bodies. Under the assumption that these forces are frequency independent, the friction moment as a function of the pendulum deflection angle can be written in the form [29] of Equation (1).

$$M_{fr} = -mgR(c + b\varphi^p)sign\left(\frac{d\varphi}{dt}\right), \tag{1}$$

where m—the pendulum mass; R—the ball radius; c, b, p—the approximation parameters determined from the experiment. Here, coefficient c "is responsible" for the adhesive component of friction, and the second part on the right-hand side of Equation (1) is "responsible" for internal friction during the deformation of contacting bodies and capillary forces.

Using Equation (1), we can calculate the specific surface energy in the contact zone as the ratio of the work $A_t(\varphi) = mgRc\varphi$ performed by the adhesion forces when the surface of the ball is separating from the surface of the contact spot during its rotation through a small angle φ when it moves away from the equilibrium position, to S:

$$\sigma = \frac{A_t}{S} = \frac{mgc}{2a}, \tag{2}$$

The parameter σ in its physical meaning and dimensions coincides with the similar parameter γ used in [3], which is referred to by the author as surface energy or surface tension.

The phenomenon of a sharp decrease in the period of the pendulum during its swings is associated with the action of the same elastic bonds between the surfaces of the contacting bodies that do not break as the pendulum swings. The dependence of the moment of these forces on the ball rotation angle, based on some physical considerations, can be written in the form [29]

$$M_{el}(\varphi) \approx 2\gamma a^2 R|\varphi|^{n+1}\left(\frac{\pi}{2} - \frac{R}{a}\varphi\right) \cdot sign(\varphi), \tag{3}$$

where a is the contact spot radius; γ, n are the approximation parameters determined from the experiment. In our opinion, the parameter γ characterizes the elastic pressure of the adhesion forces acting between the ball surface and the test surface. In [3], the author, in the case of the Van-der-Waals forces action denoted as σ, calls the similar parameter as Van-der-Waals stress.

Under the experimental conditions, when the mass center of the pendulum lies on the contact spot, and at very small values of its amplitude and angular velocity, the differential equation of swings of the pendulum practically coincides with the equation of swings of a mathematical pendulum. Using Equation (1), we have a differential equation in the form

$$I\frac{d^2\varphi}{dt^2} + mgR\varphi = -mgR(c + b\varphi^p)sign\left(\frac{d\varphi}{dt}\right), \tag{4}$$

where I—the moment of pendulum inertia relative to the mass center.

The solution of Equation (4) in a first-term approximation of the asymptotic theory of nonlinear oscillations gives the relationship of the pendulum swings amplitude α on time in an implicit form [29]:

$$t(\alpha) = -\frac{T}{4}\int_{\alpha_0}^{\alpha} \frac{d\varphi}{\frac{1}{p+1}b\varphi^p + c}, \quad (5)$$

where T—the average value of the pendulum swing period.

Using Equation (5) as the regression equation for approximating the experimental damping curves of the swing amplitudes, we can determine the numerical values of the parameters c, b, and p.

Similarly, taking into account Equation (3) for the moment of elastic forces, solving the differential equation of the pendulum swing, in the second approximation of the asymptotic theory of nonlinear oscillations, we can obtain the dependence of the pendulum swing period on the amplitude in the form (Equation (28) in [29]).

$$T(\alpha) = T_0\left[1 - \sqrt{\pi}\gamma\frac{a^2\alpha^n\,\Gamma(\frac{n}{2}+\frac{3}{2})}{mg\,\Gamma(\frac{n}{2}+2)}\left(1 - 0.55\,\frac{R\alpha}{a}\right)\right]^{-1} \quad (6)$$

Using an equation of the form Equation (6) as a regression equation for the experimental dependence $T(\alpha)$, one can find the numerical values of the parameters T_0, γ, and n, which are stable with respect to the choice of initial approximations of the values of these parameters during calculations in the process of nonlinear approximation.

The total moment of rolling resistance forces can be written as

$$M(\varphi) = M_{\text{fr}}(\varphi) + M_{\text{el}}(\varphi). \quad (7)$$

The dependence $M(\varphi)$ with the known dependence $\varphi(t)$ allows us to construct a phenomenological theory of rolling resistance of adhesion forces in the DPR mode. In particular, the dependence $M_{el}(\varphi)$ allows one to construct a skeletal curve and the dependence $M_{fr}(\varphi)$ allows one to construct a hysteresis loop around the virgin curve. Examples of such curves are shown below.

4. Pendulum Device Based on One Ball: Design and Measurement Procedure

As mentioned previously, when using a pendulum device based on one ball, it is necessary to solve the problem of instability of the pendulum swing plane. The instability effect enhances when the mass center of the pendulum is close to the contact spot, and the friction is small. If the pendulum is mounted with one support ball on a hard flat surface, it will have the ability to swing around two horizontal axes and rotate around a vertical axis.

In our experiments, it was found that in the DPR mode, after careful starting of the swings, the pendulum rotation around the vertical axis practically did not occur or quickly stopped due to the spin friction being high. In addition, it was noted, if the pendulum shape where the moments of inertia of the main horizontal central inertia axes differ significantly from each other (Figure 8), the swings with a short period around the long axis (the U axis with a minimum moment of inertia) damps earlier than swings with a large period around the short axis (V axis with a maximum moment of inertia).

Thus, in order to build a device based on a single-ball pendulum and solve the problem of holding the pendulum swing plane, it is necessary that the main moments of inertia of the pendulum around which the swings occur should differ significantly in size as explained later. In addition, the mass center of the pendulum should be on the contact spot, which eliminates both the ball slippage during its swings and the influence of base vibrations on swings [28,29]. The first condition can be fulfilled due to the pendulum design—it should have the most elongated shape. The second condition can be ensured by careful balancing. There is no spinning around the vertical axis due to proper balancing.

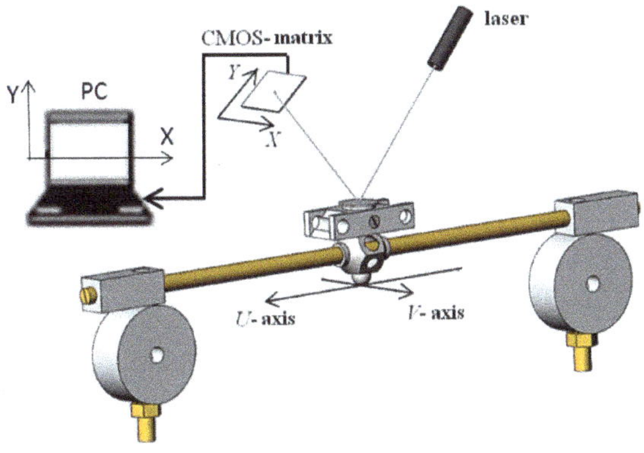

Figure 8. The pendulum design with a system for recording the pendulum deviations from the equilibrium position.

Figure 8 shows a single-ball pendulum specially designed for measurements. In this pendulum, the ratio of the moments of inertia I_v/I_u is about 25, which experimentally gave the ratio of the pendulum swing periods T_v/T_u equal to about 5. This has been achieved with extensive tests. The diameter of the ball was 12.1 mm. The mass of the pendulum was 0.406 kg. The mass center of the pendulum was near the contact spot.

To measure the amplitude and swing time of the pendulum, an optical recording system for its oscillations was used, consisting of a semiconductor laser, a focusing device (not shown in Figure 8), a mirror mounted on the pendulum, and a CMOS matrix. The computer records the signal from the matrix. This setup allows us to record the beam displacement path, reflected from the pendulum along the X and Y axes of the coordinate system of the CMOS matrix. Examples of writing $Y(t)$ and $X(t)$ of a balanced pendulum are shown in Figures 9 and 10, respectively. The Control over the balancing of the pendulum was also carried out by constructing its trajectory in the coordinates Y, X (Figure 11)—an analogue of the Lissajous figure. It is worth noting that measurements are repeatable.

If the pendulum was perfectly balanced, and the optoelectronic system was perfectly tuned and had no noise, and there was no influence of the vibrations of the basis, the short-term transverse vibrations of the pendulum should not be excited, and the records in Figures 10 and 11 should look like a horizontal and vertical line, respectively.

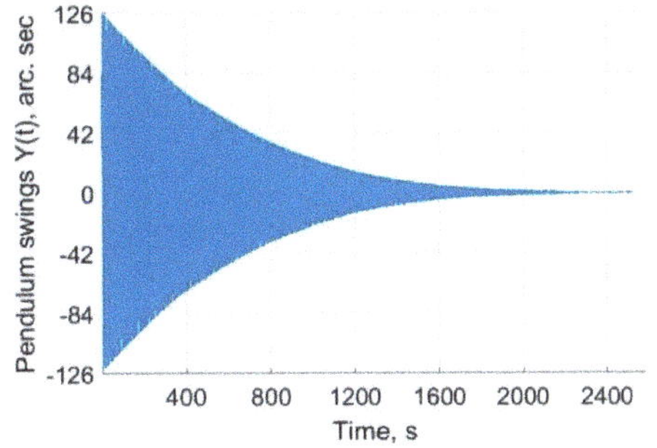

Figure 9. Recording the amplitude of long-periodic $Y(t)$ of the single-ball pendulum.

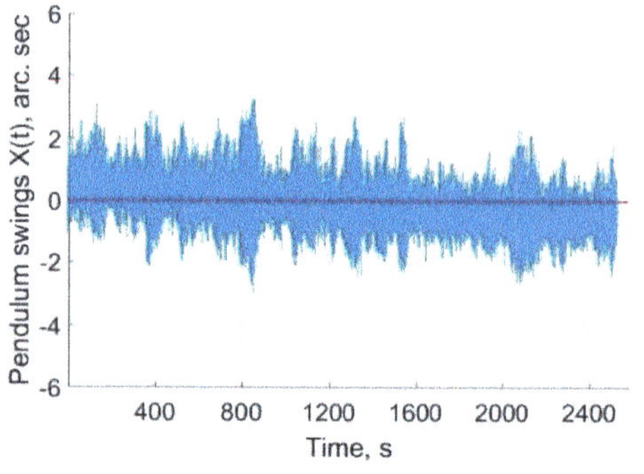

Figure 10. Recording the amplitude of short-periodic swings $X(t)$ of the single-ball pendulum.

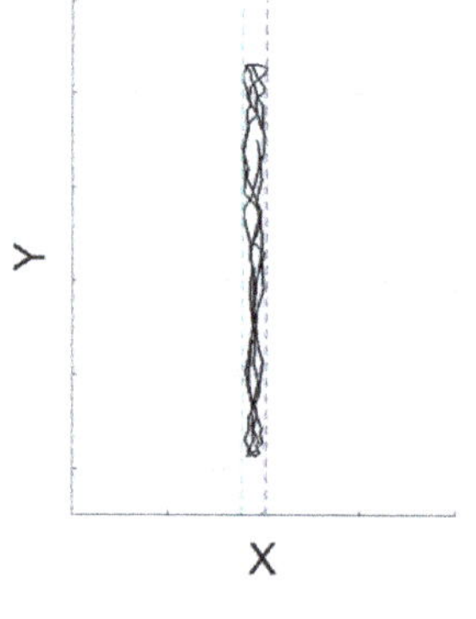

Figure 11. Trajectory record of the laser ray spot on the matrix during the first two swing periods.

5. Measurement and Calculation Results

In the experiments, a ball of radius $R = 6.05$ mm made of hard steel was used as a pendulum support. As the materials for the test samples were selected next: single-crystal silicon (roughness $Ra = 0.4$ nm, elastic modulus $E_{si} = 1.31 \cdot 10^{11}$ N/m^2, Poisson's ratio $\nu_{si} = 0.266$), hard steel ($Ra = 63$ nm, $E_{st} = 2.11 \cdot 10^{11}$ N/m^2, $\nu_{st} = 0.28$) and K8 optical glass ($Rz = 40$ nm, $E_g = 0.82 \cdot 10^{11}$ N/m^2, $\nu_g = 0.206$).

The calculated contact parameters of the hard steel ball and surfaces for testing are shown in Table 1.

Table 1. Materials contact characteristics.

Surfaces Under Test	Radius of the Contact Spot, [micrometer]	Depth of Ball Penetration into the Surfaces, [micrometer]	Contact Angle, [arcmin]
Hard steel	54	0.50	31
Silicon	59	0.60	34
Glass K8	66	0.70	38

The initial swing amplitude of the pendulum was chosen equal to $\alpha_0 \approx 6 \cdot 10^{-4}$ rad $\approx$ 124 arcsec, the final amplitude was approximately 2 arcsec, so that the maximum ball displacement is $r_{max} \approx 3.63$ μm, the minimum ball displacement is $r_{min} \approx 0{,}06$ μm. This is limited by the recording equipment's accuracy.

It is noticed that the angular and linear displacements of the ball were in the DPR zone. The results of measuring the dependences of the amplitude α on time t and the period T on amplitude α obtained for friction pairs by averaging the results of a series of three consecutive measurements for each contact pair are shown in Figures 12 and 13, respectively.

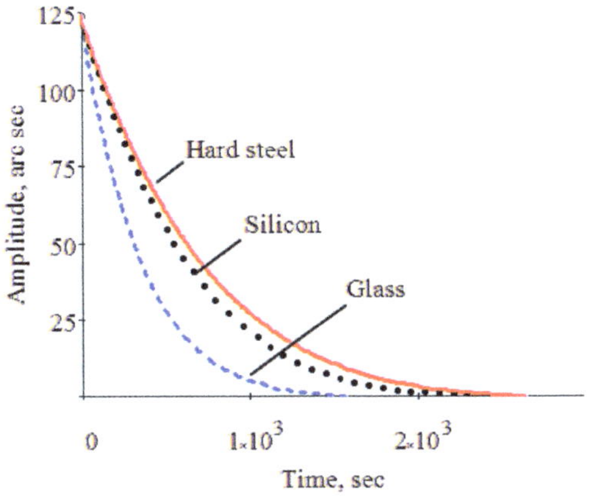

Figure 12. The curves approximation of the dependence of the pendulum swing amplitude vs. time according to Equation (5).

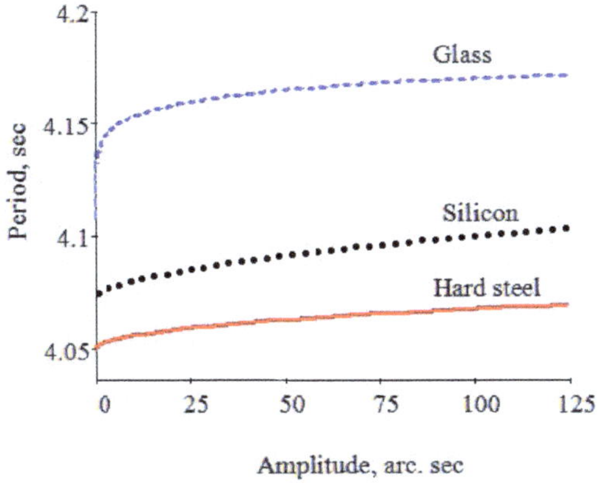

Figure 13. Curves approximation of the pendulum swing period dependence on the amplitude according to Equation (6).

The calculation results of the approximation parameters (friction parameters) of the amplitude versus time are presented in Table 2. The calculation of the average relative approximation error (ARAE) of the curves of the amplitude dependence on time, given in this table, was carried out according to Equation (8).

$$\text{ARAE} = \frac{1}{k}\sum_{i=0}^{k}\left|\frac{\alpha_i - \alpha(t_i)}{\alpha_i}\right|, \quad (8)$$

where α_i—the measured values of the amplitude, $\alpha(t_i)$—the calculated values with Formula (5).

Table 2. Values of friction parameters for several materials.

Friction Parameters (Initial Amplitude α_0 = 124 arc s)	Contact Pair		
	Hard Steel/Glass	Hard Steel/Silicon	Hard Steel/Hard Steel
p, 10^{-4}	37.330	7.060	7.961
b	0.944	0.815	0.850
c, 10^{-9}	10.510	3.840	6.601
σ, 10^{-3} J/m^2	28.900	13.200	27.300
ARAE	0.008	0.026	0.035

Note that the values of the parameter σ agree in order of magnitude with the values of "adhesive energy $w_{\text{t-s}}$ (~10^{-3} J/m^2)" obtained using an atomic force microscope (AFM) [30].

The values of the elastic interaction parameters obtained from the approximation of the reliance of the pendulum swing period on the amplitude are presented in Table 3. The calculation [31] of the average relative error of this approximation given in this table is carried out according to Equation (9).

$$\text{ARAE} = \frac{1}{k}\sum_{i=0}^{k}\left|\frac{T_i - T(\alpha_i)}{T_i}\right|, \quad (9)$$

where T_i—the measured value of the amplitude, $T(\alpha_i)$—the calculated value of the amplitude in accordance with Equation (6).

Table 3. Values of parameters for elastic rolling resistance.

Parameters of Elastic Rolling Resistance, (Initial Amplitude α_0 = 124 arc s)	Contact Pair		
	Hard Steel/Glass	Hard Steel/Silicon	Hard Steel/Hard Steel
n	0.092	0.552	0.461
T_0, s	4.071	4.072	4.049
γ, 10^6 N/m^2	28.97	290.69	95.46
ARAE	0.002	0.002	0.004

In Figure 14, graphical smoothing has been applied using least-squares smoothing according to the rule of s-nearest neighbors, in which s is selected adaptively [32].

The practical coincidence of the smoothing curve, which does not imply the use of any physical models, and the curve constructed on the basis of the dependence model $T(\alpha)$, proposed in [29,33] and leading to Equation (6), can be considered a justification for the objectivity of these models.

Graphs displayed in Table 4 show the dependences of the moments of rolling resistance forces with the angle of pendulum deviation within one swing cycle at an amplitude of 3.1 arcsec, constructed according to Equations (1), (3) and (7) and taking into account the values of the parameters obtained by approximating the experimental data for three different materials. It also shows the results of calculating the work of the adhesion forces $4A_t$ (rectangle set off in with dashed lines) and the work of dissipative forces (adhesion to tear off and internal friction) $4A$ (the area of the whole figure), calculated for one period or four deviations from the equilibrium position of the pendulum.

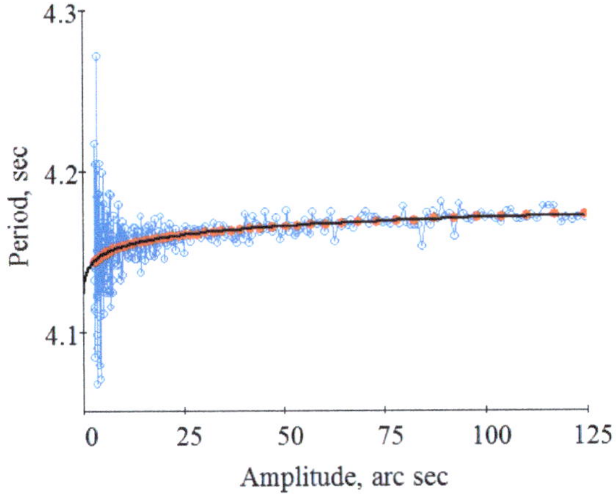

Figure 14. The dependence of the swing period on the amplitude: individual blue points—the result of measurements; individual red dots—computer smoothing by a function; black curve is an approximating curve constructed according to Equation (6).

Table 4. Dissipative motion for one cycle of the swing (*If amplitude* α *= 3.1 arc s*).

Dissipative Component Moment (nN·m) vs. Angle (arc s) Equation (1)	Elastic Component Moment (nN·m) vs. Angle (arc s) Equation (3)	The Total Moment of Resistance Moment (nN·m) vs. Angle (arc s) Equation (7)
	Glass	
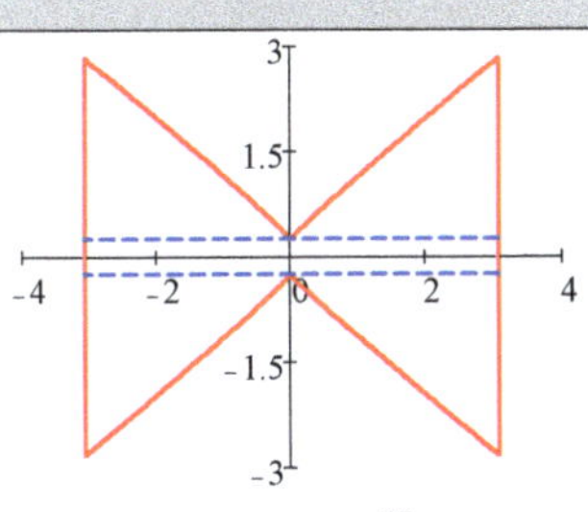 $4A_t = 15.7 \cdot 10^{-15}$ J; $4A = 98.6 \cdot 10^{-15}$ J	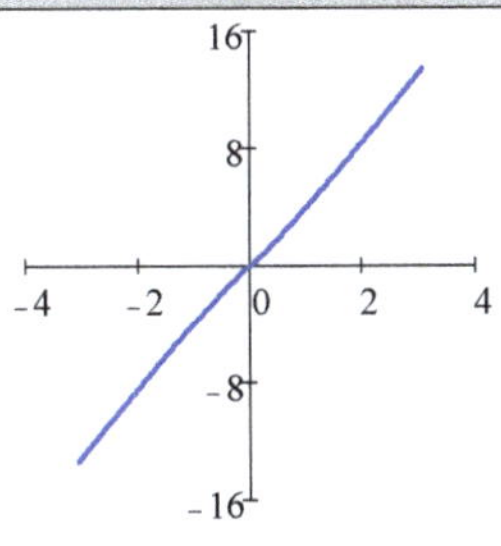	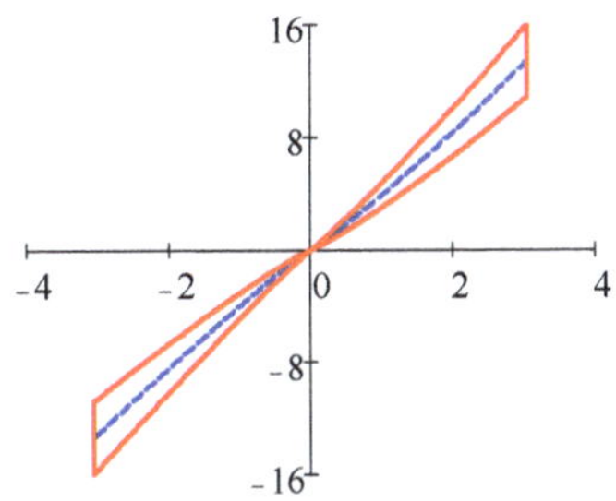
	Silicon	
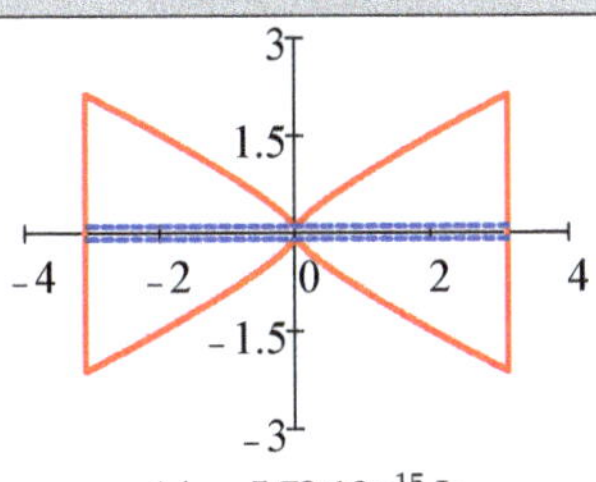 $4A_t = 5.72 \cdot 10^{-15}$ J; $4A = 75.2 \cdot 10^{-15}$ J	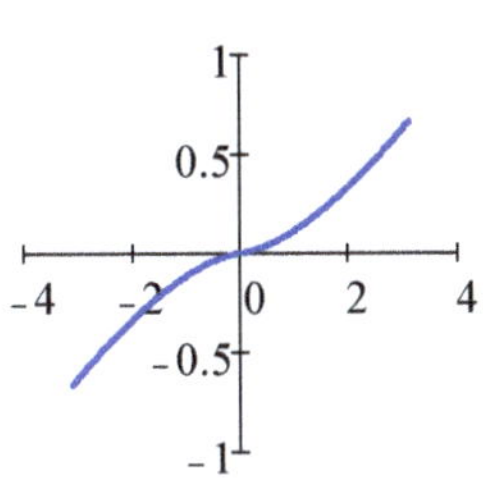	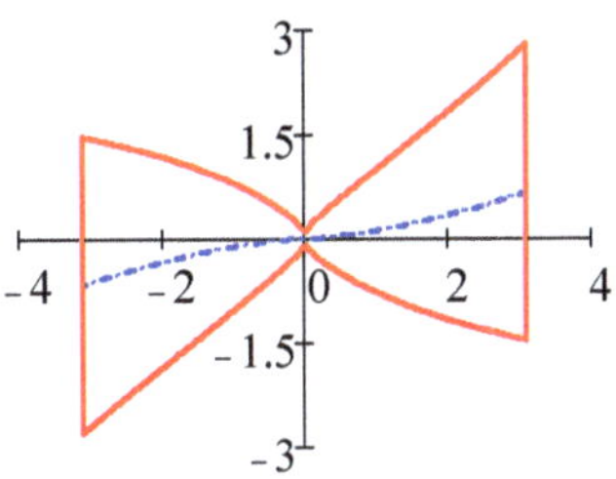

Table 4. *Cont.*

Dissipative Component Moment (nN·m) vs. Angle (arc s) Equation (1)	Elastic Component Moment (nN·m) vs. Angle (arc s) Equation (3)	The Total Moment of Resistance Moment (nN·m) vs. Angle (arc s) Equation (7)
	Hard steel	
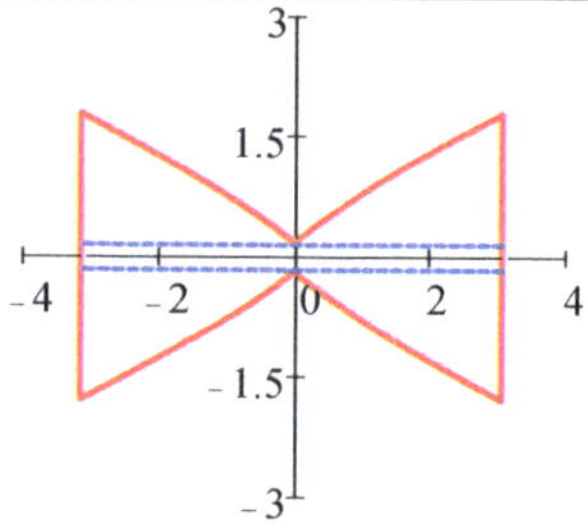 $4A_t = 9.86{\cdot}10^{-15}$ J; $4A = 64.3{\cdot}10^{-15}$ J	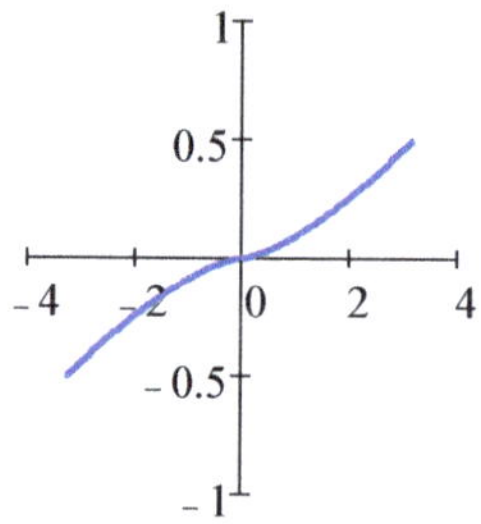	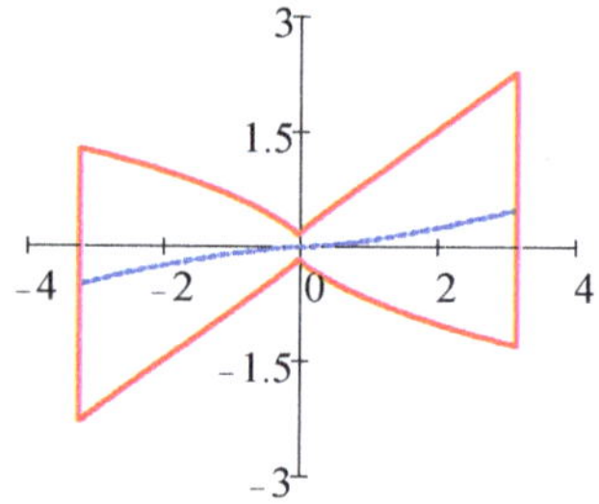

6. Conclusions

In the displacement zone where the dimensions are significantly reduced than the dimensions of the contact spot and under elastic loads, the "deep-pre-rolling" (DPR) zone, rolling resistance is determined by:

- dissipative adhesion forces, when parts of surface bodies that are in contact, leaves each other; in addition, rolling resistance is determined by the frequency-independent internal friction forces, arising during elastic deformations of contacting bodies.
- also, this rolling resistance is determined by the elastic adhesion forces that create negative pressure between the surfaces of the contacting bodies.

In the DPR zone, the spin friction around the vertical axis has a maximum value compared to the swing friction around the horizontal axes. Swing friction around horizontal axes decreases with decreasing swing speed. These friction features allowed us to build a pendulum device based on one ball with a stable swing plane. In this device, the pendulum must have a shape in which the horizontal moments of inertia differ significantly from each other reaching between 25 to 30 as a key condition. Here, the swing of the pendulum with a maximum period has a stable swing plane.

In the DPR zone, there is an effect of a sharp decrease in the swing period of the pendulum with a decrease in the swing amplitude. In this case, the rolling friction also decreases and tends to its minimum final value, determined by the work of adhesion forces on separation. In the study of rolling resistance in the DPR zone, it is necessary to measure not only the reliance of the swing amplitude on the time, but also the reliance of the swing period of the pendulum on time.

The developed phenomenological theory and measurement procedure allowed us for the first time to build a simple instrument for direct measurements with high sensitivity and accuracy of the surface energy density of the adhesion forces (or surface tension) in the case of a solid body, and the parameters of internal frequency-independent friction and the pressure generated by adhesion forces. Key applications are found in precision positioning at a small scale or telescopes at a large scale and also any trials for balancing mechanical devices around a single ball.

Author Contributions: Conceptualization, I.G. and S.M. Methodology, I.G., N.R. and S.M.; validation, I.G., N.R; formal analysis, I.G.; investigation, I.G., N.R. and S.M.; data curation, N.R.; writing—original draft preparation, I.G. and S.M.; writing—review and editing, S.M.; supervision, I.G.; project administration, I.G and S.M. All authors have read and agreed to the published version of the manuscript.

Funding: The research received no external funding.

Data Availability Statement: Research data can be shared through authors of this paper.

Acknowledgments: The authors would like to express their thanks to the support of BNTU and KFUPM through DROC and IRC-IMR center.

Conflicts of Interest: The authors declare no conflict of interest.

References

1. Vicente, F.S.; Guillamón, M.P. Use of the fatigue index to study rolling contact wear. *Wear* **2019**, *436–437*, 203036. [CrossRef]
2. Cross, R. Coulomb's Law for rolling friction. *Amer. J. Phys.* **2016**, *84*, 221–230. [CrossRef]
3. Cherepanov, G.P. The laws of rolling. *Phys. Mesomech.* **2019**, *22*, 242–254. [CrossRef]
4. Popov, V.L. *Contact Mechanics and Friction: Physical Principles and Applications*; Springer: Berlin/Heidelberg, Germany, 2017; pp. 231–253. [CrossRef]
5. Bowden, F.P.; Tabor, D. *The Friction and Lubrication of Solids*; Oxford University Press: New York, NY, USA, 1950; p. 372.
6. Johnson, K.L. *Contact Mechanics*; Cambridge University Press: Cambridge, UK, 1987; p. 52. [CrossRef]
7. Mekid, S. A non-linear model for pre-rolling friction force in precision positioning. *Proc. Inst. Mech. Eng. Part J J. Eng. Tribol.* **2004**, *218*, 305–312. [CrossRef]
8. Tan, X.; Modafe, A.; Ghodssi, R. Measurement and Modeling of Dynamic Rolling Friction in Linear Microball Bearings. *J. Dyn. Syst. Meas. Control* **2006**, *128*, 891–898. [CrossRef]
9. Amthor, A.; Zschaeck, S.; Ament, C. High Precision Position Control Using an Adaptive Friction Compensation Approach. *IEEE Trans. Autom. Control* **2009**, *55*, 274–278. [CrossRef]
10. Benditkis, R.; Necsulescu, D.S. Comments on Rolling Resistance. *J. Tribol.* **1994**, *116*, 658–660. [CrossRef]
11. van Spengen, W.M. MEMS reliability from a failure mechanisms perspective. *Microelectron. Reliab.* **2003**, *43*, 1049–1060. [CrossRef]
12. Mekid, S. Dedicated instruments for nano-engineering education: Integrated nano-manipulation and micro-nanomachining. *Int. J. Mech. Eng. Educ.* **2019**, *49*, 60–71. [CrossRef]
13. Mekid, S.; Bashmal, S. Engineering manipulation at nanoscale: Further functional specifications. *J. Eng. Des. Technol.* **2019**, *17*, 572–590. [CrossRef]
14. Mendeleev, D.I. Opytnoe issledovanie kolebaniya vesov i vozobnovlenie prototipa ili osnovnoj obrazcovoj russkoj mery massy v 1893–1898 gg. (Experimental research: Weight change and the renewal of the prototype or the main exemplary Russian measure of mass in 1893–1898.). *L. Gos. Nauch.-Tekhn. Izd-Vo Lenhimsektor* **1931**, *20*, 302. (In Russian)
15. Herbert, E.G. Some Recent developments in hardness testing. *Engineer* **1923**, *135*, 686–687.
16. Kratkoe rukovodstvo k mayatniku Gerberta dlya ispytaniya tverdosti (Herbert Pendulum Quick Guide for Hardness Testing). Moskva; Mashmetizdat: 1933. 12p. (In Russian)
17. Avdeev, B.A.; Mashgiz, M. Ispytatel'nye mashiny i pribory (Testing machines and devices). **1957**, *35*, 354. (In Russian)
18. Pod Red, L.K.; Martens, M. Tekhnicheskaya enciklopediya (Technical Encyclopedia). *Sov. Enciklopediya* **1929**, *5*, 465. (In Russian)
19. Halama, R.; Podešva, J.; Suzuki, R.; Matsubara, M.; Čech, R. Mechanics of Herbert Pendulum Hardness Tester and its Application. *Key Eng. Mater.* **2017**, *741*, 122–127. [CrossRef]
20. Matsubara, M.; Sakamoto, K. Improved Herbert Hardness Tester. *Exp. Tech.* **2011**, *36*, 73–76. [CrossRef]
21. Suzuki, R.; Kaburagi, T.; Matsubara, M.; Tashiro, T.; Koyama, T. Hardness Measurement for Metals Using Lightweight Herbert Pendulum Hardness Tester With Cylindrical Indenter. *Exp. Tech.* **2015**, *40*, 795–802. [CrossRef]
22. Suzuki, R.; Kaburagi, T.; Matsubara, M.; Setagawa, T.; Shimizu, R. Effect of test load on damping hardness in Herbert hardness test. In Proceedings of the Ibaraki District Conference, Ibaraki, Japan, 1 January 2016; p. 212. [CrossRef]
23. Kuznecov, V.D. Fizika tverdogo tela (Solid state physics). *Tomsk. Krasn. Znamya* **1937**, *5*, 448–480. (In Russian)
24. Heins, R.W.; Street, N. An Evaluation of the Rehbinder-Kuznetsov Pendulum Technique In Hardness Measurements. *Soc. Pet. Eng. J.* **1965**, *5*, 177–183. [CrossRef]
25. Pendulum Hardness Tester/Buiged Laboratory Instruments, Biuged, Guangzhou, China 2019, CO.LTD [Electronic Re-source]. Guangzhou, 2019. Mode of Access. Available online: http://www.biuged.com/En_Pr_d_gci_106_id_105.html (accessed on 7 September 2019).
26. *ISO 1522:2006*; Paints and Varnishes—Pendulum Damping Test. 2006.
27. Tomlinson, G. CVI.A molecular theory of friction. London, Edinburgh, Dublin Philos. *Mag. J. Sci.* **1929**, *7*, 905–939. [CrossRef]
28. Gilavdary, I.; Mekid, S.; Riznookaya, N. Micro-slippage effects in pre-rolling induced by a disturbed and undisturbed pendulum with spherical supports. *Proc. Inst. Mech. Eng. Part J J. Eng. Tribol.* **2013**, *228*, 46–52. [CrossRef]
29. Gilavdary, I.; Mekid, S.; Riznookaya, N. A new theory on pure pre-rolling resistance through pendulum oscillations. *Proc. Inst. Mech. Eng. Part J J. Eng. Tribol.* **2012**, *227*, 618–628. [CrossRef]
30. Szoszkiewicz, R.; Bhushan, B.; Huey, B.D.; Kulik, A.J.; Gremaud, G. Adhesion hysteresis and friction at nanometer and micrometer lengths. *J. Appl. Phys.* **2006**, *99*, 014310. [CrossRef]

31. MathSoft. In *Mathcad 12. User's Guide*; Math-Soft, Inc.: Cambridge, MA, USA, 2004; p. 163.
32. Gilavdary, I.; Mekid, S.; Riznoukaya, N. Measurement of extremely low pre-rolling resistance in dry and lubricating conditions. *Surf. Topogr. Metrol. Prop.* **2020**, *8*, 035004. [CrossRef]
33. Borawski, A.; Szpica, D.; Mieczkowski, G. Research on Tribological Features of Brake Friction Materials—Com-parison of the Results Obtained with the Pin-On-Disc and Ball-Cratering Methods. *Mechanika* **2022**, *28*, 317–322. [CrossRef]

lubricants

MDPI

Article

Efficient Sub-Modeling for Adhesive Wear in Elastic–Plastic Spherical Contacts

Minsi Li, Guo Xiang and Roman Goltsberg *

Department of Mechanical Engineering, Technion, Haifa 32000, Israel; liminsi@campus.technion.ac.il (M.L.); guo.xiang@campus.technion.ac.il (G.X.)
* Correspondence: groman@technion.ac.il

Abstract: This paper presents a novel approach for simulating adhesive wear in elastic–plastic spherical contacts using an improved finite element sub-model. Initially, a global model with a coarse mesh identifies the potential wear region under combined normal loading and tangential displacement. Subsequently, a refined mesh sub-model simulates the crack initiation and propagation until the formation of a wear particle. This refined sub-model efficiently handles a wide range of spherical radii and normal loads. An expression is derived relating the dimensionless wear volume and wear rate to the dimensionless normal load, revealing the limited effect of the sphere radius on the wear rate. The effect of the mechanical properties on the wear particle morphology is also analyzed.

Keywords: spherical contact; adhesive wear; sub-model; combined loading; wear rate

1. Introduction

Wear is the key factor causing material loss and mechanical failure between contacting surfaces, which typically occurs in various tribo-components. There are mainly five types of wear in tribology: abrasive, erosive, corrosive, fatigue and adhesive [1]. Among them, adhesive wear is the least avoidable between contacting surfaces that are subjected to strong adhesive bonds, in which the material removal occurs at the contacting asperities of the mating surfaces [2]. Although very common, an accurate prediction of adhesive wear is still a challenging question in tribology [3]. Hence, developing an effective model to reveal the mechanism of adhesive wear is of great significance.

Archard [4] replaced Holm's concept [5] for the removal of atoms with the removal of spherical wear particles of the same radius as their contact area to describe the adhesive wear process. This alternative approach was supported by experimental evidence [6], showing that the material removal is not at an atomic level.

Archard [4] proposed a model for the adhesive wear, based on a linear relation to calculate the wear volume, which was expressed as $W = KPL/H$ where W is the wear volume, P is the normal load, L is the sliding distance, H is the hardness and K is the wear coefficient.

Due to the simplicity and practicality of Archard's model in engineering applications, it became a well-accepted wear model in the following decades [7]. However, it has inherent limitations, one of which is the lack of a physical understanding of the mechanism of the wear particle formation. Therefore, Archard's model can only give a reliable prediction for cases where the wear coefficient K was already found experimentally, implying that several wear experiments imitating the correct tribological system are required to obtain wear predictions.

Indeed, a lot of work has been done to perform the integrated analysis of the fracture behavior to enable adhesive wear modeling [7]. The pioneering study performed by Hills and Ashelby [8] introduced the classical fracture mechanics to explore the wear behavior between contacting surfaces. In Ref. [8], the crack was initially generated in a pre-selected location. Subsequently, Suh and coworkers [9,10] developed the delamination theory of

Citation: Li, M.; Xiang, G.; Goltsberg, R. Efficient Sub-Modeling for Adhesive Wear in Elastic–Plastic Spherical Contacts. *Lubricants* **2023**, *11*, 228. https://doi.org/10.3390/lubricants11050228

Received: 29 April 2023
Revised: 15 May 2023
Accepted: 16 May 2023
Published: 18 May 2023

wear, in which they assumed the crack nucleation and subsurface crack propagation. Here also, the crack location was predetermined. Although the wear particle formation process can be investigated by a predetermined initial crack location, it limits the crack to propagate from this predetermined location, which may cause an unreliable prediction. Hence, for proper wear modeling, a model that can simulate the crack initiation is required.

The finite element method (FEM) is an efficient method to determine the crack initiation based on the damage mechanism concept. By introducing the Johnson–Cook strain-based ductile fracture criterion [11], Wu and Shi [12] were able to simulate a 2D crack initiation and propagation using a finite element model for a cylindrical contact. According to [12], the crack initiation was at the contact surface, which is in contradiction to the assumption made in the previously mentioned studies [8–10].

To simulate the adhesive wear of 3D spherical contacts an FE model, Zhang and Etsion [13] used the JC criterion, similar to [12], for the crack initiation. However, a different approach was used for the crack propagation in [13]. Specifically, a fracture energy criterion [14] was used, where the fracture energy, G_f, described the energy needed to open a unit area of the crack [14]. By incorporating this physical approach into the adhesive wear simulation, the crack propagation was modeled as a crack opening process rather than the same way as its generation.

In Ref. [13], the static friction and the adhesive wear of a spherical contact under combined normal and tangential loading were presented for both elastic and elastic–plastic regimes for the normal loading of a sphere with a 10 mm radius. In Zhang's study [13], an extremely refined mesh was required to obtain an accurate wear prediction, which was very time consuming, limiting the study to one material and two normal loads. However, some insights regarding the formation of the adhesive wear particle were obtained in both [12,13]. The research showed that the crack initiation started at the trailing edge of the contact surface and then, another crack was initiated at the leading edge. As the tangential displacement increased, the cracks propagated and eventually connected with each other, resulting in the formation of a wear particle. It can be inferred that investigating the adhesive wear using predetermined cracks may not provide accurate results.

In order to improve the computing efficiency of the FE wear model [13], Zhang and Etsion [15] adopted a sub-model technique which enabled them to efficiently calculate the wear volume and wear rate for different normal preloads. While this sub-model was appropriate to characterize the wear particle morphology, the model was not able to present the realistic mechanical response during the wear process, which is required to better analyze the mechanism of the wear particle formation, tangential stiffness and friction behavior.

In the present study, an improved sub-model technique is suggested which innates the advantages of both the models previously proposed by Zhang and Etsion [13,15]. Thus, the present improved sub-model is able to efficiently predict the mechanical response in a spherical contact and the wear particle morphology for different radii, normal preloads and material properties.

2. Theoretical Background

Figure 1, taken from [13], presents a schematic representation of the contact problem of a rigid flat and a homogeneous deformable sphere under combined normal and tangential loading. The contact condition between the rigid flat and the outer surface of the sphere is a full stick. The loading process consists of two stages. At the beginning of the simulation, a normal load P is applied to the rigid flat, resulting in a vertical displacement of the rigid flat, also termed as the interference and denoted by ω_0. Due to the deformation of the sphere tip, a circular contact area A of diameter d_0 is obtained. The characteristics of the contact area, friction coefficient and corresponding critical load, L_c, were studied previously by Brizmer et al. [16,17]. The normal load is usually normalized by the critical normal load at the yield inception under the full stick contact condition, $P^* = P/L_c$, given in Ref. [16].

$$L_c = \bar{L_c}\frac{\pi^3 Y}{6} C_\nu^3 \left(R\left(1-\nu^2\right)\frac{Y}{E}\right)^2 \tag{1}$$

where $\bar{L_c} = \left(8.88\nu - 10.13\left(\nu^2 + 0.089\right)\right)$ and $C_\nu = 1.234 + 1.256\nu$, and R, E, Y and ν are the radius of the sphere, Young's modulus, the yield strength and Poisson's ratio, respectively. With this normalization, the contact is elastic–plastic for $P^* > 1$. For $P^* < 1$, the contact is elastic.

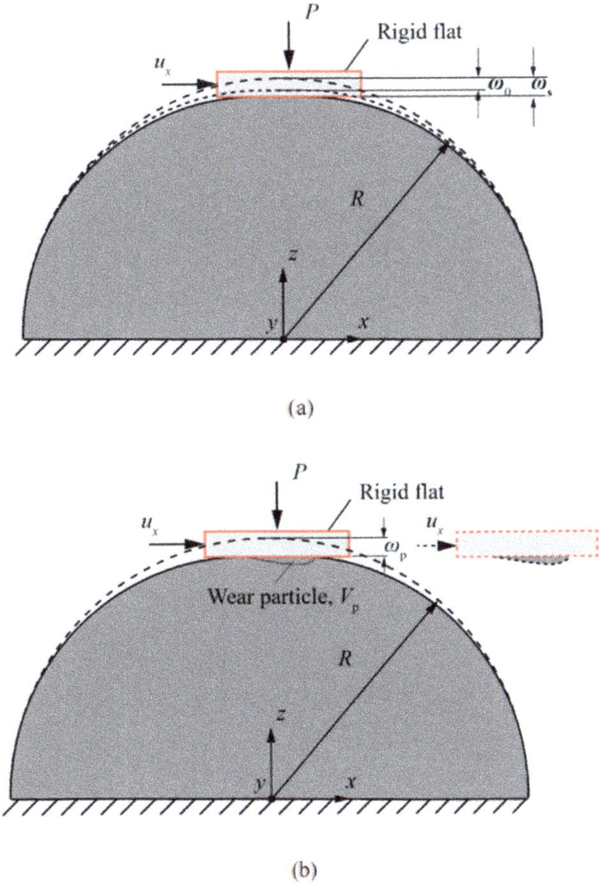

Figure 1. Adhesive wear of the spherical contact with a rigid flat (**a**) schematic and (**b**) cross section (plane y = 0) of the wear particle, taken from [13].

Following the normal preload, a tangential displacement u_x is applied to the rigid flat until a formation of the wear particle is achieved.

Similar to Ref. [13], the adhesive wear propagation is studied using the FEM. To model the fracture process, two fracture criteria are used. First, the Johnson–Cook (JC) criterion [11] is used for the damage initiation of the first crack. According to the JC criterion, the crack occurs when the strain at a material point reaches the equivalent plastic strain ε^{f} defined in [11] as the following equation.

$$\varepsilon^{f} = \left(D_1 + D_2 e^{D_3 \sigma^*}\right)\left(1 + D_4 \ln \dot{\varepsilon}^*\right)\left(1 + D_5 T^*\right) \tag{2}$$

As shown in Equation (2), the plastic strain consists of three parts, which describes the effect of the stress, the strain rate and the temperature, respectively. For a quasi-static and isothermal analysis, the effects of the strain rate and temperature are absent $(D_4 = D_5 = 0)$. Hence, Equation (2) can be reduced to the following.

$$\varepsilon^{f} = D_1 + D_2 e^{D_3 \sigma^*} \tag{3}$$

The equivalent plastic strain ε^{f} in Equation (3) depends on the stress triaxiality σ^*, which is given in the following form.

$$\sigma^* = \sigma_m / \overline{\sigma} \tag{4}$$

where σ_m is the hydrostatic stress and $\overline{\sigma}$ is the von Mises equivalent stress.

Equation (3) and first bracket of Equation (2) follow the form presented by Hancock and Mackenzie [18]. The coefficient D_1 is the crack nucleation strain for the materials that allow for a considerable plastic flow prior to the crack nucleation. D_1 will be zero otherwise. D_2 is a material constant and D_3 expresses how the equivalent plastic strain is affected by the stress triaxiality σ^*. These three parameters can be obtained from the tensile tests [19,20].

The fracture energy G_f [14] for the damage evolution is used to enable the crack propagation. The opening of a crack requires the expenditure of energy to overcome the associated stress. Therefore, the crack opening process leads to the absorption of energy. The amount of energy needed per unit of the crack area in opening the crack from zero to a given width c is calculated using the following equation [14].

$$G_f = \int_0^c \sigma d\omega \tag{5}$$

With this approach, the softening response after the damage initiation, which is characterized by the reduction in the associated stress needed to open a crack as the crack length increases, is expressed by a stress-displacement response rather than by a stress–strain response. In the FEM, this stress-displacement concept requires a definition of a characteristic width c and the characteristic width has size effect on the fracture energy. The fracture energy given by Hu and Wittmann [21] from the tensile tests is a function of the dimensionless crack size, normalized by the specimen width, which shows that the measured fracture energy is significantly influenced by the dimensionless crack size. In order to eliminate the crack size effect, size-independent fracture energy should be used. In a further study the relation for the size-independent fracture energy was given as the following equation [22].

$$G_F = \frac{K_{IC}^2}{E} \tag{6}$$

where K_{IC} is the size-independent specific fracture energy. This equation enables the extraction of fracture energy without experiments and is free of the crack size effect.

In this study, the ABAQUS/explicit 2020 platform was used to predict the adhesive wear behavior. In the given wear modeling problem, this fracture energy is associated with the length of the mesh size in ABAQUS [23] since, in order to open a crack, the crack needs to go through an entire element at the crack tip. The sensitivity of the mesh was evaluated in [15] to determine the impact of the mesh size on the wear volume. It was shown in [15] that below a certain mesh size, a further reduction had a minor effect on the results but significantly increased the computation time.

Using the sub-model technique presented in [15], the boundary conditions of all the surfaces can be obtained from the deformation distribution of the global model. This means that all the surfaces are pre-deformed in the same way as they are in the global model. However, the rigid flat is not included in this sub-model, meaning that the normal load and tangential displacement cannot be calculated. Hence, it is impossible to capture the frictional behavior at the contacting surface. Moreover, although the less efficient global model presented in Ref. [13] was able to capture the friction behavior, as the predefined normal load decreases, a much smaller mesh size is needed, which requires a longer computation time. Therefore, a more efficient sub-model that can overcome the above limitations is suggested hereafter.

3. FE Model

The solution process consisted of two steps. First, a simulation using the global model without element deletion was performed to derive the size and deformation distributions for the sub-model, which is similar to the process described in [15]. In the second step, an improved sub-model was used with the obtained deformation distribution and element deletion to simulate the wear and friction behavior.

Figure 2 presents the global model with different sections consisting of different mesh densities and the location and assembly of the improved sub-model. The mesh design in the global model was the same as the one described in Ref. [13]. As shown in Figure 2a, three different sections are shown. Section I in red was where the potential wear particle was expected and had the finest mesh. Section II and III highlighted in yellow and blue, respectively, had a gradually increasing mesh size with an increasing distance from section I.

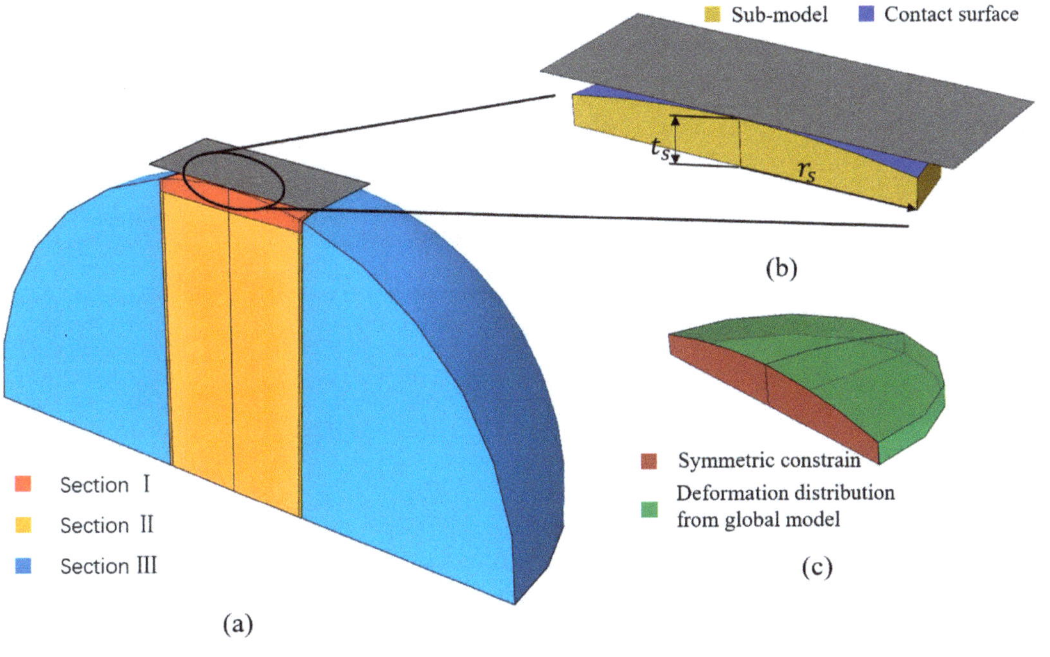

Figure 2. Global model with a mesh design, the improved sub-model assembly and boundary condition: (**a**) global model with a mesh design, (**b**) location of the improved sub-model with sub-model parameters, (**c**) boundary conditions.

Using the sub-model technique [15], only a part of section I was used for the second solution step. This sub-model had a refined mesh compared to the mesh in the global model for better accuracy. As shown in Figure 2b, a plate-like sub-model highlighted in yellow with definitions of its thickness t_s and radius r_s in contact with a rigid flat is shown. The contact surface between the rigid flat and the improved sub-model is highlighted in blue.

The thickness t_s and the radius r_s of the improved sub-model were chosen based on the location of the maximum stress in section I of the global model. Due to the fact that the investigated material model is assumed to be perfectly elastic–plastic, these two parameters indicate the size of the obtained yield region. The thickness t_s and the radius r_s are defined as the vertical and tangential distances from the tip of the sphere to where the yield region is included.

These two parameters were slightly larger than those in [15], as their values were selected based on the location of the maximum strain found in the global model. Furthermore, using the geometry of the sub-model similar to [15] will result in the crack propagating beyond the sub-model boundary. This happens, since the element deletion was not included in the deformation distribution applied on all the surfaces of the sub-model in [15]. Thus, the interference in [15] was reduced and the thickness of the sub-model necessary to prevent the crack from penetrating the sub-model boundary was smaller.

Investigation on the effect of these two parameters, t_s and r_s, was performed by separately conducting simulations with twice the thickness and radius. Although the differences in the wear volume and static friction coefficient were less than 7%, the computational time almost doubled. This indicated that the sub-model's size determined by the maximum stress location was sufficient to effectively capture the wear process.

Figure 2c shows the boundary conditions applied to the improved sub-model. The contact surface was left free of constraints. A symmetric constraint was imposed on the symmetrical surface, highlighted in red, while the deformation distribution from the global model was applied to the side and bottom of the surface, highlighted in green. The deformation distribution from the global model serves two purposes: primarily, to produce contact between the plate-like sub-model and the rigid flat and to eliminate the effects caused by the geometry change, as the stress field depends on the geometry and the improved sub-model is not in a spherical shape.

It should be mentioned that an alternative solution was attempted, where instead of using the deformation distribution from the global model, fixed boundary conditions were used. However, these boundary conditions resulted in a significantly different crack initiation and propagation compared to Ref. [13]. As a result of the fixed boundary condition on the bottom, the sub-model was unable to deform in the normal loading direction, which led to a 50% reduction in the interference. Although the tangential behavior was primarily dictated by the behavior of the contact surface, it was also heavily influenced by the fixed boundary condition on the side, as this condition significantly altered the stress field. Moreover, the comparison presented in the results section demonstrates that the friction behavior obtained using the improved sub-model, which incorporated the deformation distribution on both the bottom and side, closely matched the results obtained in [13] using the global model. Therefore, the deformation distribution boundary conditions are necessary for accuracy and efficiency.

Finally, a full stick contact condition was defined between the rigid flat and the sphere. After the crack initiates, one more interaction must be defined, as there is a newly formed surface due to the failed elements deletion from the mesh. Eventually, a wear particle may form between the rigid flat and the slip interface, which is the interface between the newly formed wear particle and the bulk of the sphere. The interaction property applied on the slip interface is assumed to be frictionless to simplify and simulate. Thus, a shear crack cannot transmit shear or normal tensile stresses [24].

The elements with an aspect ratio close to one were used to reduce the mesh sensitively to the crack propagating direction; representing the uniform mesh as was also shown in [15]. In the present study, the mesh independence check was performed by increasing the mesh density and guaranteeing the relative error within a small, predetermined tolerance (10%) for all the normal loads and sphere radii. For instance, using R = 10 [mm] and P^* = 100, changing the mesh size from 0.01 mm to 0.005 mm resulted in a change of the results below 9%. However, the calculating time of the latter was approximately twice that of the former. Therefore, the mesh size of 0.01 mm was used for R = 10 [mm] and P^* = 100.

4. Results and Discussion

The adhesive wear and friction behaviors of the spherical contact under combined loading were studied by carrying out wear simulations using the improved sub-model for the sphere radius ranging from 5 [mm] to 50 [mm], with dimensionless normal loads P^* ranging from 15 to 150. It was found in the current model that the tangential displacement $u_x = 9\omega_0$ was sufficient to achieve a wear particle for all the different simulations. It was also observed in [13] with the global model. Hence, to ensure the full formation of the wear particle, a value of $u_x = 12\omega_0$ was used for all the simulations. Similarly, in Refs. [13,15], the sphere material was aluminum 2024 T351, as this particular material was thoroughly studied both numerically and experimentally [25–28]. The material properties of aluminum 2024 T351 are: $E = 74$ Gpa, $v = 0.33$, $Y = 325$ Mpa and a density of $\rho = 2780\ \mathrm{Kg/m^3}$. The JC criterion coefficients of the damage initiation for aluminum 2024 T351 are (Ref. [28]): $D_1 = 0.13$, $D_2 = 0.13$ and $D_3 = -1.5$. For the damage evolution, the fracture energy was $G_f = 20\ \mathrm{KJ/m^2}$ [28].

In the following, the verification of the improved sub-model is presented, followed by the investigation of the tangential mechanical response and the wear particle morphology for the different normal loads and sphere radii. The effect of changing the different mechanical properties ($D_1, D_2, D_3, G_f, E/Y$) on the obtained results is also discussed.

4.1. Verification of the Improved Sub-Model

The results from Ref. [13] were used to verify the adequacy of the present model. Figure 3 shows the five instants for the fracture evolution obtained by the improved sub-model, corresponding to the five instants obtained in Ref. [13] for the $R = 10$ [mm] and $P^* = 100$ case with the definition of the wear particle parameters. The black color represents the free edge formed during the fracture evolution process; the deeper blue color represents the crack formed below the contact surface. Figure 4 shows the friction behavior and the interference for the same case, as shown in Figure 3. Both the dimensionless tangential force Q/P and the dimensionless interference ω/ω_0 are plotted with respect to the dimensionless tangential displacement u_x/ω_0. The five instants of the fracture evolution are marked in vertical dash line.

For instant I_A, the crack initially appeared on the trailing edge of the contact surface. The term 'trailing edge' refers to the edge located in the direction opposite to the tangential displacement u_x, as indicated by the arrow shown in Figure 3a pointing in the u_x direction. With an increasing tangential displacement, the crack spread along the edge of the growing contact interface, and at instant I_B, the tangential force reached its maximum, resulting in a sliding inception. At instant I_C, the first crack at the leading edge under the contact surface appeared. With the fracture extending below the contact surface, the fracture from the trailing edge linked with the one from the leading edge at instant I_D, as shown in Figure 3d. A region of un-failed elements that were not deleted according to the JC and fracture energy criteria was surrounded by the linked cracks. Eventually, at instant I_E the un-failed region vanished leading to the wear particle creation. These five stages for the fracture evolution using the improved sub-model matched well with the results given in [13]. The differences in the corresponding tangential displacement for the five instants, as shown in Figure 4, were 4%, 6.8%, 5.4%, 2.3% and 1.5%, respectively.

Along with the crack initiation and propagation, the tangential force behavior of the improved sub-model corresponded well to the results from [13]. The tangential stiffness reduced with the increasing tangential displacement u_x until instant I_B, at which the tangential stiffness completely vanished. The static friction coefficient represented by Q_{max}/P was 0.36 for the improved sub-model and 0.35 in [13]. The relative error of the static friction coefficient between these two models was less than 5%.

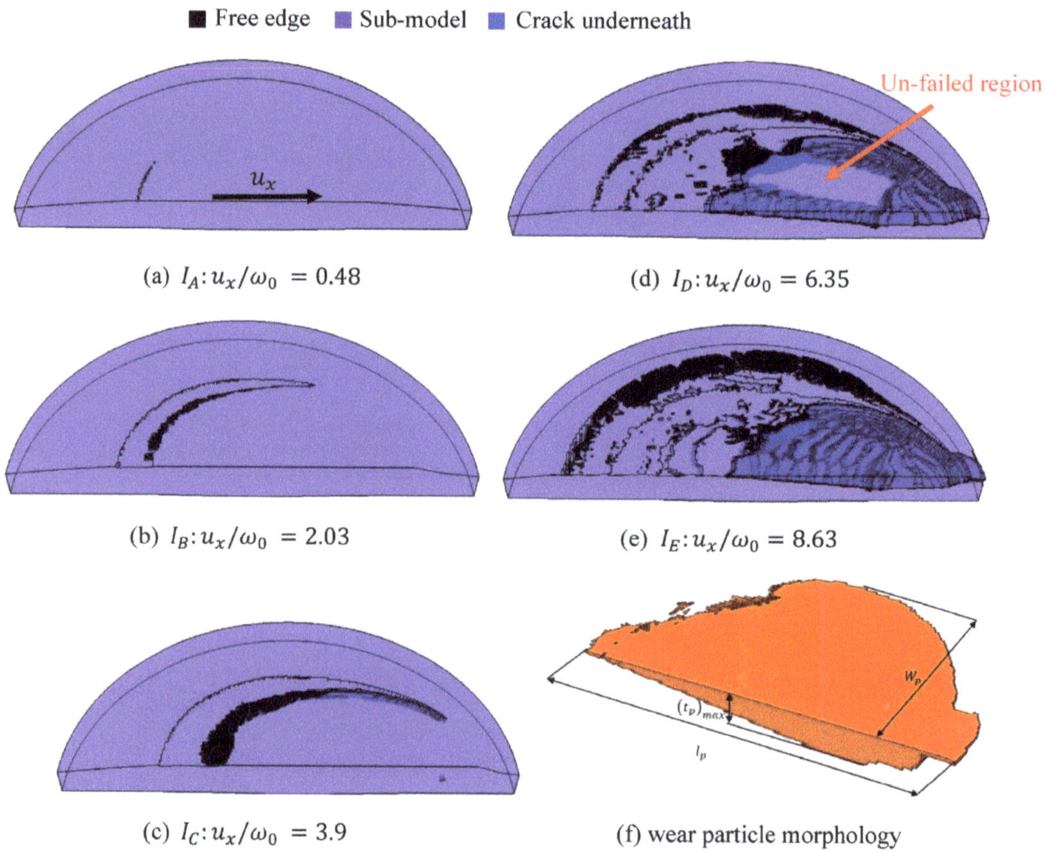

Figure 3. Evolution of the fracture, from instants I_A to I_E for $R = 10$ [mm], $P^* = 100$.

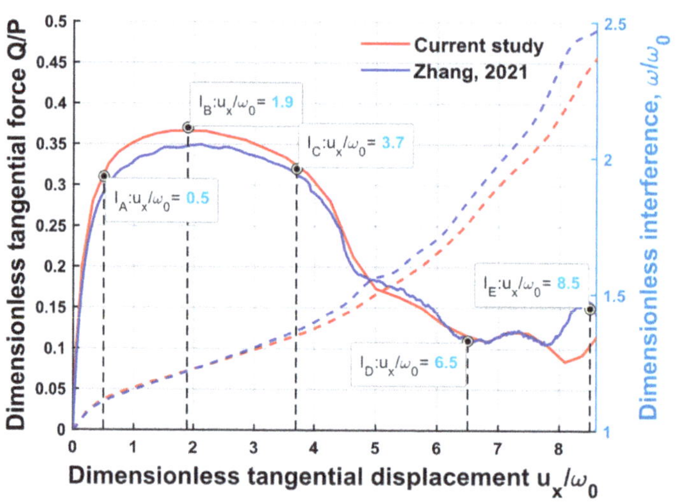

Figure 4. Dimensionless tangential force Q/P and dimensionless interference ω/ω_0 vs. dimensionless tangential displacement u_x/ω_0 for $R = 10$ [mm], $P^* = 100$, comparing with Ref. [13].

Regarding the comparison between the dimensionless interference shown in Figure 4, the difference between [13] and the current result increased with the dimensionless displacement, which was due to the element deletion, as the element size used in the improved sub-model was larger than in [13]. However, the largest difference was less than 7%.

The wear particles formed in [13] and the one formed using the improved sub-model were similar in shape, having the shape shown in Figure 3f. The maximum thickness and length of the wear particle in [13] were $(t_p)_{max} = 0.012R$ and $l_p = 0.16R$, with a particle volume of $V_p = 4.3e^{-2}$ $[\text{mm}^3]$. The corresponding value using the improved sub-model were $(t_p)_{max} = 0.0133R, l_P = 0.155R$ and $V_p = 4.36e^{-2}$ $[\text{mm}^3]$. The differences between these three parameters were less than 10%, especially for the wear particle volume having a 2% difference. An additional parameter was also denoted in Figure 3f, which was the wear particle half width W_p, being $W_p = 0.063R$ for this case.

To conclude, the results predicted by the improved sub-model were in a good agreement with the predictions made in [13], validating the improved sub-model for predicting the tangential behavior wear particle geometry. These results verified that the new sub-model can provide a reliable prediction for the wear morphology and the mechanical response, thereby validating the effectiveness of the present model.

The computation time was reduced from 90 h to 30 h, compared to the model used in [13]. A further reduction in the computation time can be achieved by reasonably increasing the loading rate of the simulation, both in the normal and tangential directions. Moreover, as shown in Figure 3, it can be observed that some non-contact and non-wear regions existed, which had no effect on the wear predictions when the tangential loading was terminated, indicating that the current sub-model simulating area can be further reduced for the purpose of a parametric study. Therefore, using a smaller sub-model simulating region compared to the present one, which fully covers the potential wear and contact regions and reasonably increases the loading rate, can further reduce the simulating time. For R = 10 [mm] and P^* = 100, using the above accelerating techniques reduced the computing time from 30 h to 15 h, and the relative errors of the wear particle volume and the static friction coefficient compared to the current improved sub-model were less than 3%.

4.2. The Effect of the Normal Load P^*

Table 1 summarizes the parameters used in the sub-model to study the effect of the normal load on the wear volume and the tangential load. The elements r_1 and t_1 are the radius and thickness of section I in the global model (Figure 2a), respectively. The elements a_1 and a_s are the mesh sizes of section I in the global model and in the improved sub-model, respectively. Superscript $*$ is used to indicate the dimensionless parameter, normalized by the sphere radius R.

Table 1. Geometrical and mesh parameters used in the global and improved sub-model.

P^*	$r_1^*(r_1/R)$	$t_1^*(t_1/R)$	$a_1^*(a_1/R)$	$r_s^*(r_s/R)$	$t_s^*(t_s/R)$	$a_s^*(a_s/R)$
15	0.1	0.015	0.0015	0.07	0.008	5×10^{-4}
20 to 30	0.1	0.018	0.002	0.08	0.012	6×10^{-4}
50	0.13	0.022	0.002	0.1	0.02	7×10^{-4}
75	0.2	0.05	0.003	0.14	0.026	0.001
100	0.2	0.05	0.004	0.15	0.038	0.001
150	0.25	0.07	0.004	0.19	0.05	0.002

For accuracy, the mesh size a_1 should be decreased as the applied normal load decreases. On the other hand, in order to reduce the number of elements and the computation time, the radius and thickness of section I in the global model were reduced as well since decreasing the normal load resulted in a smaller potential region for the wear particle.

The parameters used in the sub-model were adjusted according to the stress field in the global model. As the simulations were conducted for the different radii, ranging from 5 to 50 mm, the parameters were normalized using the sphere radius R.

Figure 5 presents the results for the dimensionless tangential load Q/P and the dimensionless interference ω/ω_0 vs. the dimensionless tangential displacement u_x/ω_0 for different dimensionless normal loads P^* and different radii R. Figure 5 demonstrates that the transient dimensionless tangential forces and interferences exhibited a similar trend for the given material properties under different sphere radii and normal loads.

(**a**)

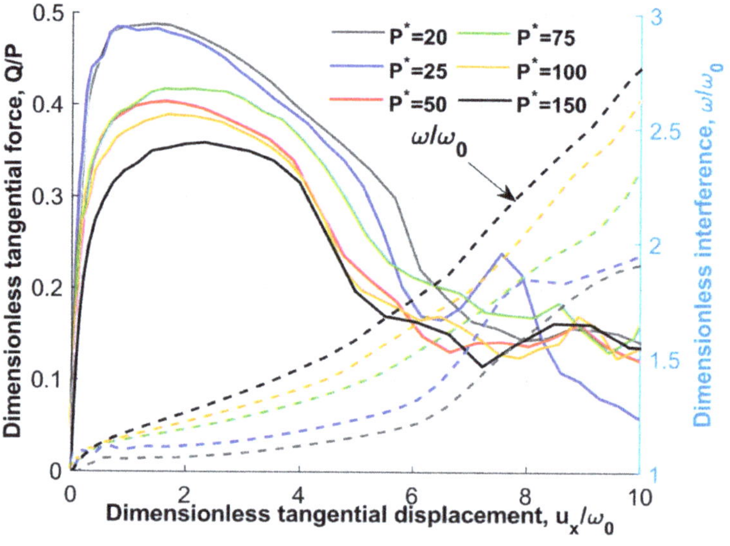

(**b**)

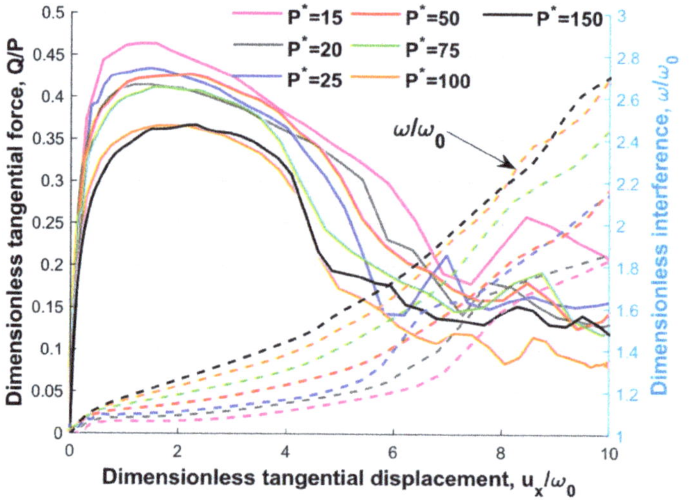

Figure 5. *Cont.*

(c)

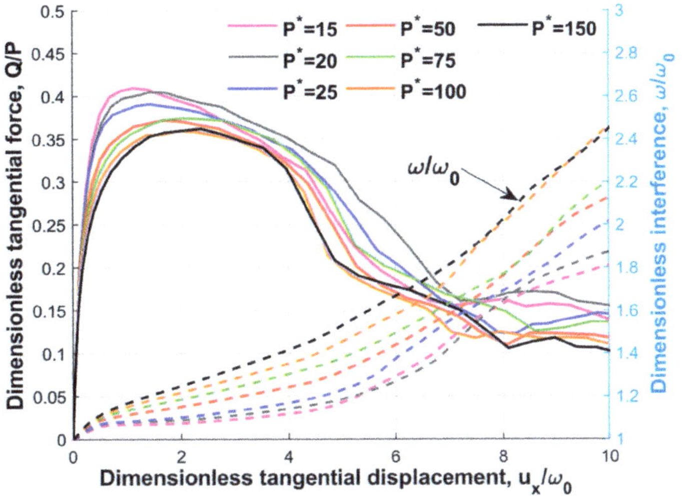

Figure 5. Dimensionless tangential load Q/P and dimensionless interference ω/ω_0 vs. dimensionless tangential displacement u_x/ω_0 for different normal loads and radii R, (**a**) R = 5 [mm], (**b**) R = 10 [mm], (**c**) R = 50 [mm].

The dimensionless tangential force curves exhibited a certain non-monotonic behavior, particularly following the formation of the wear particles. These non-monotonic fluctuations were caused by the inherent numerical error. More specifically, the deletion and recontacting of the elements on the newly generated surface during the tangential loading led to the observed non-monotonic fluctuations. While it was impossible to entirely eliminate these fluctuations, they were reduced by implementing a more refined mesh. Similar fluctuations were also reported in Ref. [13]. Figure 4 provides a comparison that demonstrates that the fluctuation levels were within an acceptable range when compared to the previously published results.

As shown in Figure 5, the curves for the larger sphere radius were more consolidated. This behavior was caused by the dimensionless sub-model parameters (r_s^*, t_s^*). Although these two parameters allowed for the wear particle formation, the sub-model geometry might not have been large enough to exclude the influence of the applied boundary conditions. A few simulations of R = 5 [mm] were carried out using larger geometry, and the friction behavior was close to the friction behavior of R = 50 [mm]. It should be noted that the wear volume and wear particle parameters had a slight difference, below 10%. Since, as will be shown, the dimensionless results were independent of the radius, it was more beneficial to look at the larger radii as long as the adhesion, which was associated with nano and micron scale, was not considered.

Furthermore, it can be observed that the dimensionless interference during tangential loading increased with the increase in the dimensionless normal load. This can be explained by the junction growth under the combined loading [29].

Figure 6 illustrates the maximum dimensionless tangential forces, which indicate the static friction coefficient corresponding to the different sphere radii plotted against the dimensionless normal load. In general, the static friction coefficient decreased with the increasing dimensionless normal load, which showed a consistent result with [17]. A comparison between the results presented in [17] is also shown in percentages by the dashed lines. The differences for the small normal load, $15 < P^* < 25$, were lower than 10%; however, for the larger normal loads, $50 < P^* < 150$, the differences became larger, but were no more than 35% with respect to [17].

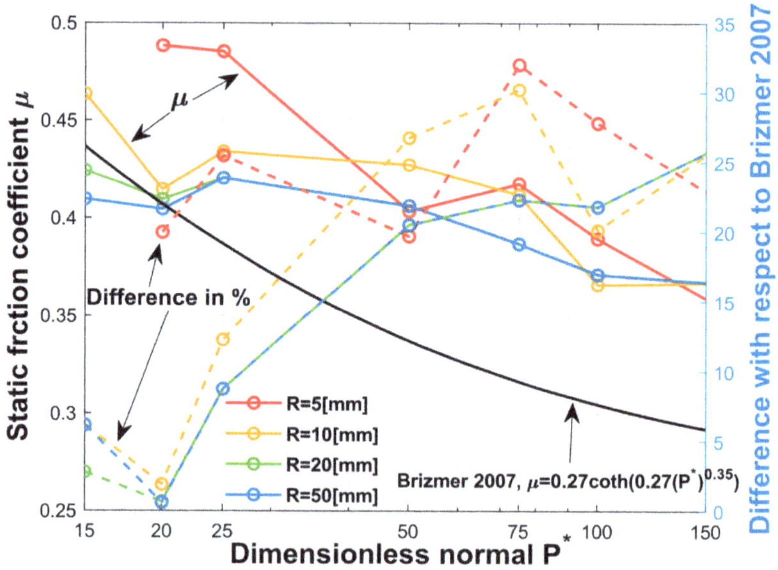

Figure 6. Static friction μ vs. dimensionless normal load P^* for the different radii and comparison with Ref. [17].

As shown in Figure 7, the maximum tangential force and the corresponding normal load are shown for the different sphere radii, where the different colors represent the different sphere radii.

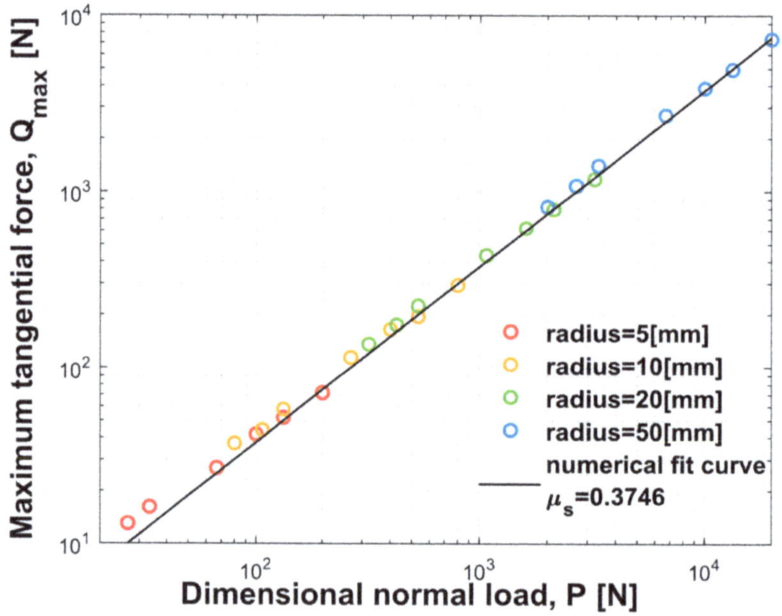

Figure 7. Maximum tangential force vs. normal load for the different radii and different normal loads.

A numerical fitted curve, in the black solid line, with $Q_{max} = \mu_s * P$ resulted in $\mu_s = 0.37$ for all the simulations conducted in the current study, regardless of the sphere radius. Interestingly, in Rabinowicz [30], for 50 separate experiments with different material combinations, a lower limit for the friction coefficient, approx. 0.37, was reported.

4.3. Wear Characteristics

The effects of the normal load intensity and the sphere radius on the wear particle morphology and wear rate were also analyzed. The geometry of a typical wear particle is presented in Figure 3f and the obtained results for the wear particle length l_p, thickness t_p and half width W_p are shown in Figure 8, in both dimensional and dimensionless manners. As shown in Figure 8, the dimensional and dimensionless parameters are indicated by the solid and the dashed lines, respectively. The same normalization as in [15] was adopted in the current study. All three wear particle parameters were normalized using the diameter of the contact area after the normal loading, d_0, which is a function of the radius and the interference in the following formula [16].

$$d_0 = 2\sqrt{\omega_0 R} \tag{7}$$

It can be seen in Figure 8 that all the dimensional parameters of the wear particle increased as P^* increased, indicating that increasing the normal load led to an increase in the wear particle size.

A consolidation of the dimensionless results can be observed. With the increases in the dimensionless normal load, the dimensionless wear particle length remained constant at a value of one. The behavior of the dimensionless wear particle length in this study was in contrast to that reported in [15], where l_p^* was observed to increase with the dimensionless normal load caused by the junction growth. This difference can be attributed to the deletion of elements on the trailing edge of the contact surface in the current sub-model. Following the initiation of the crack, the element adjacent to the crack, as shown in Figure 3a, continued to be deleted from the wear particle. This deletion of the adjacent element ultimately resulted in a reduction in the wear particle length. Although some deletion occurred due to the small thickness at the trailing edge, it didn't affect the prediction of the wear volume.

As shown in Figure 8b, the dimensionless wear particle thickness increased with the dimensionless normal load, and this behavior corelates well with [15]. The numerical curve fitting shown in the black solid line provided a function t_p^* with respect to the P^* with a goodness of $R^2 = 0.97$.

$$t_p^* = 7.1e^{-4}P^* + 0.021 \tag{8}$$

It can be observed that the predicted thickness was slightly higher compared to the results in Ref. [15]. The differences in t_p were caused by the differences in the obtained interferences in both sub-models. Including the rigid flat in the sub-model resulted in a slightly increased interference. As the interference increased, so did the size of the plastic region [31], forcing the crack to propagate more in the vertical direction and resulting in a thicker wear particle.

As shown in Figure 8c, the dimensionless wear particle half width was almost constant with a value of $W_p^* = 0.5$, meaning half of the contact diameter d_0. This was due to the growth of the junctions primarily affecting the contact area in the tangential loading direction, which was perpendicular to the width direction of the wear particle. As a result, the wear particle width remained unchanged.

The wear volume V_p is defined as the volume of the wear particle. The wear rate w represents the wear volume per unit of tangential displacement, and it could be used for surface wear evaluation.

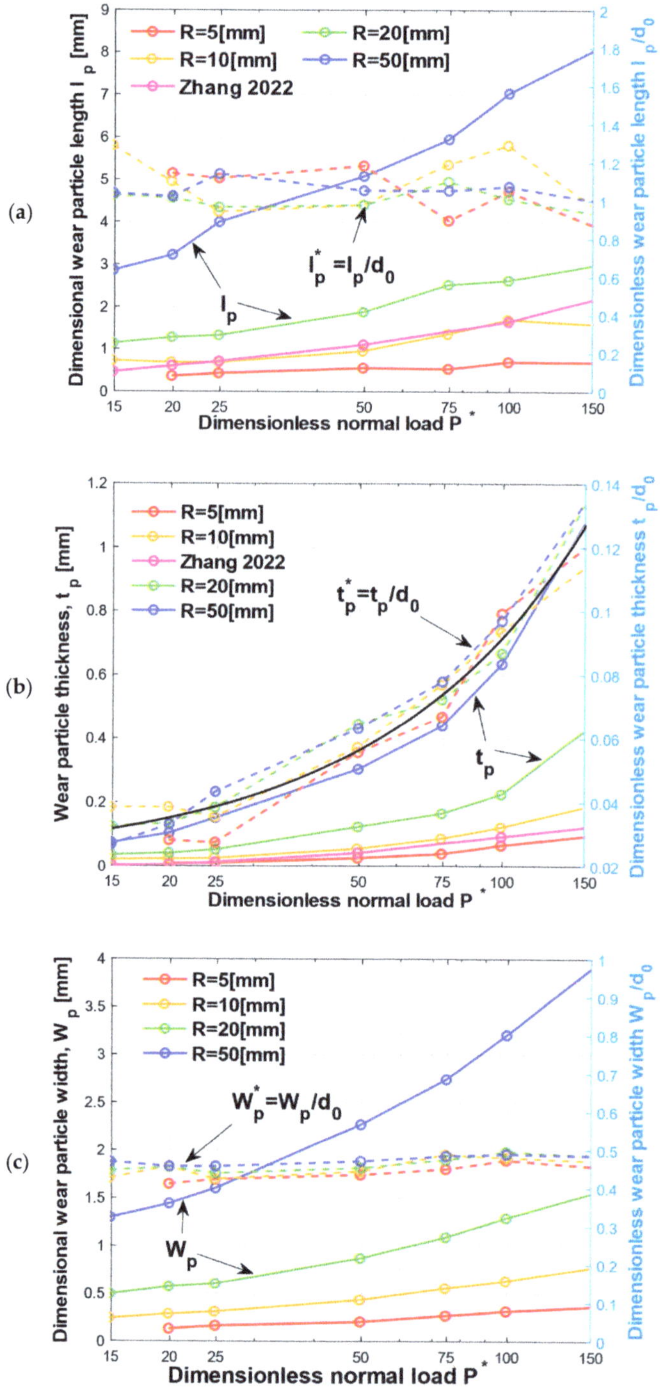

Figure 8. Wear particle geometrical characteristics: (**a**) wear particle length, (**b**) wear particle thickness, (**c**) wear particle half width, presented in both dimensional and dimensionless manners, comparing with Ref [15].

The wear rate is given by the following equation.

$$w = V_p / s \tag{9}$$

where s is the sliding distance when the wear particle is fully detached from the sphere and is calculated as the following.

$$s = u_p + l_p \tag{10}$$

where u_p is the tangential displacement of the rigid flat when the wear particle is formed. It is noteworthy that the simulation ran until the tangential displacement of u_p and not s was achieved. With this method, the computational time can be reduced while ensuring a more realistic result. Obtaining a complete detachment of the wear particles by defining the sliding distance for the simulation required an extensive computational time and led to an unphysical wear volume. This was because the element deletion driven by the JC criterion and fracture energy continued after the wear particle was fully formed (at tangential displacement u_p). This led to an additional element removal from the wear particle during the detachment. Consequently, the wear volume for the complete detachment was smaller than the wear particle volume V_p when it was just formed. The previous studies on the adhesive wear also utilized the same concept [13,15,32].

Figure 9 presents the results of the wear volume V_p and the wear rate w for the various dimensionless normal loads P^* and different sphere radii R. The wear volume was normalized by the hemisphere volume, $V_p^* = V_p / V_0$, where $V_0 = 2\pi R^3/3$. The wear rate was normalized term by term, where the nominator was normalized by the hemisphere volume and the denominator was normalized by ω_0.

$$w^* = V_p^* / \left(\left(u_p + l_p \right) / \omega_0 \right) \tag{11}$$

This normalization led to a general relation between the dimensionless wear rate w^* and the dimensionless normal load P^*, allowing for the prediction of the wear rate and the wear volume independent of the radius for aluminum 2024 T351. The only parameter required to obtain the dimensionless wear rate was ω_0, which can be calculated by following equation [33].

$$P^* = (\omega^*)^{3/2} \left(1 - \exp\left(\frac{1}{1 - (\omega^*)^{\beta}} \right) \right) \tag{12}$$

where β is the linear function of Poisson's ratio, $\beta = 0.174 + 0.08\nu$, and ω^* is normalized by the critical interference δ_c for the stick contact condition, given by [33].

$$\delta_c / \omega_c = 6.82\nu - 7.83\left(\nu^2 + 0.0586 \right) \tag{13}$$

where ω_c is the critical interference at the yield inception for a slip contact condition [33].

$$\omega_c = \left(C_v \frac{\pi (1 - \nu^2)}{2} \left(\frac{Y}{E} \right) \right)^2 R \tag{14}$$

Figure 9a shows that the wear volume increased as P^* and R increased. The dimensionless wear volume curves were consolidated. In order to demonstrate the relation between V_p^* and P^*, a numerical curve fitting with a goodness of $R^2 = 0.997$ was given as following equation.

$$V_p^* = 3.819 \times 10^{-9} (P^*)^2 \tag{15}$$

By using Equation (15) and the normalization $V_p^* = V_p / V_0$, the dimensional wear volume can be calculated. The obtained result corelated well with the result in [15], with a difference less than 9%.

Figure 9b shows a similar behavior for the wear volume. The relation between w^* and P^* were given by a numerical fitted curve with a goodness of R^2 = 0.989.

$$w^* = 6.39\times10^{-12}(P^*)^{2.5} \tag{16}$$

By substituting Equations (12)–(14) into Equation (16), the dimensional wear rate can be calculated. The presented model deals with the creation of a wear particle from a well-defined spherical asperity. The other shapes of the contacting asperities or post-abrasive asperities were not examined in the present study.

(a)

(b)

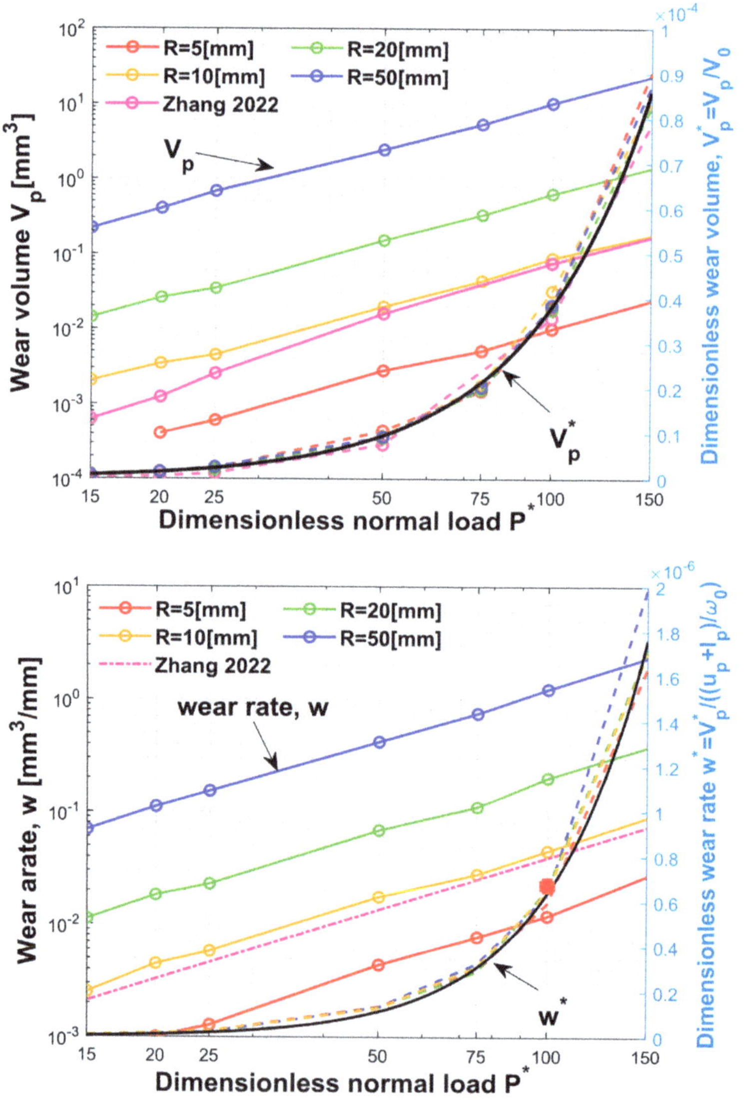

Figure 9. (**a**) Wear volume V_P and dimensionless wear volume $V_p^* = V_p/V_0$; (**b**) wear rate $w = V_p/s$ and dimensionless wear rate $w^* = V_p^*/((u_p + l_p)/\omega_0)$.

Since there is an exchange of asperities in in the contact between rough surfaces in relative motion during the tangential displacement, and assuming that all these asperities are spherical, the wear rate for the entire rough surface was derived from the uniformed dimensionless result of one asperity obtained in this study.

The power-law dependency on the dimensionless normal load for the wear rate reported in [15] is given by the following equation.

$$w = 3.26\times10^{-5}(P^*)^{1.54} \tag{17}$$

The dimensionless relation obtained in the current study had the power of 2.5 in P^* term. The reason for this difference was the normalization used for the dimensionless wear rate. The normalization used for the dimensionless wear rate was $w^* = (V_P/V_0)/(s/\omega_0)$. The denominator was normalized by ω_0, which was affected by the dimensionless normal load, as shown in Equation (12). The numerator was normalized by V_0, which was a constant. Hence, in Equation (16), a higher power on P^* term was obtained.

To verify the expression obtained above, a simulation using the sphere radius, $R = 15$ [mm], was carried out for a dimensionless normal $P^* = 100$. The dimensionless wear rate was 6.754×10^{-7}, and the relative difference between the simulation result and the predicted results from Equation (16) was approx. 2.5%.

In Ref. [15], a transition region was observed for the wear rates. Three regimes of the wear rate with different exponents of P^* were identified, which were $P^* \leq 20, 20 < P^* \leq 30$ and $P^* \geq 30$, representing the mild wear, transition region and severe wear, respectively. However, with the improved sub-model, this transition region was not observed. The power-law dependency on the dimensionless normal load for the wear rate w in the current study is given by following equation.

$$\begin{cases} w = 4.6\times10^{-5}(P^*)^{1.5}, \ P^* < 30 \\ w = 2.74\times10^{-5}(P^*)^{1.6}, \ P^* > 30 \end{cases} \tag{18}$$

The change in the power of P^* was not as significant between the two regions compared to what was reported in [15]. This difference was due to the different boundary condition at the contact surface of the sub-model. The deformation distribution used on the contact surface in [15] did not include the crack initiation and propagation, resulting in a smaller interference for the same normal load in the transition region. For instance, for $R = 10$ [mm] with $P^* = 25$, the dimensionless interference when the wear particle was formed $\omega_p^* = \omega_p/\omega_0$ was 1.2 in [15], while in the current study $\omega_p^* = 2.1$.

4.4. Parametric Study for the Material Properties

The results presented above were limited to a specific material, aluminum 2024 T351. To obtain a more general understanding of how the material properties affected the wear particle formation and mechanical behavior, a parametric study was utilized. The initiation and propagation of a crack was influenced by the three coefficients D_1, D_2 and D_3 in the JC failure model and by the fracture energy G_f.

The range of the studied coefficients D_1 and D_2 was from 0.05 to 0.2, and D_3 ranging from -1 to -2. These ranges covered the changes of approx. 50% of the original values for the aluminum analyzed in the current study. However, different materials can have values outside this range, for example FeCoNiCr high entropy alloy [34] had values of $D_1 = 0.004$, $D_2 = 0.665$, and $D_3 = -1.5$ and free-cutting steel 50SiB8 [35] had values of $D_1 = 0.0733$, $D_2 = 0.7204$, and $D_3 = -1.5643$.

The effect of D_1 on the friction behavior and wear volume V_p was very small, and the maximum difference of the static friction coefficient and wear volume was less than 5% and 6%, respectively, for the range of the tested D_1 values. D_2 had a relatively large influence on the wear volume. The maximum change in the wear volume was 30% for the smallest value of D_2.

For D_1 and D_2, the most affected result was the wear particle thickness t_p. As these two parameters decreased, t_p increased. As shown in Equation (3), D_1 and D_2 influenced the equivalent plastic strain in the same way. As D_1 and D_2 decreased, ε_f decreased. For the spherical contact under normal loading, the stress triaxiality σ^* had a negative value as the element was being compressed, and it increased with the distance from the contact surface. This implies that the equivalent plastic strain for the crack generation decreased with an increasing distance from the contact surface. By reducing ε_f, the crack propagation in the vertical direction became higher. Hence, the wear particle thickness increased as D_1 and D_2 decreased.

For the influence of D_3, the most affected values were u_p and V_p. A maximum increases in u_p about 60% was observed, compared to the original values of u_p and V_p for the smallest value of D_3. As D_3 decreased, the equivalent plastic strain increased for the negative value of σ^*, meaning that a larger strain is need for the element deletion, which explains the increase in u_p. The wear volume change was caused by an additional element deletion during the larger tangential displacement., which may have resulted in fewer physical results.

Figure 10 displays the friction behavior and dimensionless interference for various values of the fracture energy. While the JC parameters were more available in the literature, the same was not observed for the fracture energy. Hence, for the current parametric study on G_f, a range spanning two orders of magnitude around the fracture energy of aluminum was used. It is important to note that the fracture energy values shown in Figure 10 were normalized by the fracture energy of aluminum 2024 T351.

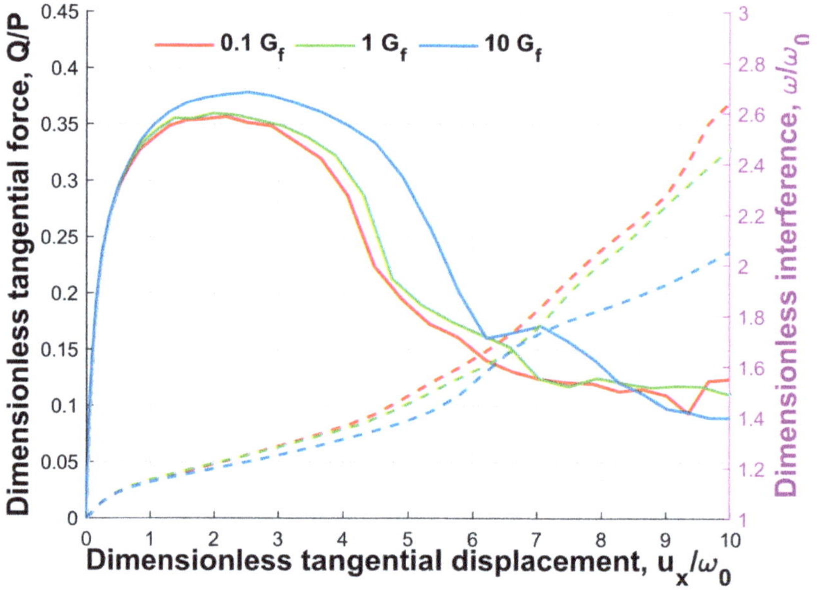

Figure 10. Dimensionless tangential load Q/P and dimensionless interference ω/ω_0 vs. dimensionless tangential displacement u_x/ω_0 for the different fracture energies.

It can be observed from Figure 10 that the static friction coefficient increased with the fracture energy. As shown in Figure 3, the maximum value of the dimensionless tangential force, or the static friction coefficient, was achieved after the crack initiation. This suggests that the fracture energy had an impact on the static friction coefficient. Since the force required to open a crack increased with the fracture energy, the static friction increased as well. However, it should be noted that the influence of the fracture energy was relatively small, with a maximum difference of only 6.2% observed between the two orders of magnitude of the fracture energy.

In contrast, the dimensionless interference decreased with the fracture energy. The change in the interference consisted of two contributors, namely the junction growth and element deletion. With an increase in the fracture energy, the number of elements deleted decreased, leading to a reduction in the dimensionless interference.

While changing D_1, D_2, D_3 and G_f caused some changes to the wear particle morphology, the wear particle formation process for the different parameters was similar to those presented in Figure 3.

However, for a different material, Ti-6Al-4V, a difference in the crack propagation behavior was observed, as presented in Figure 11. The material properties were [36]: E = 110 Gpa, ν = 0.33, Y = 880 Mpa and a density of ρ = 4430 Kg/m^3. The JC criterion [36] and the fracture energy [37] were: $D_1 = -0.09$, $D_2 = 0.25$, $D_3 = -0.5$ and $G_f = 47$ KJ/m^2.

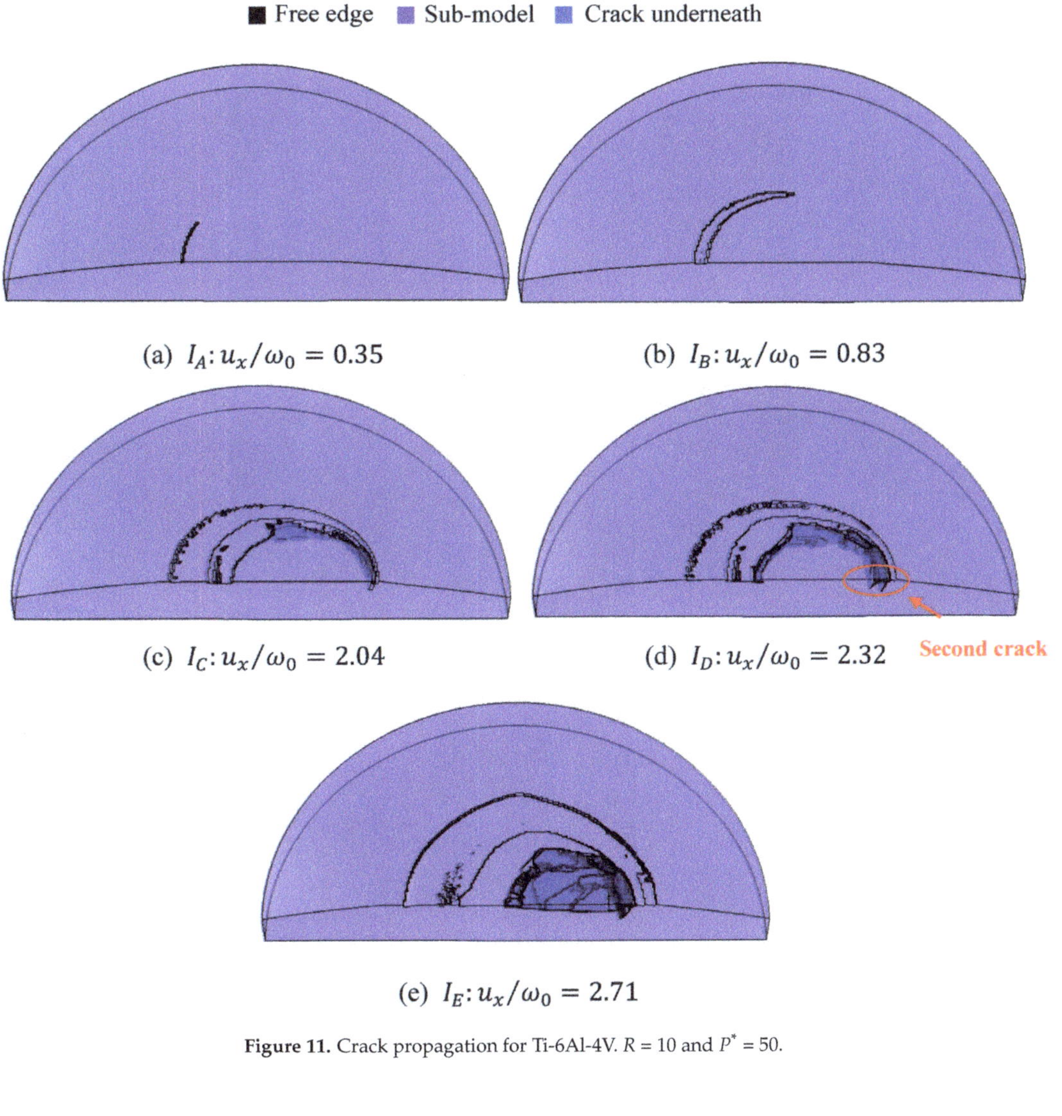

Figure 11. Crack propagation for Ti-6Al-4V. R = 10 and P^* = 50.

As shown in Figure 11, at instant I_A, a crack was generated on the trailing edge of the contact surface. At instant I_B, the tangential force reached its maximum. These two stages are similar to Figure 3. However, at instant I_C, instead of a crack generated below the leading edge of the contact surface, the crack propagated along the edge of the contact surface and reached the leading edge. Interestingly, at I_D, a second crack was generated within the contact surface surrounded by the first crack. As the tangential displacement increased, the second crack linked with the crack at the trailing edge of the contact surface and led to the formation of a wear particle at instant I_E.

Since it was not observed that changing the JC coefficients or fracture energy led to such changes in the wear particle formation process, it was assumed that this new crack propagation may have been caused by the change in E/Y. For aluminum 2024 T351 $E/Y = 227$ and for Ti-6Al-4V, this value was $E/Y = 125$.

5. Conclusions

An efficient FE model was presented for adhesive spherical contact under combined normal loading and tangential displacement. The model consisted of a global model that helped determine the potential fracture location and a sub-model with a refined mesh for the precise prediction of the wear particle formation. The improved sub-model included the rigid flat and two criteria were employed to simulate the crack initiation and propagation. Thus, the present sub-model inherited the efficiency of the sub-model introduced in the literature and the capability to predict the tangential mechanical response.

The current study primarily focused on the effects of the sphere radius and a range of relatively high normal loads on the wear particle evolution. However, the developed method can be extended to the wear simulations under various parameter combinations, including the materials, operating conditions, etc. The main contributions of the present study were the unified dimensionless relation between the dimensionless wear rate and the dimensionless normal load for the sphere radius varying from 5 to 50 [mm], as well as the insights gained into how the JC parameters and material properties (E/Y) affect the friction and wear behaviors. The obtained relations for the wear volume and wear rate may be generalized for the wider range of mechanical properties with the presented efficient sub-model.

Furthermore, the present method offers the possibility of extending the single spherical wear model to the adhesive wear prediction for rough surfaces and for fretting wear modeling. Nevertheless, the modeling of the adhesive wear can be improved by introducing a physical approach for the crack propagation, which is not based on element deletion. Such an approach would provide a more realistic and physically based simulation of the crack propagation and wear particle formation.

Author Contributions: Conceptualization, M.L. and R.G.; writing—original draft preparation, M.L.; writing—review and editing, G.X. and R.G.; supervision, R.G. All authors have read and agreed to the published version of the manuscript.

Funding: This research received no external funding.

Data Availability Statement: Not applicable.

Acknowledgments: The present study is supported by the Israel Academy of Sciences and Humanities and the Council for Higher Education Excellence Fellowship Program for International Postdoctoral Researcher. This paper is part of the IEA AMT IA technical activities.

Conflicts of Interest: The authors declare that they have no known competing financial interests or personal relationships that could have appeared to influence the work reported in this paper.

Nomenclature

a	mesh size
c	characteristic width of crack
d	asperity radius
E	Young's modulus
G_F	size-independent fracture energy
G_f	fracture energy
H	hardness
K	Archard's wear coefficient
L_c	critical load at yield inception under full stick contact condition
L	sliding distance for Archard's wear model
l_p	length of wear particle
P	normal load
P^*	dimensionless normal load, $P^* = P/L_c$
R	sphere radius
r	section I radius
s	sliding distance
t	section I thickness
t_p	thickness of wear particle
u_p	tangential displacement when wear particle is formed
u_x	tangential displacement
V_0	volume of original hemisphere, $V_0 = 2/3\pi R^3$
V_p	volume of wear particle
W	Archard's wear volume
W_p	wear particle width
w	wear rate
Y	yield strength
υ	Poisson's ratio
ω	interference
ω_p	interference when the wear particle is formed
Subscripts	
0	at normal preloading
p	wear particle
s	sub-model
1	global model
Superscript	
$*$	dimensionless

References

1. Devaraju, A. A critical review on different types of wear of materials. *Int. J. Mech. Eng. Technol.* **2015**, *6*, 77–83.
2. Burwell, J.T.; Strang, C.D. On the empirical law of adhesive wear. *J. Appl. Phys.* **1952**, *23*, 18–28. [CrossRef]
3. Molinari, J.-F.; Aghababaei, R.; Brink, T.; Frérot, L.; Milanese, E. Adhesive wear mechanisms uncovered by atomistic simulations. *Friction* **2018**, *6*, 245–259. [CrossRef]
4. Archard, J. Contact and rubbing of flat surfaces. *J. Appl. Phys.* **1953**, *24*, 981–988. [CrossRef]
5. Holm, R. *Electric Contacts: Theory and Application*; Springer Science & Business Media: New York, NY, USA, 2013.
6. Bowden, F.P.; Tabor, D. *The Friction and Lubrication of Solids*; Oxford University Press: Oxford, UK, 2001.
7. Zhang, H.; Goltsberg, R.; Etsion, I. Modeling Adhesive Wear in Asperity and Rough Surface Contacts: A Review. *Materials* **2022**, *15*, 6855. [CrossRef]
8. Hills, D.A.; Ashelby, D.W. On the application of fracture mechanics to wear. *Wear* **1979**, *54*, 321–330. [CrossRef]
9. Fleming, J.R.; Suh, N.P. Mechanics of crack propagation in delamination wear. *Wear* **1977**, *44*, 39–56. [CrossRef]
10. Suh, N.P. The delamination theory of wear. *Wear* **1973**, *25*, 111–124. [CrossRef]
11. Johnson, G.R.; Cook, W.H. Fracture characteristics of three metals subjected to various strains, strain rates, temperatures and pressures. *Eng. Fract. Mech.* **1985**, *21*, 31–48. [CrossRef]
12. Wu, A.; Shi, X. Numerical investigation of adhesive wear and static friction based on the ductile fracture of junction. *J. Appl. Mech.* **2013**, *80*, 041032. [CrossRef]
13. Zhang, H.; Etsion, I. Evolution of adhesive wear and friction in elastic-plastic spherical contact. *Wear* **2021**, *478*, 203915. [CrossRef]
14. Hillerborg, A.; Modéer, M.; Petersson, P.-E. Analysis of crack formation and crack growth in concrete by means of fracture mechanics and finite elements. *Cem. Concr. Res.* **1976**, *6*, 773–781. [CrossRef]

15. Zhang, H.; Etsion, I. An advanced efficient model for adhesive wear in elastic—Plastic spherical contact. *Friction* **2022**, *10*, 1276–1284. [CrossRef]
16. Brizmer, V.; Kligerman, Y.; Etsion, I. The effect of contact conditions and material properties on the elasticity terminus of a spherical contact. *Int. J. Solids Struct.* **2006**, *43*, 5736–5749. [CrossRef]
17. Brizmer, V.; Kligerman, Y.; Etsion, I. Elastic–plastic spherical contact under combined normal and tangential loading in full stick. *Tribol. Lett.* **2007**, *25*, 61–70. [CrossRef]
18. Hancock, J.W.; Mackenzie, A.C. On the mechanisms of ductile failure in high-strength steels subjected to multi-axial stress-states. *J. Mech. Phys. Solids* **1976**, *24*, 147–160. [CrossRef]
19. Liu, X.; Ma, H.; Fan, F. Modified Johnson–Cook model of SWRH82B steel under different manufacturing and cold-drawing conditions. *J. Constr. Steel. Res.* **2021**, *186*, 106894. [CrossRef]
20. Murugesan, M.; Jung, D.W. Johnson Cook material and failure model parameters estimation of AISI-1045 medium carbon steel for metal forming applications. *Materials* **2019**, *12*, 609. [CrossRef]
21. Hu, X.; Wittmann, F. Size effect on toughness induced by crack close to free surface. *Eng. Fract. Mech.* **2000**, *65*, 209–221. [CrossRef]
22. Hu, X.-Z. An asymptotic approach to size effect on fracture toughness and fracture energy of composites. *Eng. Fract. Mech.* **2002**, *69*, 555–564. [CrossRef]
23. Systèmes, D. *Abaqus 6.9 User's Guide and Theoretical Manual*; Hibbitt, Karlsson & Sorensen, Inc.: Providence, RI, USA, 2009.
24. Nahshon, K.; Pontin, M.; Evans, A.; Hutchinson, J.; Zok, F. Dynamic shear rupture of steel plates. *J. Mech. Mater. Struct.* **2007**, *2*, 2049–2066. [CrossRef]
25. Mabrouki, T.; Girardin, F.; Asad, M.; Rigal, J.-F. Numerical and experimental study of dry cutting for an aeronautic aluminium alloy (A2024-T351). *Int. J. Mach. Tools Manuf.* **2008**, *48*, 1187–1197. [CrossRef]
26. Wierzbicki, T.; Bao, Y.; Lee, Y.-W.; Bai, Y. Calibration and evaluation of seven fracture models. *Int. J. Mech. Sci.* **2005**, *47*, 719–743. [CrossRef]
27. Abdelhafeez, A.M.; Soo, S.L.; Aspinwall, D.; Dowson, A.; Arnold, D. A coupled Eulerian Lagrangian finite element model of drilling titanium and aluminium alloys. *SAE Int. J. Aerosp.* **2016**, *9*, 198. [CrossRef]
28. Asad, M.; Mabrouki, T.; Ijaz, H.; Khan, M.A.; Saleem, W. On the turning modeling and simulation: 2D and 3D FEM approaches. *Mech. Ind.* **2014**, *15*, 427–434. [CrossRef]
29. Brizmer, V.; Kligerman, Y.; Etsion, I. A model for junction growth of a spherical contact under full stick condition. *J. Tribol.* **2007**, *129*, 783–790. [CrossRef]
30. Rabinowicz, E.; Tanner, R.I. Friction and wear of materials. *J. Appl. Mech.* **1966**, *33*, 479. [CrossRef]
31. Kogut, L.; Etsion, I. Elastic-plastic contact analysis of a sphere and a rigid flat. *J. Appl. Mech.* **2002**, *69*, 657–662. [CrossRef]
32. Zhang, J.; Alpas, A.T. Transition between mild and severe wear in aluminium alloys. *Acta Mater.* **1997**, *45*, 513–528. [CrossRef]
33. Brizmer, V.; Zait, Y.; Kligerman, Y.; Etsion, I. The effect of contact conditions and material properties on elastic-plastic spherical contact. *J. Mech. Mater. Struct.* **2006**, *1*, 865–879. [CrossRef]
34. Bobbili, R.; Madhu, V. A modified Johnson-Cook model for FeCoNiCr high entropy alloy over a wide range of strain rates. *Mater. Lett.* **2018**, *218*, 103–105. [CrossRef]
35. Gerstgrasser, M.; Smolenicki, D.; Akbari, M.; Klippel, H.; Roelofs, H.; Cadoni, E.; Wegener, K. Analysis of two parameter identification methods for original and modified Johnson-Cook fracture strains, including numerical comparison and validation of a new blue-brittle dependent fracture model for free-cutting steel 50SiB8. *Theor. Appl. Fract. Mech.* **2021**, *112*, 102905. [CrossRef]
36. Zhang, Y.; Outeiro, J.C.; Mabrouki, T. On the selection of Johnson-Cook constitutive model parameters for Ti-6Al-4 V using three types of numerical models of orthogonal cutting. *Procedia CIRP* **2015**, *31*, 112–117. [CrossRef]
37. Shrestha, S.; El Rassi, J.; Kannan, M.; Morscher, G.; Gyekenyesi, A.L.; Scott-Emuakpor, O.E. Fracture toughness and fatigue crack growth rate properties of AM repaired Ti–6Al–4V by Direct Energy Deposition. *Mater. Sci. Eng. A* **2021**, *823*, 141701. [CrossRef]

lubricants

MDPI

Article

A Multi-Scale Investigation to Predict the Dynamic Instabilities Induced by Frictional Contact

Farouk Maaboudallah * and Noureddine Atalla *

Department of Mechanical Engineering, Groupe d'Acoustique de l'Université de Sherbrooke (GAUS), Sherbrooke, QC J1K 2R1, Canada
* Correspondence: farouk.maaboudallah@usherbrooke.ca (F.M.); noureddine.atalla@usherbrooke.ca (N.A.)

Abstract: We propose a new variational formulation to model and predict friction-induced vibrations. The multi-scale computational framework exploits the results of (i) the roughness measurements and (ii) the micro-scale contact simulations, using the boundary element method, to enrich the contact zone of the macroscopic finite element model of rubbing systems with nominally flat contact boundaries. The resulting finite elements at the contact interface of the macroscopic model include (i) a modified normal gap and (ii) a micro-scale description of the contact law (i.e., pressure gap) derived by solving the frictionless contact problem on a rough surface indenting a rigid half-plane. The method is applied to a disc brake system to show its robustness in comparison with classical deterministic formulations. With respect to the traditional complex eigenvalues analysis, the proposed multi-scale approach shows that the inclusion of roughness significantly improves the results at low frequencies. In this panorama, any improvement of dynamic instabilities predictions should be based on an uncertainty analysis incorporating roughness combined with other parameters such as friction coefficient and shear moduli of the pads, rather than on roughness itself.

Keywords: dynamic instabilities; frictional system; multi-scale framework; contact mechanics; roughness

Citation: Maaboudallah, F.; Atalla, N. A Multi-Scale Investigation to Predict the Dynamic Instabilities Induced by Frictional Contact. *Lubricants* **2023**, *11*, 344. https://doi.org/10.3390/lubricants11080344

Received: 13 June 2023
Revised: 3 August 2023
Accepted: 8 August 2023
Published: 11 August 2023

1. Introduction

The frictional contact problem plays a major role in engineering applications, whether in studying the fretting fatigue in dovetail blade roots in aeronautics [1], valve systems for nuclear cooling [2] or braking systems in the automotive industry [3–5]. The mechanical behavior of frictional systems mentioned above is micro-scale dependent. Their performance is conditioned by numerous microscopic phenomena, such as roughness, progressive damage of asperities as well as resulting debris [6,7]. For instance, it was shown in [8] that the macroscopic behavior of the braking system is strongly influenced by the microscopic contact properties, which involves the roughness of the sliding interfaces. Hence, it is important to review and to overcome the assumption of perfectly flat surfaces by taking into account the statistical characteristics of rough surfaces.

In general, the modeling of rough surfaces requires the knowledge of two functions: (i) the height distribution function (HDF) [9] and (ii) the spatial function, also known by the autocorrelation function (ACF) or its Fourier transform, which is called the power spectral density (PSD) function [10]. The first one characterizes the roughness through the average height parameters [11,12], while the second describes the spacial arrangement and the variation of the asperities, which often depicts a fractal behavior [13,14]. Most rough surfaces have a self-affine property, which means that the profile remains similar under different magnification [13,15]. The latter can be decoded by a PSD having a power-law shape as a function of the wavenumber (or wavelength) with a potential plateau characterized by a longer wavelength, namely the long-distance roll-off wavevector [16–18].

With regard to the mechanical contact models, the first contribution goes back a century with the widely used Hertz theory [19,20]. This theory assumes perfect smooth contacting surfaces, and hence, it is not valid for rough surfaces since it failed to give a reasonable estimation for the real contact area and for the pressure-gap behavior. The strong deviation from the perfect contact assumption led researchers to develop more realistic models to take into account the roughness. The first asperity-based models, pioneered by research work [9,21], are based on a statistical method and the random process theory introduced for the first time in [10,22]. The concept of their approach is to transform the problem of two contacting rough surfaces to a deformable rough one in contact with a rigid plane. Then, the roughness is modeled by a number of isolated spherical asperities with the same radius, and their height is described randomly following a given probability density function. The asperity-based models provide a good approximate solution for the evolution of the contact area especially in the case of a small load [23]. However, their main drawback is the omission of interactions between the asperities, which can lead to an overestimation of the contact pressure. With the development of advanced analytical models [24–27], interactions between asperities can be taken into account. The coalescence of asperities is another effect that has to be considered for a realistic representation. In fact, the contact spots do not grow independently in reality. They merge and, therefore, contact patches develop. Afferante et al. [28] proposed a suitable approach to address the coalescence of two contacting spots. The concept of their approach is that the asperities with overlapping contacts spots are eliminated and corrected with a single contact patch, namely the equivalent asperity. It should be noted that the contact area of the new asperity includes the area of the suppressed contact spots. In parallel to the aforementioned analytical models, Person proposed an ingenious theory [13,14,29] to tackle the contact problem for rough surfaces. The fundamental concept of the fractal approach is to solve the contact problem at different scale by introducing the evolution of contact pressure probability density. The author mentioned in [14] that the latter satisfies a diffusion-like equation with an appropriate boundary conditions. Persson's approach has been criticized on the pretext that the proof of the diffusion-like equation is not rigorous, since it is derived assuming full contact and used with the boundary condition on its solution to model the partial contact. In general, it provides a solution that is quantitatively inconsistent with the available solutions, which leads to somewhat smaller contact areas and, thus, greater pressures than those found in the literature [30,31].

The development of computing facilities has enabled the emergence of accurate numerical models that, unlike previous analytical models, are free of assumptions. A couple of numerical techniques can be distinguished: (i) the finite element method (FEM) [2,32], (ii) the boundary element method (BEM) [33] and (iii) Green's function molecular dynamics (GFMD) [34,35]. The first class of methods is the most used and the most accurate technique. It is based on the optimization of the variational formulation of the contact problem [36,37]. In general, it is applied to a representative volume element (RVE) of the rough surface and requires intensive computing, since the element size should be small enough to capture the asperity waviness and the roughness [2]. The last two classes are more attractive for modeling rough contacts. This is essentially due to the fact that only the rough surface needs to be discretized, and not the bulk, as required by FEM, which allows increasing the mesh density and, hence, perform more accurate studies. Despite these advances, BEM and GFMD are based on the fundamental elasticity theory. The generalization of these approaches to take into account non-linearities is sometimes possible but is not an easy task. For a detailed review, see [32,38].

In the industrial context, FEM is mainly used to model and predict the behavior of frictional systems. In most applications, the contact interfaces are assumed to be flat and have dimensions generally much larger than the microscopic scale of the roughness. Therefore, explicit integration of the roughness in these interfaces is not possible, since it generates extremely dense meshes which, in turn, will increase the computation time. In consequence, authors in literature abandon the classical path of explicitly modeling surface roughness by proposing multi-scale embedded strategies to enhance the computational cost of the direct FEM. For instance, the authors in [39–41] suggested integrating the micro-mechanical contact laws in the sliding interfaces. The fundamental concept behind this approach is to derive and then integrate the constitutive contact equations using the homogenization process, which describes the microscopic behavior within the contact zone. Bonari et al. [42] also developed a novel multi-scale formulation that takes advantage of micro-scale simulations to enrich the contact interfaces of macroscopic FE models. The developed method allows taking into account any desirable topography to model efficiently and within a reasonable CPU time the frictional system. In the same spirit, Paggi et Reinoso [43] introduced a new idea that takes into account the shape of the roughness to correct the normal separation in macroscopic contact interfaces. Tison et al. [3,44] investigated the use of rough contact to model the dynamic behavior of a disc brake system. They used a multi-scale framework based on the random field theory [45,46]. The integration of the latter within the complex eigenvalues analysis (CEA) provides a remarkable prediction, which allowed for better correlation with the experimental results.

Regarding the dynamic instabilities that represent the core of this paper, it is well accepted that the frictional contact influences drastically the global dynamic behavior of a rubbing system [47,48]. Taking into account the contribution of the frictional contact to the system's overall stiffness matrix, dynamic instabilities are predicted using two methods: (i) transient analysis and (ii) CEA. The latter is the most widely used in the industry, as it offers a very good compromise between computation time and accuracy. Indeed, CEA approach uses the mode coupling theory to evaluate the stability of the rubbing system. Thus, it will be unstable if and only if at least one predicted vibratory mode has a strictly positive real part [49]. Since the frictional contact problem is closely linked to dynamic instabilities, the contribution of the roughness should be taken into account and investigated in detail for a potential enhancement of the deterministic CEA. As pointed above, the classic FEM is time consuming since the mesh should be fine enough to capture all the physics of the rough surface. A novel strategy has therefore to be found to fulfill two main criteria: (i) an accurate prediction of the dynamic instabilities and (ii) a reasonable computational cost. In this light, the main contribution of this paper is to suggest a simple alternative approach, highly flexible, that allows accurately and reliably predicting the dynamic instabilities of an industrial braking system. The proposed multi-scale finite element formulation, presented in detail in Section 2, abandons the traditional path of explicitly introducing roughness by means of the asperities in the contacting interfaces. In fact, it considers that the contact interfaces of a large-scale FE model are perfectly flat and smooth. But each element (contact patch) is enriched by adding two contributions, namely (i) the pressure-gap contact law and (ii) the contact element activation. The first one is derived using BEM on a generated rough surface and is assigned to each patch. The use of micro-scale contact law, for each patch, will modify the penalty method and will introduce the roughness in an implicit manner. The second enhancement is introduced within the normal gap function. Indeed, a critical gap is added to the classic formulation. It is derived from HDF resulting from the topography measurement. The embedded strategy is then applied to a disc brake system in Section 3 to investigate the prediction of dynamic instabilities.

2. Governing Equation

The first section introduces the basic governing equations for the non-linear contact problem. After recalling the general framework of the boundary value problems with constraints, the variational formulation of the weak problem will be presented for the case of frictionless contact problems of two deformable solids. This is the first step of a complex eigenvalues analysis (CEA) where the equilibrium position, u_e, is determined. Finally, we will present the principle of our approach, which consists of integrating roughness in macroscopic FE models in order to predict dynamic instabilities of a frictional system in a more realistic way.

2.1. General Framework

In the framework of continuum mechanics, a solid is modeled by an open domain Ω, which is assumed to be bounded and regular. The boundary of Ω is divided in two parts: (i) the first one denoted $\partial\Omega_u$ where the Dirichlet boundary conditions are applied and (ii) the second one denoted $\partial\Omega_f$ where the Neumann boundary conditions are applied. So, without any loss of generality, the formulation of the problem when two deformable bodies Ω_1 and Ω_2 (see Figure 1), with $\Omega = \Omega_1 \cup \Omega_2$, come in contact at a single contact zone $\partial\Omega_c$ is given by a system of elliptical partial differential equations that can be categorized as follows:

- static balance of momentum equation,

$$\begin{cases} \nabla \cdot \underline{\underline{\sigma}} &= 0 & \text{in } \Omega = \Omega_1 \cup \Omega_2 \\ \underline{\underline{\sigma}}.\underline{n} &= \underline{\sigma_0} & \text{at } \partial\Omega_f \end{cases}, \tag{1}$$

- compatibility equations,

$$\begin{cases} \underline{\underline{\varepsilon}}(\underline{u}) &= \frac{1}{2}(\nabla \cdot \underline{u} + \nabla \cdot \underline{u}^T) & \text{in } \Omega = \Omega_1 \cup \Omega_2 \\ \underline{u} &= \underline{u_0} & \text{at } \partial\Omega_u \end{cases}, \tag{2}$$

- constitutive relation,

$$\underline{\underline{\sigma}} = {}^4\underline{\underline{\mathbb{C}}} : \underline{\underline{\varepsilon}}(\underline{u}) \text{ in } \Omega = \Omega_1 \cup \Omega_2, \tag{3}$$

where $\underline{\underline{\sigma}}$ is the Cauchy stress tensor and $\underline{\underline{\varepsilon}}$ is the so-called small strain tensor, which is related to the displacement field $\underline{u}$. $\underline{\sigma_0}$ and $\underline{u_0}$ are the stress and the displacement field imposed on the boundaries $\partial\Omega_f$ and $\partial\Omega_u$, respectively. ${}^4\underline{\underline{\mathbb{C}}}$ is the Hook fourth-order tensor for the considered domain.

In the case of a frictionless contact between two bodies, a relevant complimentary conditions should be formulated. The latter defines the geometrical and the mechanical state of the two contacting surfaces. The complimentary conditions are called Hertz–Signorini–Moreau conditions and can be written as

$$g \geq 0, \;\; \sigma_n \leq 0, \;\; g.\sigma_n = 0, \;\; \underline{\sigma_t} = 0 \text{ at } \partial\Omega_c = \Gamma_c^1 \cup \Gamma_c^2, \tag{4}$$

where σ_n and $\underline{\sigma_t}$ refer to the normal and the tangential contact stresses, respectively. g refers to the gap function between the slave and master surfaces. Hence, the Hertz–Signorini–Moreau conditions can be read as the non-penetration–non-adhesion conditions. In other words, if the two bodies Ω_1 and Ω_2 are in contact, then the gap function g is equal to 0 and the normal contact stress σ_n is below 0; otherwise, $g > 0$ and $\sigma_n = 0$.

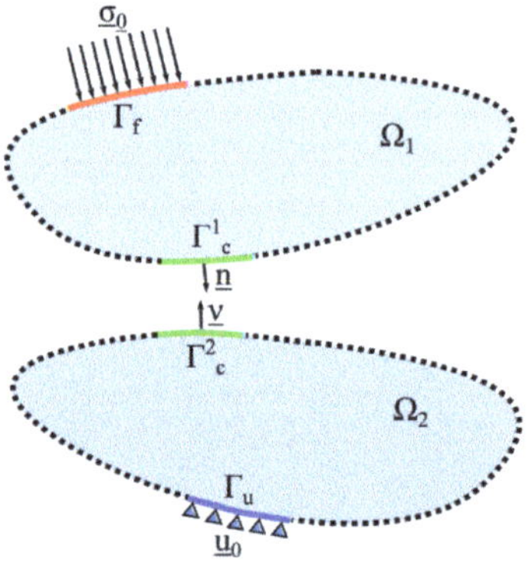

Figure 1. Contact between two deformable solids Ω_1 and Ω_2. Ω_1 refers to the slave body and Ω_2 is the master one. The active contact interface is $\Gamma_c = \Gamma_c^1 = \Gamma_c^2$.

2.2. *Variation Formulation of the Frictionless Contact Problem*

The statement of the weak formulation for the contact problem is obtained by multiplying the strong form in Equation (1) by a test function $\underline{v}$ and integrating over the domain Ω, and we write

$$\int_{\Omega} \nabla \cdot \underline{\underline{\sigma}} \cdot \underline{v}\, d\Omega = 0 \;\; \forall \underline{v} \in \mathbb{V}. \tag{5}$$

By performing integration and applying the divergence theorem, Equation (5) becomes

$$\int_{\partial\Omega} \underline{n} \cdot \underline{\underline{\sigma}} \cdot \underline{v}\, d\Gamma - \int_{\Omega} \underline{\underline{\sigma}} : \nabla \underline{v}\, d\Omega = 0, \tag{6}$$

where : is the contraction operation of two second-order tensors and $\underline{n}$ denotes the outward normal at the slave surface Γ_c^1. We have $\underline{n} = -\underline{\nu}$ where $\underline{\nu}$ is the outward normal at master surface Γ_c^2 (see Figure 1).

Note that by introducing the traction vector $\underline{\sigma}_0$, the contact boundary conditions and the virtual displacement $\delta\underline{u}$ in the test function $\underline{v}$, the first term in Equation (6) may be rewritten as the sum of two main parts: (i) the frictionless contact contribution and (ii) the Neumann boundary condition contribution. Symbolically, we write

$$\int_{\partial\Omega_c} \underline{n} \cdot \underline{\underline{\sigma}} \cdot \delta(\underline{\rho} - \underline{r})\, d\Gamma + \int_{\partial\Omega_f} \underline{\sigma}_0 \cdot \delta\underline{u}\, d\Gamma - \int_{\Omega} \underline{\underline{\sigma}} : \delta\nabla\underline{u}\, d\Omega = 0, \tag{7}$$

where $\underline{\rho}$ and $\underline{r}$ are the displacement vectors of master and slave points, respectively. Furthermore, the quantity $\underline{r} - \underline{\rho}$ is the gap vector describing the position of the slave point $\underline{r}$ and his projection into the master surface $\underline{\rho}$.

In the case of a normal gap, $g_n\underline{n} = \underline{r} - \underline{\rho}$, the variational form in Equation (7) may be written as follows:

$$\int_{\partial\Gamma_c^1} \sigma_n \delta g_n\, d\Gamma - \int_{\partial\Omega_f} \underline{\sigma}_0 \cdot \delta\underline{u}\, d\Gamma + \int_{\Omega} \underline{\underline{\sigma}} : \delta\nabla\underline{u}\, d\Omega = 0. \tag{8}$$

It should be noted that the mathematical constraints of the strong and weak forms are not the same. For the strong form in Equation (1), the Cauchy tensor is required to be smooth enough; i.e., $\underline{\underline{\sigma}} \in \mathcal{C}^1(\Omega)$, which is not the case for the weak form in Equation (6). The order of differentiation in the weak formulation is lower than the strong one. In fact, the order of differentiability in the weak integral form is distributed between the Cauchy stress tensor and the test function.

Finally, the balance of virtual work, called also the weak form, for the frictionless contact problem can be formulated as follows:

Find $\underline{u} \in \mathbb{U} = \{\underline{u} \in \mathbb{H}^1(\Omega) \mid \underline{u} = \underline{u}_0 \text{ on } \partial\Omega_u\}$such as

$$\begin{cases} \displaystyle\int_{\Gamma_c^1} \sigma_n \delta g_n\, d\Gamma - \int_{\partial\Omega_f} \underline{\sigma}_0 \cdot \delta\underline{u}\, d\Gamma + \int_{\Omega} \underline{\underline{\sigma}} : \delta\nabla\underline{u}\, d\Omega = 0 \\ \mathbb{V} = \{\delta\underline{u} \in \mathbb{H}^1(\Omega) \mid \delta\underline{u} = 0 \text{ on } \partial\Omega_u\} \\ \mathbb{C} = \left\{\delta\underline{u} \in \mathbb{V} \mid (\underline{r} + \delta\underline{r} - \underline{\rho} + \delta\underline{\rho}) \cdot \underline{n} \geq -g_{n0}\right\} \end{cases}, \tag{9}$$

where $\mathbb{H}^1(\Omega)$ denotes the Hilbert space of the first order, $\delta\underline{r}$ and $\delta\underline{\rho}$ refer to $\delta\underline{u}$ but in the contact interface. And, finally, g_{n0} represents the initial gap between contacting surfaces.

2.3. *Overview of the Embedded Computational Strategy to Include the Roughness on Macro-Scale Model*

Under the framework of the finite element method, the formulation in Equation (9) will be solved using the most popular scheme: the penalty method. The goal as mentioned above is to find the equilibrium position due to the frictionless contact (see Section 1 in [4] for more detail).

To approximate the Hertz–Signorini–Moreau condition, the normal contact stress arising at the contact interface will be expressed as a function of the gap using the following approximation:

$$\sigma_n(g) = \varepsilon_n(< -g >) = \begin{cases} 0 & g > 0, \text{there is no contact} \\ \varepsilon_n(-g) & g \leq 0, \text{there is contact} \end{cases}, \quad (10)$$

where ε_n is a non-positive, continuous and strictly decreasing function defining the evolution of the contact stresses, σ_n, as a function of the gap, g, between the master and the slave surfaces. In the case of the linear penalty method, the function ε_n will simply denote the classic contact stiffness (i.e., penalty coefficient). Finally, $< \cdot >$ refers to $\max\{\cdot, 0\}$.

The penalty method can be seen as an approximation of the contact constraints. It leads to a small penetration between the slave and master surfaces (see the case when $g < 0$ in Equation (10)). It should be noted that the definition in Equation (4) will be met if the contact stress is higher for a small penetration. In other words, the penalty method does not restrict the penetration between the contacting surfaces but it resists it. If it is deeper (i.e., $g < 0$), the value of $\varepsilon_n(-g)$ will be higher and, hence, the real contact stress (i.e., reaction) will appear.

Under the last assumption, the virtual work in Equation (9) can be divided into the classic solid mechanics part, δW_s, and the contribution from the contact problem, namely δW_c. The weak formulation of the system is obtained using the penalty method. We have

$$\delta W_c + \delta W_s = 0, \quad (11)$$

where

$$\delta W_c = \int_{\Gamma_c^1} \varepsilon_n(< -g_n >)\delta g_n \, d\Gamma, \quad (12)$$

and

$$\delta W_s = \int_{\Omega} \underline{\underline{\sigma}} : \delta \nabla \underline{u} \, d\Omega - \int_{\partial\Omega_f} \underline{\sigma_0} \cdot \delta \underline{u} \, d\Gamma. \quad (13)$$

In this paper, the the potential energy of the contact interaction will be formulated as follows:

$$\Pi_c = \frac{1}{2} \int_{\Gamma_c^1} \varepsilon_n(< -(g_n - g_n^{\text{cr}}) >)^2 \, d\Gamma, \quad (14)$$

where g_n^{cr} is called the critic gap. It is a value that will be defined for each contact element and from which the latter will be activated. The first enrichment introduced in Equation (14) will allow for a non-homogeneous activation of the contact status for all nodes of the contact interface.

Considering Node-To-Surface (N2S) discretization, the slave surface Γ_c^1 can be presented by nodes denoted by $\underline{r}^i$ and the master one, Γ_c^2, can be discretized as set of segments denoted by $\Gamma_{c,j}^2$. Let us assume that we have N slave points and M master segments. One can observe from Figure 2 that for all slave nodes $\underline{r}^i$, one or more master segments $\Gamma_{c,j}^2$ can be determined using the normal projection. Hence, the i-th contact element can be defined by a combination of one slave node $\underline{r}^i$ with its correspondent master segment $\Gamma_{c,j}^2$. By using the last definition and the fact that we have n_c contact finite elements, the discretized form of Equation (14) can be rewritten as follows:

$$\begin{cases} \Pi_c \approx \frac{1}{2} \sum_{i=1}^{n_c} \int_{\Gamma_{c,i}^2} \varepsilon_n^i(< -(g_n^i - g_n^{\text{cr},i}) >)^2 \, d\Gamma \\ \varepsilon_n^i \neq \varepsilon_n^j \quad \forall i \neq j \end{cases}, \quad (15)$$

where ε_n^i denotes the contact law for the i-th contact element. It pairs up the normal contact stress with the gap distance. Note that the contact law from one element to another is not the same. Each contact element will have its own law constituting the second enrichment.

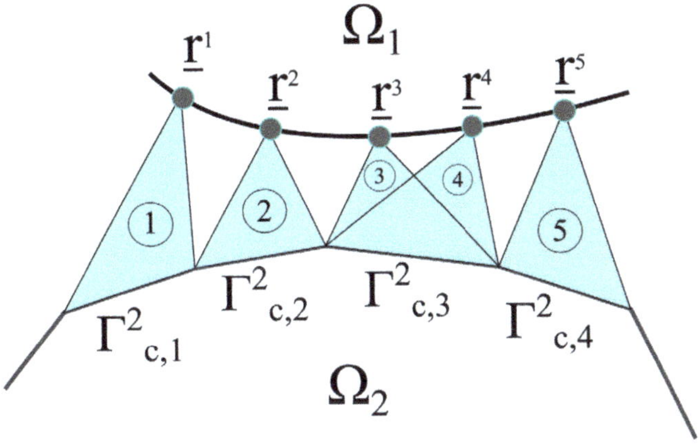

Figure 2. Construction of the contact finite element using Node-To-Surface (N2S) discretization. The contact element contains only one slave node attached to a master surface. The case presented in this figure contains $n_c = 5$ contact elements.

The idea behind the proposed strategy is to integrate the heterogeneity observed in the asperity scale into macroscopic models. In general, the present methodology incorporates three main steps: (i) The first one consists in characterizing the rough surface (the objective of Section 3.1). (ii) The second one aims to solve, at the roughness scale, the frictionless contact problem. This step is important as it allows deriving the pressure-gap's contact law by taking into account the roughness. Finally, (iii) the third one consists in introducing the height distribution as well as the local behavior of the contact stiffness to enrich the gap, equivalently the contact detection, and the penalty function. The overall procedure is summarized in Figure 3.

In the first step (Figure 3a), the rough surface is characterized for example by using a confocal laser scanning microscopy (CLSM) technique in order to measure the height z of each point. Since the roughness is random [10], the height of the rough surface is considered, in this work, as a random variable (RV) with an independent Cartesian coordinates. The measured realization of the RV will be used to estimate the HDF denoted by $P(z)$. The latter will be introduced in the third step into the macroscopic FE model through the gap function. Practically, the statistics of the rough surface will be transmitted to the FE model using the new form of the normal gap in Equation (14) or (15). Hence, we write

$$P(z) \equiv P(\Delta), \tag{16}$$

where $\Delta = g_n - g_n^{\text{cr}}$ is a RV. The realization Δ_i^j of Δ will be assigned for each contact element i.

In a second step (Figure 3b), several rough surfaces will be generated numerically based on the statistics and the spectral analysis of the measured surface. The objective is to apply the BEM in order to solve microscopically the contact problem of the rough surfaces. The purpose of this micro-scale simulation is to compute the micro-mechanical behavior of the contact interface involving the roughness. The latter is modeled by the pressure-gap law for each generated rough surface and injected into the weak form through the function ε_n of Equation (10). It is agreed that this requires intensive computation on several rough surfaces, especially if advanced BEM solvers are used to take into account interfacial or material non-linearities such as adhesion or plasticity. Here, the crucial goal is to derive a set of local contact laws, by taking into account the roughness, for each contact element (patch).

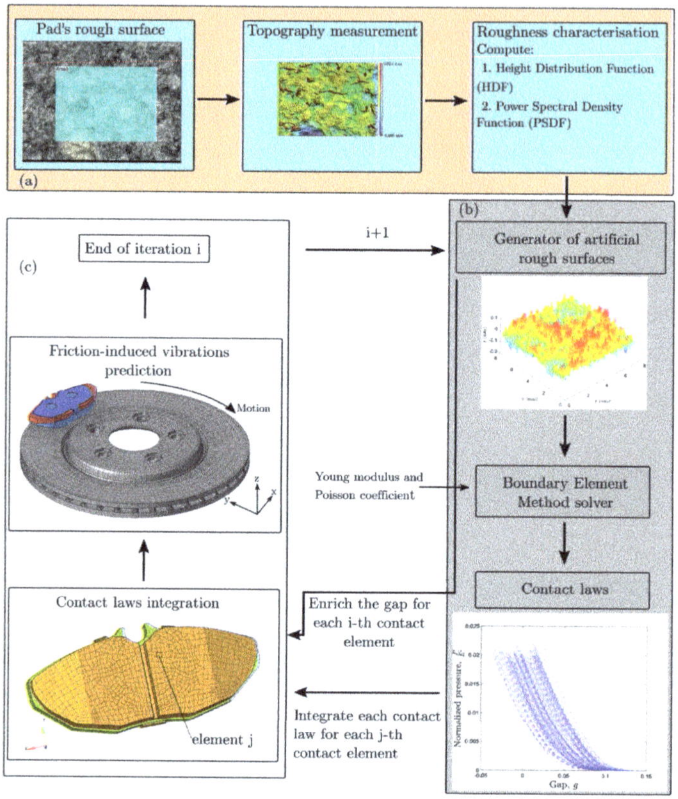

Figure 3. The flowchart of the computational framework: (**a**) The first step aims to measure and then characterize the observed roughness. The functions deduced from the characterization, namely the height distribution function and the power spectral density function, will be used in the second step (**b**) to numerically generate rough surfaces similar to what has been measured. In this step (**b**), BEM solver is used to solve the low-scale contact problem. The results of the BEM and the characterization will enrich the contact laws and the separation between master and slave surfaces. In the third step (**c**), the friction-induced vibrations of an automotive disc brake system are predicted.

Finally, in the last step (Figure 3c), the contact laws as well as the enriched gap function will be assigned to each patch. The integration procedure is the same for both parameters. Indeed, assuming that the contact interface is smooth and flat, the sampled results resulting from the BEM simulations and the measured surface characterization will be embedded within the contact elements. Following the logic of the embedded approach, the roughness will not be represented explicitly in the FE model. It will be included implicitly through the new formulation of the contact gap and the modified penalty function. Next, the enriched FE model can be solved using CEA to compute the dynamic instabilities. In general, the procedure falls into a fully parallelizable loop. After the surface characterization, steps 2 and 3 (Figure 3b,c) can be repeated as many times as needed in order to obtain a representative statistical prediction.

3. Application to a Disc Brake System

In this section, the previously proposed computational strategy is implemented to predict the dynamic behavior of an automotive disc brake system. First, the pad surface is measured by a CLSM. The objective is to characterize the pad surface topography by means of (i) the height distribution function (HDF) and (ii) the power spectral density

(PSD) function. Second, a batch of artificial surfaces will be randomly generated based on the previously measured functions (HDF and PSD). The aim is to apply BEM, developed in [50], to predict the contact laws, i.e., the evolution of the load as a function of the gap. These elements will be integrated, toward the end, into a macroscopic FE model of a disc brake to reproduce the heterogeneity observed in the contact interfaces in order to predict the dynamic instabilities induced by friction.

3.1. Pad Surface Characterization

The pad's surface is measured experimentally through a CLSM. Figure 4 shows an example of the measured topography over a scan area of 1.8×1.4 mm^2 with a resolution of 256×256 pixels. The measured topography (see Figure 4b) appears irregular and depicts characteristics of randomness. Hence, the random process theory, widely applied to analyze roughness, can be used to model accurately the probability function of the surface heights also known by the HDF. The latter holds only the out-of-plan information. To complete the characterization, the PSD is used to describe the spatial arrangement in the plane.

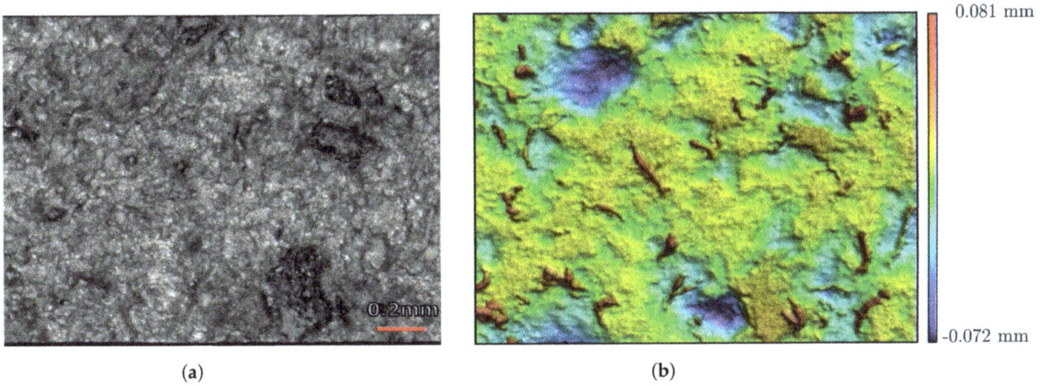

Figure 4. Confocal (**a**) scan and (**b**) measurements on the surface of a disc braking system pad.

Figure 5a illustrates the computed 2D-PSD of the pad's surface using the developments in our previous research work [18,38]. It shows clearly that the pad's roughness is highly isotropic. Such a result allows computing the radial PSD (averaged) in order to study the effect of resolution and scan length. Moreover, it can be seen, in Figure 5b, that the behavior of the measured PSD is not sensitive to resolution or scan length. It conserves its shape despite the change of the measurement parameters. With regard to the PSD behavior, Figure 5b depicts roughly two linear regions in the log–log plot. It starts from the lower frequency, which is inversely correlated to the scan length $q_L = \frac{2\pi}{L}$, up to a high measured frequency, related to the short-distance cut-off wavevector, defined by $q_s = \frac{2\pi}{\Delta}$ where Δ refers to the sampling length. In fact, the radial PSD has a shape similar to the bi-fractal surfaces defined in [51]. The difference lies in the definition of the slopes of the linear regions. In fact, bi-fractal surfaces have a theoretical PSD form that can be defined as follows:

$$\mathrm{PSD}(q) = \begin{cases} C_0\left(\frac{q}{q_L}\right)^{c_1} & \text{if } q_L < q < q_c \\ C_1\left(\frac{q}{q_c}\right)^{c_2} & \text{if } q_c \leq q < q_s \\ 0 & \text{else} \end{cases}, \tag{17}$$

where the exponents $c_i, i \in \{1,2\}$ (the slope of the linear regions in log–log plot) are related to the Hurst exponent H_i as follows: $c_i = -2(H_i + 1)$ with $0 < H_i < 1$. In our case, the two regions have a slope of -2 and -4.5, which correspondents to Hurst exponents of $H_1 = 0$ and $H_2 = 1.25$, respectively. The first region defines a borderline case of fractality, while the second is completely outside the fractal or the self-affine framework. However, the model defined in Equation (17) will be used to characterize the PSD of the pad's surface. The parameters c_1, c_2, q_L, q_c, C_0 and C_1 are computed using a suitable optimization scheme on the radially averaged PSD exactly as defined in Equation (22) of [18].

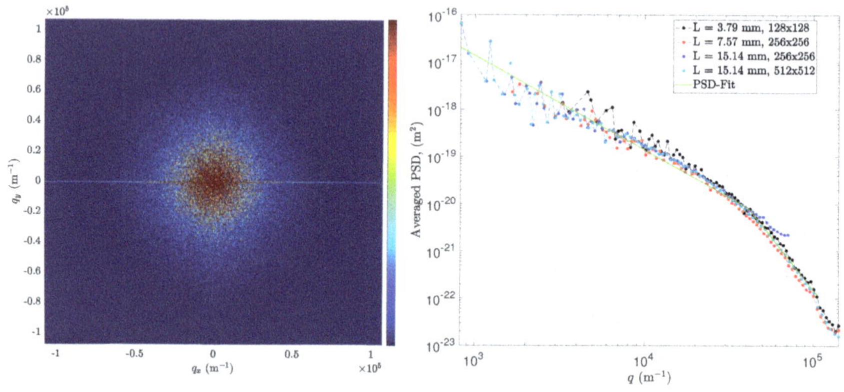

(a) (b)

Figure 5. Power spectral density functions (PSD) of the pad's rough surface: (**a**) 2D-PSD of the 15×15 mm^2 with a resolution of 512×512 pixels. (**b**) The effect of scan length and resolution on the averaged 2D-PSD.

Figure 6 complements the information given by the PSD. It presents the HDF of the measured pad's surface in a normalized manner and compares it with the centered and reduced normal law. As a first observation, one can see that the surface is approximately Gaussian. A notable difference is observed for the higher heights which makes the HDF asymmetric. This behavior was expected since automotive disc brake pads are usually compacted and ground to correct height. From Figure 6, the roughness of the pad can be defined by means of several parameters. For example, the mean (or median) plan $\bar{z} = 0.0256$ mm, the root mean square (rms) roughness $\sigma = 0.0331$ mm, the skewness $s = -0.2410$ and the kurtosis $\kappa = 3.6549$.

Note that in the following, the fitted PSD and HDF will be taken into account to numerically generate rough surfaces having the same spatial arrangement and roughness as that measured experimentally. The numerical procedure to generate artificial rough surface is inspired by Wu's algorithm [52]. The latter is adapted to take into account PSDs with shapes similar to what has been defined in Equation (17).

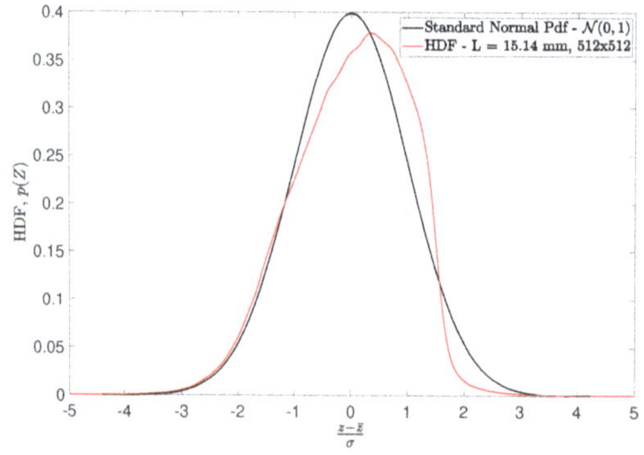

Figure 6. Comparison between the normalized height distribution function (HDF) of the pad's rough surface (red curve) and the standard Gaussian distribution (back curve).

3.2. Contact Law of Pad/Disc Interfaces

In the previous section, the surface of the pad has been fully described by two quantities: (i) PSD and (ii) HDF. These will act as an input data to artificially generate a batch of random rough surfaces. The artificial surfaces are intended to mimic the roughness behavior of the automotive disc brake pads so that they can be used to compute contact laws, i.e., load-separation curves. An example of an artificial rough surface is presented in Figure 7a. It has the same spectral and statistical specifications as the measured pad's surface, namely (i) PSD properties (compare Figure 7b with Figure 5a) and (ii) the roughness parameters (the mean plan and the rms roughness).

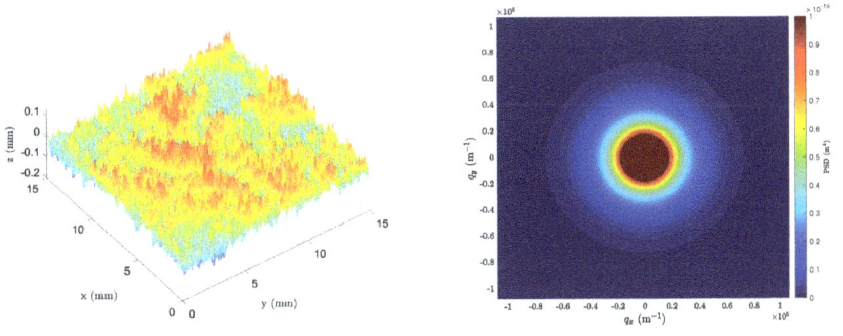

(**a**) (**b**)

Figure 7. 3D view of the artificial rough surface generated with a resolution of 512 pixels in both directions (on the left—(**a**)) and its 2D-PSD (on the right—(**b**)).

After generating batch of artificial random surfaces, the next step is to compute the evolution of contact stresses against contact kinematics for each generated pad's topography. At this stage, many models can be used. The study of the latter has been addressed in our previous research [38], where authors tried to explain the philosophy behind each theory with a comparative study. The concern here is to apply directly the result of [38] in order to solve frictionless rough contact and thus, obtain the evolution of contact stresses against the surface separation, commonly known as the gap. More precisely, BEM solver, introduced by Bemporad and Paggi [50], will be used to solve the contact problem between the artificial rough surfaces and the rigid surface of the disc. The choice of using BEM is motivated by the fact that it offers the best trade-off between FEM and semi-analytical models. Indeed, it is free of any kind of assumption unlike semi-analytical models, and it allows discretizing only the rough surface without the bulk, which can save considerably the CPU cost.

As explained above, the curve load-separation for each generated rough surface is obtained using BEM code developed in [50]. BEM simulation is performed on a generated rough surface of $2.825 \times 2.825\ \text{mm}^2$ with a resolution of 128 pixels in both directions. This numerical setup is chosen because (i) BEM solvers show strong convergence on meshes containing at least 128×128 elements [38] and (ii) the single finite element (patch) of the pad's surface has a averaged area equal to $2.825 \times 2.825\ \text{mm}^2$ (see the element j defined in Figure 3c). Since the pad's material involves an orthotropic behavior, only Young's modulus, $E_z = 1983.3$ MPa, in the most stressed direction, in this case the z-direction, and the Poisson's ratio $\nu_{xz} = 0.4054$ are taken into account as inputs for BEM simulations. It should be noted that these micro-scale simulations assume a homogeneous and linearly elastic rough surface. Therefore, geometric (i.e., large deformations), material (i.e., plasticity) and contact interface non-linearities (i.e., adhesion and friction) are not considered in this computational framework.

Figure 8 shows BEM results for 100 generated rough surfaces. It depicts the evolution of the normalized contact pressure, $\frac{P}{E^*}$ (where E^* is the composite modulus) versus the separation. The latter is defined as a distance between the imposed displacement of the rigid surface (i.e., the surface of the disc) and the median plan, $\bar{z}$, of the rough surface. Each simulation, performed on a generated rough surface of 2.825×2.825 mm^2, demonstrates a non-linear mechanical behavior of the considered set of asperities. For a small displacement, the separation between the two surfaces is large, which implies that only few asperities will be in contact, thus generating a low contact pressure. As long as the imposed displacement increases (synonymous to a decrease of the separation between the rigid surface and the midplane), the real contact area evolves, leading to an increase in contact pressure and, hence, the contact stiffness. This behavior, i.e., pressure-gap law, will be integrated into each patch (i.e., finite element) of the pad's surface of the macroscopic FE model. Following this logic and according to the strategy defined in Figure 3, the macroscopic FE model of a disc brake system will have flat and smooth contact interfaces including the finite element patches. It should be noted that each patch will be driven by a contact law (BEM solution on single generated rough surface), instead of the classic penalty coefficient. Moreover, the gap of each patch is enriched by a threshold according to Equation (16). The new gap formulation will benefit from the HDF of Figure 6 to add a contribution (i.e., threshold) modeling the randomness of the measured height z. This enrichment is intended to ensure the activation of the contact elements in a non-uniform way just like what happens in surfaces with low-scale asperities.

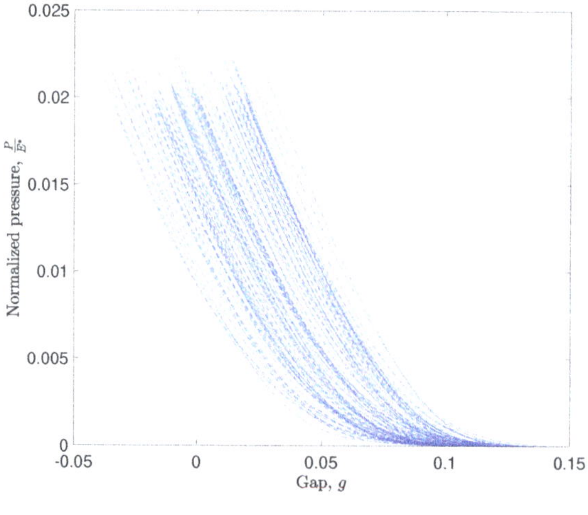

Figure 8. Evolution of the normalized contact pressure, $\frac{P}{E^*}$, as a function of the gap, g for 100 generated rough surfaces. The contact pressure is defined as $P = \frac{F}{L^2}$, where F denotes the contact load and L^2 refers to the finite element path. The gap or the surface separation is defined as a distance between the rotating disc (rigid flat surface) and the median rough surface, $\bar{z}$.

3.3. Application to Dynamic Instabilities Prediction

The multi-scale approach is implemented to predict dynamic instabilities of a disc brake system model composed from three main structural subsystems (see Figure 3c): two pads and a rotating disc. The pad subsystem is broken down into three elements, namely the lining material, the backplate (lining support) and the shim that tends to reduce vibrations. The braking action is represented by a 5 bar pressure applied to the pads on a circular surface. This pressure replaces the action of the hydraulic system which tends to move the piston to squeeze pads against the rotating disc. The current FE model of the braking system has been used to conduct several numerical tests in our previous work [4,5].

The applied multi-scale analysis is divided into four major steps that are part of parallelized loop. After the pad's topography characterization, the loop is started by:

1. **Artificial rough surface generation**: The purpose of this step is to use the characterized HDF and PSD to generate a batch of artificial rough surfaces similar to the measured topography.
2. **Micro-scale contact simulations**: Here, BEM solver introduced in [38,50] is used to solve the elastic contact problem considering artificial surface asperities and a rigid half-plane. The resolution is performed on the whole batch of the generated rough surfaces. Solving the micro-scale contact problem minimizes the convex quadratic program (QP) (see Equation (50) in [38]). The obtained results are the contact load (or the contact pressure) and the separation between the two contacting bodies.
3. **Enrichment of the contact element**: The objective of the third step is to assign each micro-scale contact law (obtained in the second step) to each flat patch (i.e., contact element). Moreover, the gap between slave and master nodes of the macro-scale FE model (disc brake system FE model) is modified by adding a threshold to each gap in order to activate the contact elements in a non-uniform manner. As mentioned above, the added threshold depends closely on the measured HDF.
4. **Complex eigenvalues analysis**: At this stage, the traditional CEA is performed. The beginning of the last step starts with a quasi-static analysis. Its goal is to solve, progressively as the load increases, the frictional contact using the enriched contact elements. At the end of this step, the well-known complex modal analysis is performed to compute both complex eigenvalues and eigenvectors and identify the unstable modes (those with negative damping). For more details, see Section 2.1 in [4].

In the studied example, 300 iterations are used to compute the dynamic instabilities of the braking system.

The result of the multi-scale approach is presented in Figure 9 (blue marker) where the negative damping, a ratio between the real and the imaginary part of the complex eigenvalues, is plotted as a function of the frequency. This figure shows only dynamic instabilities, which means unstable modes with a positive real part (i.e., negative damping below zero). In order to compare the robustness and limitations of the multi-scale approach, the result of FEM considering perfect contact with a pad/disc friction coefficient of 0.5 (red square marker) as well the results of a stochastic model based on a FAST-FEM solver (black marker) developed by Maaboudallah et al. [4] are added in the same figure. It is known [3,44] that stochastic modeling leads to better correlation with experimental data. At first sight, it can be seen that the introduced contact elements enrichment affects the dynamic instability predictions. In particular, the unstable modes predicted by FAST-FEM in the low frequency range are reproduced by the multi-scale approach. However, the latter fails to reproduce the instabilities predicted by FAST-FE or even by standard finite elements with perfect contact at high frequencies. Despite these observations, some unstable family modes around 13 and 15 kHz are roughly highlighted by the multi-scale approach. This unconventional behavior tends to demonstrate that the roughness affects mainly low frequencies. In other words, the addition of the roughness contribution, under the framework mentioned above, can be seen as a necessary but not sufficient condition. Indeed, a robust prediction of dynamic instabilities requires the integration of the roughness but also the randomness of the most sensitive variables like the friction coefficient of pad/disc interface and the normal Young's modulus of the pad.

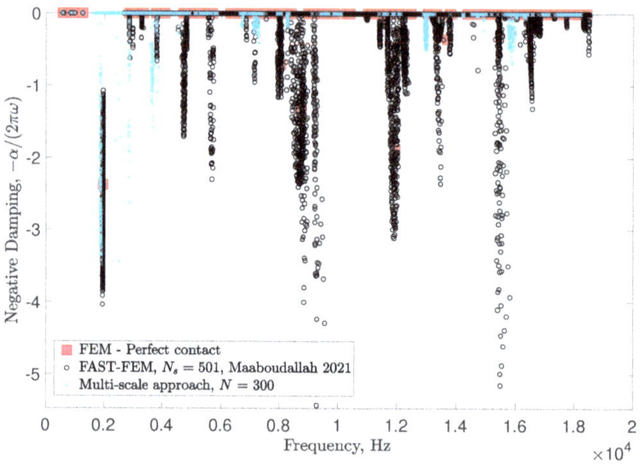

Figure 9. Prediction of dynamic instabilities: comparison between (i) the proposed multi-scale approach considering 300 iterations, (ii) FAST-FEM solver [4] and (iii) the traditional FEM using perfect contact.

4. Conclusions

In this paper, we have presented a multi-scale computational method to predict friction-induced vibrations. The method uses micro-mechanic contact simulations on characterized rough surface to enrich the macro-scale finite element of a rubbing system. The idea behind the proposed approach lies in four essential points:

1. Roughness characterization using the power spectral density function and the height distribution function;
2. Micro-scale contact simulations on the characterization roughness using the boundary element method;
3. Enrichment of the contact finite element using (i) the micro-scale contact laws (obtained from BEM) and (ii) the modified gap functions (obtained from HDF);
4. Performing the complex eigenvalue analysis on the updated stiffness matrix.

The multi-scale method was applied to a scale model of a braking system. The results show that the proposed approach is more accurate than the classical one using a perfect contact law, based on a comparison with a stochastic model accounting for the uncertainties on several parameters such as friction. In addition, it is found that roughness has an effect on the predictions of dynamic instabilities. Indeed, taking into account the roughness brings out new unstable modes in the low frequencies. However, the effect of the integrated roughness is not clear in the high frequencies, since many stochastically predicted modes were not reproduced by the multi-scale strategy. Results of the investigation demonstrate that roughness is only one element among others that must be taken into account for a robust and accurate prediction of the dynamic instabilities. An uncertainty analysis must be conducted by varying the most important parameters including roughness to correctly model and predict the dynamic instabilities of frictional systems.

Author Contributions: Conceptualization, F.M.; methodology, F.M.; software, F.M. and N.A.; validation, F.M.; formal analysis, F.M.; investigation, F.M.; resources, N.A.; writing—original draft preparation, F.M.; writing—review and editing, F.M. and N.A.; visualization, F.M.; supervision, N.A.; project administration, F.M. and N.A. All authors have read and agreed to the published version of the manuscript.

Funding: This research received no external funding.

Acknowledgments: The authors acknowledge the financial support of Natural Sciences and Engineering Research Council of Canada (NSERC) and Fiat Chrysler Automobile Canada (FCA). This research was enabled in part by support provided by Compute Canada, Compute Quebec and SHARCNET. The first author would like to warmly thank Vladislav Yastrebov and Marco Paggi for their immense help and extremely relevant explanations during the workshops organized at Mines-Paristech France and IMT-Lucca Italy, respectively. This research work could not have been done without their help.

Conflicts of Interest: The authors declare no conflict of interest.

Abbreviations

The following abbreviations are used in this manuscript:

HDF	height distribution function
ACF	autocorrelation function
PSD	power spectral density
FEM	finite element method
BEM	boundary element method
GFMD	Green's function molecular dynamics
RVE	representative volume element
CEA	complex eigenvalues analysis
N2S	Node-to-Surface
CLSM	confocal laser scanning microscopy
RV	random variable
RMS	root mean square
FAST	Fourier sensitivity amplitude test

References

1. Sun, S.; Li, L.; Yue, Z.; Yang, W.; Zhao, Z.; Cao, R.; Li, S. Experimental and numerical investigation on fretting fatigue behavior of Nickel-based single crystal superalloy at high temperature. *Mech. Mater.* **2020**, *150*, 103595. [CrossRef]
2. Yastrebov, V.A.; Durand, J.; Proudhon, H.; Cailletaud, G. Rough surface contact analysis by means of the Finite Element Method and of a new reduced model. *Comptes Rendus Mécanique* **2011**, *339*, 473–490. [CrossRef]
3. Tison, T.; Heussaff, A.; Massa, F.; Turpin, I.; Nunes, R.F. Improvement in the predictivity of squeal simulations: Uncertainty and robustness. *J. Sound Vib.* **2014**, *333*, 3394–3412. [CrossRef]
4. Maaboudallah, F.; Atalla, N. An efficient numerical strategy to predict the dynamic instabilities of a rubbing system: Application to an automobile disc brake system. *Comput. Mech.* **2021**, *67*, 1465–1483. [CrossRef]
5. Maaboudallah, F.; Atalla, N. A "data-driven uncertainty" computational method to model and predict instabilities of a frictional system. *Adv. Model. Simul. Eng. Sci.* **2023**, *10*, 3. [CrossRef]
6. Zhou, Q.; Luo, D.; Hua, D.; Ye, W.; Li, S.; Zou, Q.; Chen, Z.; Wang, H. Design and characterization of metallic glass/graphene multilayer with excellent nanowear properties. *Friction* **2022**, *10*, 1913–1926. [CrossRef]
7. Ren, Y.; Huang, Z.; Wang, Y.; Zhou, Q.; Yang, T.; Li, Q.; Jia, Q.; Wang, H. Friction-induced rapid amorphization in a wear-resistant (CoCrNi)88Mo12 dual-phase medium-entropy alloy at cryogenic temperature. *Compos. Part B Eng.* **2023**, *263*, 110833. [CrossRef]
8. Hetzler, H.; Willner, K. On the influence of contact tribology on brake squeal. *Tribol. Int.* **2012**, *46*, 237–246. [CrossRef]
9. Greenwood, J.A.; Williamson, J.B.P.; Bowden, F.P. Contact of nominally flat surfaces. *Proc. R. Soc. London. Ser. A Math. Phys. Sci.* **1966**, *295*, 300–319. [CrossRef]
10. Nayak, P.R. Random Process Model of Rough Surfaces. *J. Lubr. Technol.* **1971**, *93*, 398–407. [CrossRef]
11. Abbott, E.J.; Firestone, F.A. Specifying Surface Quality—A Method Based on Accurate Measurement and Comparison. *J. Mech. Eng.* **1933**, *55*, 569–572.
12. Bhushan, B. Solid Surface Characterization. In *Introduction to Tribology*; John Wiley & Sons, Ltd.: Hoboken, NJ, USA, 2013; Chapter 2, pp. 9–89. [CrossRef]
13. Persson, B.N.J.; Albohr, O.; Tartaglino, U.; Volokitin, A.I.; Tosatti, E. On the nature of surface roughness with application to contact mechanics, sealing, rubber friction and adhesion. *J. Phys. Condens. Matter* **2004**, *17*, R1–R62. [CrossRef]
14. Persson, B.N.J. Contact mechanics for randomly rough surfaces. *Surf. Sci. Rep.* **2006**, *61*, 201–227. [CrossRef]
15. Majumdar, A.; Bhushan, B. Fractal Model of Elastic-Plastic Contact Between Rough Surfaces. *J. Tribol.* **1991**, *113*, 1–11. [CrossRef]
16. Dodds, C.J.; Robson, J.D. The description of road surface roughness. *J. Sound Vib.* **1973**, *31*, 175–183. [CrossRef]
17. Majumdar, A.; Tien, C.L. Fractal characterization and simulation of rough surfaces. *Wear* **1990**, *136*, 313–327. [CrossRef]
18. Najah, M.; Maaboudallah, F.; Boucherit, M.; Ferguson, M.; Fréchette, L.; Charlebois, S.; Boone, F.; Ecoffey, S. Spectral analysis of the topography parameters for isotropic Gaussian rough surfaces applied to gold coating. *Tribol. Int.* **2022**, *165*, 107339. [CrossRef]
19. Hertz, H. Ueber die Berührung fester elastischer Körper. *J. Für Die Reine Und Angew. Math.* **1882**, *1882*, 156–171. [CrossRef]

20. Johnson, K.L. *Contact Mechanics*; Cambridge University Press: Cambridge, UK, 1985. [CrossRef]
21. Archard, J.F.; Allibone, T.E. Elastic deformation and the laws of friction. *Proc. R. Soc. London. Ser. A. Math. Phys. Sci.* **1957**, *243*, 190–205. [CrossRef]
22. Longuet-Higgins, M.S.; Deacon, G.E.R. Statistical properties of an isotropic random surface. *Philos. Trans. R. Soc. London. Ser. Math. Phys. Sci.* **1957**, *250*, 157–174. [CrossRef]
23. Yastrebov, V.A.; Anciaux, G.; Molinari, J.F. The role of the roughness spectral breadth in elastic contact of rough surfaces. *J. Mech. Phys. Solids* **2017**, *107*, 469–493. [CrossRef]
24. Ciavarella, M.; Delfine, V.; Demelio, G. A "re-vitalized" Greenwood and Williamson model of elastic contact between fractal surfaces. *J. Mech. Phys. Solids* **2006**, *54*, 2569–2591. [CrossRef]
25. Ciavarella, M.; Greenwood, J.A.; Paggi, M. Inclusion of "interaction" in the Greenwood and Williamson contact theory. *Wear* **2008**, *265*, 729–734. [CrossRef]
26. Paggi, M.; Ciavarella, M. The coefficient of proportionality κ between real contact area and load, with new asperity models. *Wear* **2010**, *268*, 1020–1029. [CrossRef]
27. Zhang, S.; Song, H.; Sandfeld, S.; Liu, X.; Wei, Y.G. Discrete GreenwoodWilliamson Modeling of Rough Surface Contact Accounting for Three-Dimensional Sinusoidal Asperities and Asperity Interaction. *J. Tribol.* **2019**, *141*, 121401. [CrossRef]
28. Afferrante, L.; Carbone, G.; Demelio, G. Interacting and coalescing Hertzian asperities: A new multiasperity contact model. *Wear* **2012**, *278–279*, 28–33. [CrossRef]
29. Persson, B.N.J. Theory of rubber friction and contact mechanics. *J. Chem. Phys.* **2001**, *115*, 3840–3861. [CrossRef]
30. Hyun, S.; Pei, L.; Molinari, J.F.; Robbins, M.O. Finite-element analysis of contact between elastic self-affine surfaces. *Phys. Rev. E* **2004**, *70*, 026117. [CrossRef]
31. Yang, C.; Persson, B.N.J. Contact mechanics: Contact area and interfacial separation from small contact to full contact. *J. Phys. Condens. Matter* **2008**, *20*, 215214. [CrossRef]
32. Yastrebov, V.A.; Anciaux, G.; Molinari, J.F. From infinitesimal to full contact between rough surfaces: Evolution of the contact area. *Int. J. Solids Struct.* **2015**, *52*, 83–102. [CrossRef]
33. Polonsky, I.A.; Keer, L.M. A numerical method for solving rough contact problems based on the multi-level multi-summation and conjugate gradient techniques. *Wear* **1999**, *231*, 206–219. [CrossRef]
34. Campañá, C.; Müser, M.H. Practical Green's function approach to the simulation of elastic semi-infinite solids. *Phys. Rev. B* **2006**, *74*, 075420. [CrossRef]
35. Campañá, C.; Müser, M.H. Contact mechanics of real vs. randomly rough surfaces: A Green's function molecular dynamics study. *Europhys. Lett. Assoc.* **2007**, *77*, 38005. [CrossRef]
36. Wriggers, P. Finite element algorithms for contact problems. *Arch. Comput. Methods Eng.* **1995**, *2*, 1–49. [CrossRef]
37. Yastrebov, V.A. *Numerical Methods in Contact Mechanics*, 1 ed.; John Wiley & Sons, Ltd.: Hoboken, NJ, USA, 2013. [CrossRef]
38. Maaboudallah, F.; Najah, M.; Atalla, N. *A Review on the Contact Mechanics Modeling of Rough Surfaces in the Elastic Regime: Fundamentals, Theories, and Numerical Implementations*; IntechOpen: Rijeka, Croatia, 2022; Chapter 5. [CrossRef]
39. Wriggers, P.; Reinelt, J. Multi-scale approach for frictional contact of elastomers on rough rigid surfaces. *Comput. Methods Appl. Mech. Eng.* **2009**, *198*, 1996–2008. [CrossRef]
40. Wriggers, P.; Nettingsmeier, J. Homogenization and Multi-Scale Approaches for Contact Problems. In *Computational Contact Mechanics*; Wriggers, P., Laursen, T.A., Eds.; CISM International Centre for Mechanical Sciences; Springer: Berlin/Heidelberg, Germany, 2007; pp. 129–161. [CrossRef]
41. Waddad, Y.; Magnier, V.; Dufrénoy, P.; De Saxcé, G. A multiscale method for frictionless contact mechanics of rough surfaces. *Tribol. Int.* **2016**, *96*, 109–121. [CrossRef]
42. Bonari, J.; Marulli, M.R.; Hagmeyer, N.; Mayr, M.; Popp, A.; Paggi, M. A multi-scale FEM-BEM formulation for contact mechanics between rough surfaces. *Comput. Mech.* **2020**, *65*, 731–749. [CrossRef]
43. Paggi, M.; Reinoso, J. A variational approach with embedded roughness for adhesive contact problems. *Mech. Adv. Mater. Struct.* **2020**, *27*, 1731–1747. [CrossRef]
44. Renault, A.; Massa, F.; Lallemand, B.; Tison, T. Experimental investigations for uncertainty quantification in brake squeal analysis. *J. Sound Vib.* **2016**, *367*, 37–55. [CrossRef]
45. Fenton, G.A.; Griffiths, D. Random Field Generation and the Local Average Subdivision Method. In *Probabilistic Methods in Geotechnical Engineering*; Griffiths, D.V., Fenton, G.A., Eds.; CISM Courses and Lectures; Springer: Vienna, Austria, 2007; pp. 201–223.
46. Li, C.C.; Der Kiureghian, A. Optimal Discretization of Random Fields. *J. Eng. Mech.* **1993**, *119*, 1136–1154. [CrossRef]
47. Zhang, Z.; Oberst, S.; Lai, J.C.S. On the potential of uncertainty analysis for prediction of brake squeal propensity. *J. Sound Vib.* **2016**, *377*, 123–132. [CrossRef]
48. Magnier, V.; Brunel, J.F.; Dufrénoy, P. Impact of contact stiffness heterogeneities on friction-induced vibration. *Int. J. Solids Struct.* **2014**, *51*, 1662–1669. [CrossRef]
49. Maaboudallah, F.; Atalla, N. Beyond the main order sensitivity analysis for a frictional system: Is the eXtended FAST algorithm applicable? *Nonlinear Dyn.* **2022**, 111, 5593–5614. [CrossRef]
50. Bemporad, A.; Paggi, M. Optimization algorithms for the solution of the frictionless normal contact between rough surfaces. *Int. J. Solids Struct.* **2015**, *69–70*, 94–105. [CrossRef]

51. Borri, C.; Paggi, M. Topology simulation and contact mechanics of bifractal rough surfaces. *Proc. Inst. Mech. Eng. Part J J. Eng. Tribol.* **2016**, *230*, 1345–1358. [CrossRef]
52. Wu, J.J. Simulation of rough surfaces with FFT. *Tribol. Int.* **2000**, *33*, 47–58. [CrossRef]

lubricants

MDPI

Article

A Mixed Lubrication Model of Piston Rings on Cylinder Liner Contacts Considering Temperature-Dependent Shear Thinning and Elastic–Plastic Contact

Nolan Ryan Chu [1], Robert L Jackson [1,*], Hamed Ghaednia [2] and Arup Gangopadhyay [3,†]

1 Department of Mechanical Engineering, Auburn University, Auburn, AL 36849, USA; nrc0011@auburn.edu
2 Gehring L.P., Farmington Hills, MI 48335, USA
3 Ford Motor Company, Dearborn, MI 48124, USA
* Correspondence: jacksr7@auburn.edu
† Retired.

Abstract: This work develops a numerical methodology for predicting the performance of an automotive piston ring system by considering contact and lubrication mechanics. The rough surface contact mechanics and lubrication occurs on a scale much smaller than the size of the piston rings and therefore the key aspect of the model is an algorithm that simultaneously solves the multiple mechanisms at different scales. The finite element method will be used to model the mechanical deformations of the piston ring surfaces at large scales. The quasi-steady state model includes heat generation due to solid and viscous friction. This heat generation will then be used to predict the temperature rise and thermal effects in the lubricant and component. A statistical rough surface method that renders asperities as elastic–plastic wavy surfaces predicts the solid contact area. The modified Reynolds equation will be solved to consider the effects of mixed hydrodynamic lubrication while using flow factors formulated for actual piston and ring surfaces. The lubricant viscosity depends both on temperature and shear rate. This will allow for the regimes of boundary, mixed, and full-film lubrication to be considered. The model predicts friction for various loads and speeds that are then compared to experimental measurements. Although the contacts operate mostly in the mixed lubrication regime, the model and experiments show changes in friction with load, speed, and temperature.

Keywords: rough surface elastic-plastic contact; sinusoidal asperity; boundary lubrication; flow factors; shear thinning; thermal heating; automotive

Citation: Chu, N.R.; Jackson, R.L.; Ghaednia, H.; Gangopadhyay, A. A Mixed Lubrication Model of Piston Rings on Cylinder Liner Contacts Considering Temperature-Dependent Shear Thinning and Elastic–Plastic Contact. *Lubricants* **2023**, *11*, 208. https://doi.org/10.3390/lubricants11050208

Received: 20 January 2023
Revised: 23 March 2023
Accepted: 29 March 2023
Published: 7 May 2023

1. Introduction

Combustion engines are still the most common source of power for vehicles at 97% of the vehicle market in 2021 [1]. Electric vehicles will displace some combustion vehicles, but it will take many years for them to overtake combustion vehicles in number. The rate at which this is predicted to occur varies drastically depending on the organization making the prediction (between 20% and 90% of vehicles by 2050) [2]. The adoption of electric vehicles will also be slower in developing countries [1]. Certain applications and types of vehicles, such as heavy trucks, will also mostly consist of combustion powered vehicles for a longer period. Therefore, combustion will arguably be a significant part of the market for many years.

Piston assembly is also the largest source of friction loss in combustion engines [3]. Therefore, there has been a great deal of work on reducing the friction in the piston ring–cylinder liner interface using such things and coatings and textures, as will be discussed briefly below. To evaluate possible technologies, it is also advantageous to have models capable of predicting the friction. Therefore, this work aims at creating a model that compares well to experimental measurements of a piston ring to cylinder liner interface.

The model will consist of a rough surface contact and hydrodynamic lubrication modules and therefore a brief background of these areas is also provided. While many models for the individual modules exist, there are few models that combine all these aspects at once. This work creates a model that combines wavy elastic–plastic rough surface contact and hydrodynamic lubrication and evaluates three variants, two of which introduce frictional heating.

1.1. Prior Works on the Piston Ring–Cylinder Liner Interface

Many methods and techniques have been used in the past to predict the frictional losses of a piston ring–cylinder liner interface in an engine [4–20]. The piston ring–cylinder liner interface system includes several variables such as surface topography, transient lubrication, and exhaust flow rate that interact with each other. Prior works have generally focused on only one variable at a time. Furuhama and Sumi's [4] analysis of compression ring linings is one of the first investigations on the subject. Ma et al. [5] analyzed lubricant transport and found that cylinder liner surface and ring movement significantly influence tribological behavior. Akalin and Newaz [6] analyzed the mixed lubrication regime using the Reynolds equation with flow factors. They found that hydrodynamic lubrication occurs during most parts of the stroke, but the friction coefficient increases greatly at top and bottom dead center when the sliding speed is too low for the lubricant to support much of the load. Their analysis did not calculate flow factors for a specific surface or consider elastic–plastic asperity contact, nor did they perform experiments to validate their model. Jeng [7] analyzed the lubrication conditions at the ring contact surface. Furuhama and Sasaki [8] derived a new technique to measure friction forces for small engines. Previously, this was a difficult task because they could not be isolated from much larger gas and inertia forces. Taking into account torsion, film thickness changes, and ring wear, Tian [9] studied piston ring dynamics numerically and experimentally. Harigaya et al. and Rahmani et al. investigated temperature effects on the friction and piston ring lubrication [10,11].

In recent years, several studies have focused on improving internal combustion engine efficiency. Morris et al. [12] optimized the piston ring to minimize energy losses, incidence of asperity contact, and ring mass. Bewsher et al. [13] applied atomic force microscopy to measure the boundary asperity shear strength and thus calculate localized values of frictional losses on real engine components. Howell-Smith et al. [14] tested lubricant coatings and surface textures for friction reduction. They found that surface modifications of the liner at top dead center (TDC) reduces friction by creating additional lubricant reservoirs there. In turn, this increases the power output of the engine by up to 4%. Li et al. [15] found that laser finishing could reduce the friction coefficient and weight loss of an Al-Si alloy cylinder liner by removing the aluminum layer and exposing rounded edges of silicon particles.

Efficiency can also be improved by changing the surface texture. Senatore et al. [16] studied a bronze coating with different surface textures and found that an appropriate texture geometry improves the friction coefficient and wear. Wang et al. [17] tested the effects of dimples on brass discs. They found that only a small dimple pattern reduced the friction—for large dimples, the friction coefficient actually increased. Kligerman et al. [18] developed an analytical model for partial laser surface texturing to reduce the friction in the piston ring–cylinder liner system. They found an optimal percentage of the textured portion and dimple depth depending on operating conditions. Spencer [19] developed simulations to evaluate a cross-hatched cylinder liner to reduce oil consumption, wear, and friction. Lu and Wood [20] observed an 82% reduction in piston ring–cylinder liner friction when texture grooves were normal to the sliding direction. Abril et al. [21] studied the effects of dimples and the honing groove in the cylinder liner. A slight increase in dimple density increased the minimum film thickness and reduced the friction force. Comparable increases in minimum film thickness could be obtained with deeper, larger dimples. Their honing groove analysis found that a 15 degree increase in honing angle reduced the friction coefficient by more than 14%. However, friction increased when the honing groove density was too high.

1.2. Rough Surface Contact

Contact between rough surfaces is a ubiquitous problem that can be applied to numerous phenomena such as friction, wear, and contact resistance. This work employs a statistical model in which mathematical parameters describing the surface are used to calculate probabilities and determine the contact area and load. This model was initially developed by Greenwood and Williamson [22] (GW model). They considered the interaction between a perfectly flat, rigid plane and a plane covered with spherical asperities of varying heights. They assumed that asperities behave independently of each other, and that deformation is limited to the asperities. However, they only assumed elastic contact, so other models were subsequently derived when yielding occurs at larger loads. Jackson and Green (JG) [23] derived a statistical elastic–plastic deformation model in which they established the load required for plastic deformation. As contact pressure increases, the internal stress within asperities increases as well. This results in yielding and plastic deformation. The JG model, while it includes varying fully plastic pressure not captured by most other models, is limited to small deformations where the contact radius is no more than 41% of the radius of curvature. Note that wavy asperities also result in a varying fully plastic pressure with load and can also be incorporated into statistical rough surface contact models [24].

Statistical models are reliable and easily implemented, but shortcomings exist. Those previously described assume a homogenous radius of curvature over an entire region, neglect the effects of different scales of features, and do not couple the deformation between asperities and the substrate. Bush et al. [25] developed a statistical model that accounts for variable asperity radius, but they still assumed negligible adjacent or lateral asperity interaction. Ciavarella et al. developed a model that accounts for lateral asperity interaction [26]. Afferrante et al. followed up with a coalescing asperity model, while Vakis expanded it below the mean asperity height [27,28]. These works are similar to the wavy asperity model used in this work that includes lateral asperity interaction. A recent work compared spherical and wavy asperity-based statistical models to a deterministic prediction. The wavy asperity model compared best and will therefore be implemented here [29].

1.3. Hydrodymanic Lubrication

To calculate the hydrodynamic load in modeling viscous flow of lubricant between the cylinder wall and the piston ring, the modified Reynolds Equation is used. It is a second-order partial differential equation derived from the Navier–Stokes equations assuming a Newtonian fluid, negligible inertia and body forces, negligible pressure variation across the film, laminar flow, and negligible curvature [30]. It can take many forms, depending on the physical mechanisms involved in the system.

Flow Factors are a method to determine roughness effects on lubrication flow in any of the three regimes: full film lubrication, mixed lubrication, and boundary lubrication. These regimes are depicted in Figure 1.

Boundary lubrication, which is characterized by high surface abrasion and wear, is on the left side. On the right side, the lubricant separates the surfaces sufficiently such that no solid contact occurs. The lubrication regimes can be categorized by the Stribeck curve, a plot of friction coefficient against the dimensionless bearing number, shown in Figure 2. It is used to determine transitions between flow regimes.

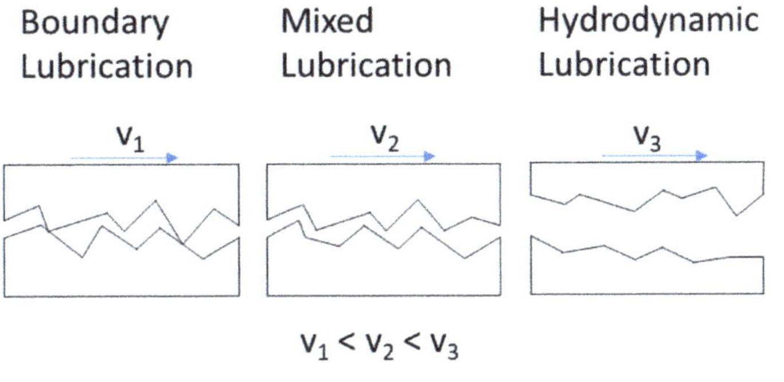

Figure 1. Lubrication Regimes.

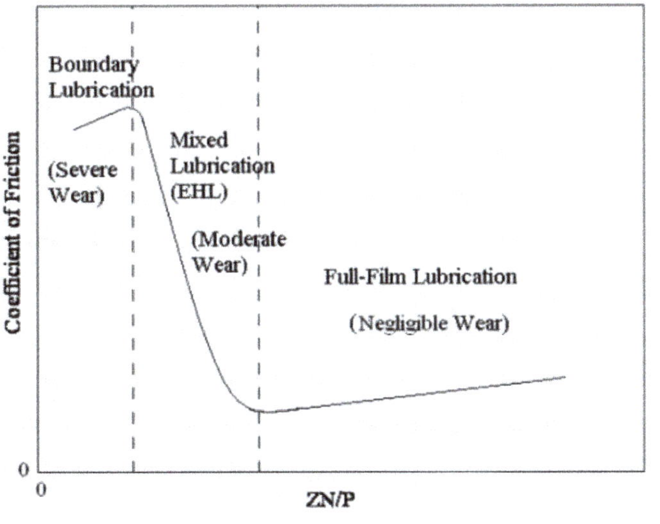

Figure 2. Sample Stribeck Curve [31].

Lubricants interact at both the macroscopic and the microscopic scales; the latter is especially important for the surfaces. Using the Reynolds equation to determine the pressure at each asperity is possible akin to a deterministic contact model, but it becomes computationally unfeasible for a sufficiently large surface resolution. Flow past individual asperities is too computationally and numerically difficult to model, so the simpler method of flow factors added to the Reynolds equation is used instead. Patir and Cheng [32] were the first to determine the effects of surface roughness on flow between three-dimensional surfaces. They derived statistical flow factors added to the Reynolds equation as follows:

$$\frac{\partial}{\partial x}\left(\varphi_x \frac{\rho h^2}{12\mu}\frac{\partial p}{dx}\right) + \frac{\partial}{\partial z}\left(\varphi_z \frac{\rho h^2}{12\mu}\frac{\partial p}{dz}\right) = \frac{\overline{U}}{2}\frac{\partial \rho h}{\partial x} + \frac{\partial \rho h}{\partial t} + \frac{\partial \varphi_s}{\partial x} \quad (1)$$

This modified Reynolds equation produces a more accurate solution that accounts for microscopic surface features. In this equation, φ_z and φ_x measure the flow resistance across asperities in the flow direction and the transverse direction, respectively, while φ_s measures lubricant transport due to shear effects. The flow factors depend on the film thickness, the RMS surface roughness, and the Peklenik number, γ. This number can be calculated from auto-correlation functions derived from the surface topography [33].

The flow factors calculated by Patir and Cheng were based off a statistically generated surface whose asperities were purely transverse, isotropic, or purely longitudinal. However, their flow factors are not totally accurate for a real surface that is not perfectly Gaussian. Other researchers tried methods to improve upon Patir and Cheng's work to find a more accurate model for specific cases. Wilson and Marsault derived an alternate form of the Reynolds equation applicable for high contact area ratios [34]. Peeken et al. investigated flow factors for sintered bearing surfaces [35]. Hu and Zheng considered different boundary conditions and numerical methods to calculate flow factors but still considered theoretical surfaces [36]. Morales-Espejel derived a transformation to calculate flow factors for a non-Gaussian surface from their counterparts for a Gaussian surface [37]. Sahlin et al. devised a new way to calculate flow factors that accounts for contact mechanics and used measured surfaces to do so [38]. Their results agreed with those of Patir and Cheng for longitudinal asperities but differed substantially for cross-hatched surfaces. Others applied various numerical and analytical methods [39–41]. The methodology used here is similar to work by Leighton et al. [42], who derived surface-specific flow factors for a piston ring–cylinder liner interface.

Previous studies of surface roughness effects on lubrication flow was mostly limited to stochastic concepts such as those first introduced by Tzeng and Saibel [43]. Patir and Cheng [24,44] derived a new method based on numerically solving the Reynolds equation over a randomly generated surface and calculating an average equation from flow quantities. Their method assumes that rough surface heights are a perfect Gaussian distribution though. This work uses the flow factors Locker et al. [45] derived for an actual cross-hatched cylinder wall by combining stochastic concepts with Peklinik's signal processing theory [33].

2. Model Methodology

Figure 3 shows the overview of the iterative axisymmetric model of the piston ring–cylinder liner interface. While a full 3D model would be ideal, that was deemed to be too computationally expensive due to the number of iterations and finite element analyses required. The model is also pseudo-steady state, meaning it does not consider transient or dynamic effects, but it does consider different sliding speeds, U (see Figure 4). As shown in Figure 3, the problem is solved iteratively by updating the forces, deformations, and surface separation with each iteration. On the inner radius of the ring, the applied load, P, (50, 100, or 150 N) was converted to a pressure using the area on the inside. This mimics the loads applied in the experiment and those experienced in an actual engine (see Figure 3). Combining the applied load on the inside with the contact and fluid pressures on the outside gives an equation that relates net radial force and surface separation. The location of the ring was numerically solved for a net zero radial force. The contact and hydrodynamic forces that solve the equilibrium equation are written in an Abaqus™ input file that creates the mesh shown later. The input file is then run in Abaqus™, and the toolbox abaqus2matlab is used to transfer the displacements back to MATLAB™. These displacements are then used to alter the piston ring profile, and the process is repeated until convergence is reached. The problem is considered converged when the difference in the forces between iterations is less than 0.5%. Additional details are provided in the following sections.

A profilometer was used to measure the surface height of a 1 mm^2 sample area from a cylinder liner and a piston ring. The diamond stylus employed has a tip radius of 2 μm and a resolution of 1 nm. Due to the profilometer's sensitivity, it was operated on a self-leveling pneumatic vibration isolation table to ensure steadiness. The surface profiles are shown in Figures 5–8. As shown in Figures 5 and 6, the cylinder liner surface has a cross-hatched finishing. The ring surface (Figures 7 and 8) also possesses some dimples by design. All surface parameters for the lubrication and the statistical rough surface contact modules were calculated from these measurements. Additional details about the parameter calculations are provided later.

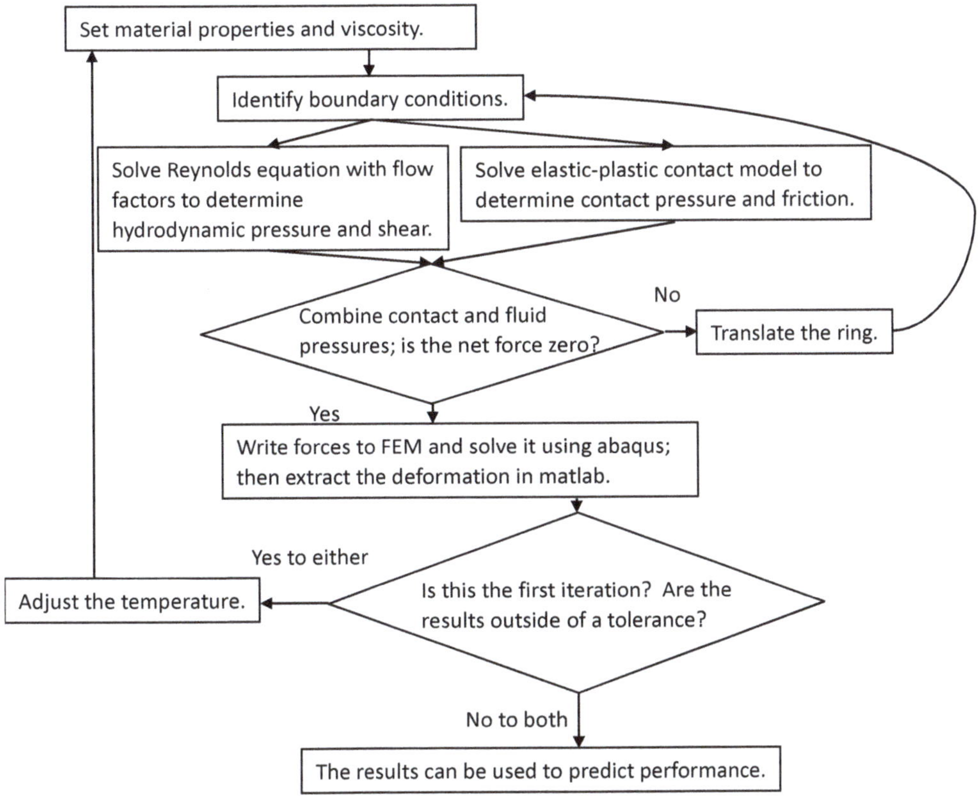

Figure 3. Model Flowchart.

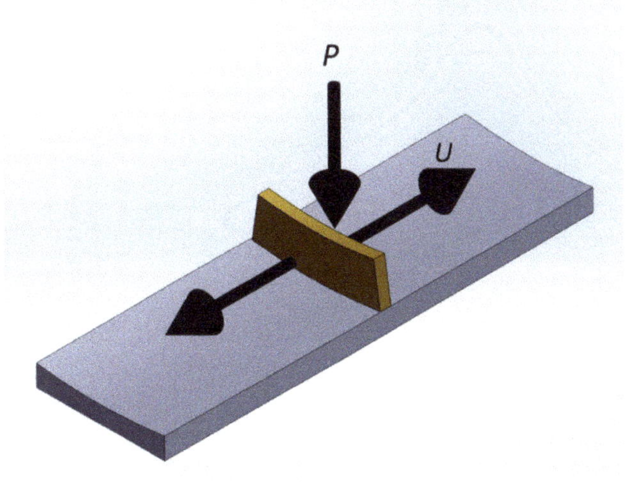

Figure 4. Schematic of piston ring and cylinder liner interface.

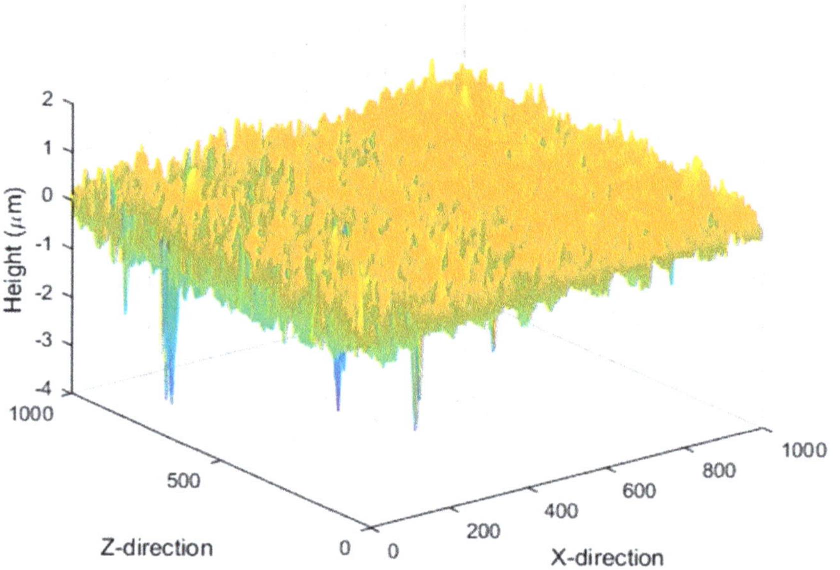

Figure 5. Isometric view of cylinder liner (x and z direction dimensions are in μm). Yellow indicates taller height and blue lower values.

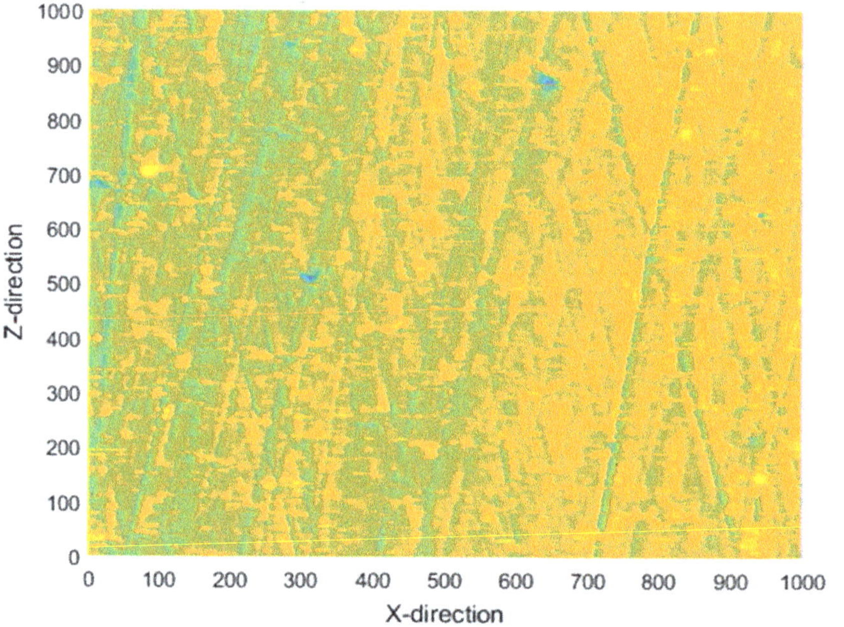

Figure 6. Top view of cylinder liner surface (x and z direction dimensions are in μm). Yellow indicates taller height and blue lower values.

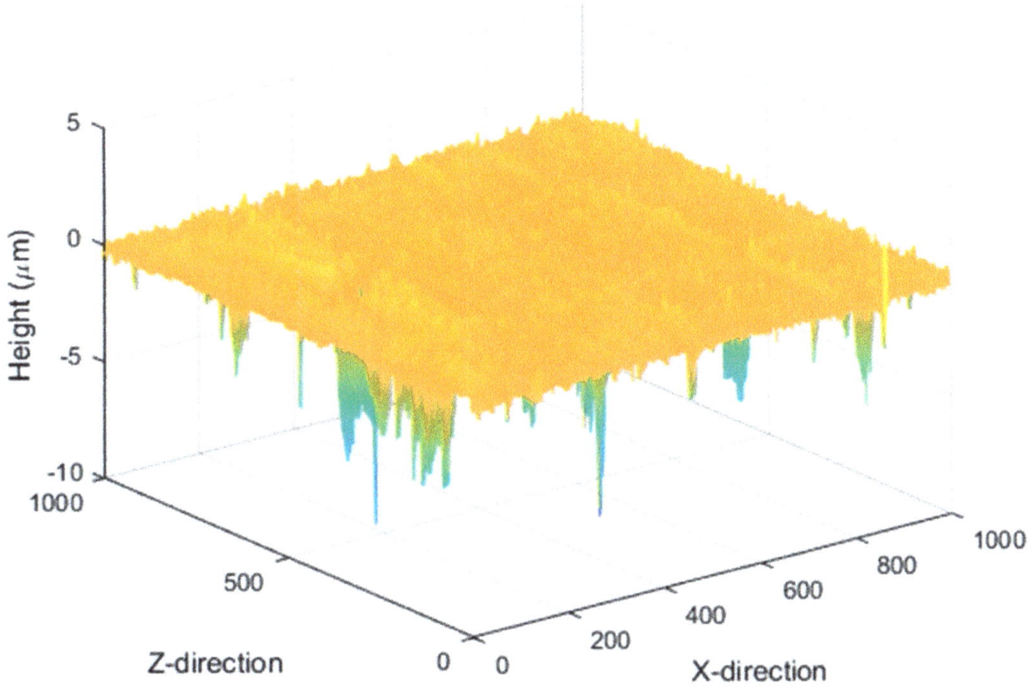

Figure 7. Isometric view of piston ring (x and z direction dimensions are in μm). Yellow indicates taller height and blue lower values.

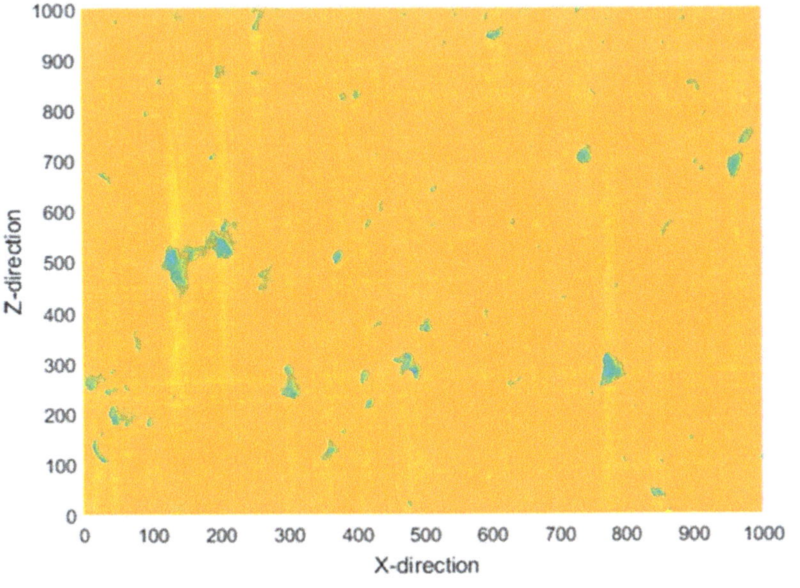

Figure 8. Top view of piston ring surface (x and z direction dimensions are in μm). Yellow indicates taller height and blue lower values.

2.1. Rough Surface Contact

This work uses the Greenwood–Williamson statistical model [22] as a base for predicting the effects of rough surface contact. The equations to find the total contact load and area are

$$P(h) = A_n \eta \int_h^{\infty} \overline{P}(z-h)\phi(z)dz \tag{2}$$

$$A(d) = A_n \eta \int_d^{\infty} \overline{A}(z-h) \cdot \phi(z) \cdot dz \tag{3}$$

P is the total contact force, A_n is the nominal area of contact (neglecting roughness), h is the mean surface separation, η is the areal asperity density, and ϕ is the asperity height distribution. Asperities are assumed to be homogenous and evenly distributed; their RMS (root mean square) height is σ_s. To calculate the statistical parameters, asperities were manually counted by scanning the surface profile and identifying points whose height was higher than any of the eight surrounding points. The radius of curvature of each asperity was calculated in two orthogonal directions, using

$$r_x = \frac{\left[1 + \frac{dh}{dx}\right]^{\frac{3}{2}}}{\frac{d^2h}{dx^2}} \tag{4}$$

and

$$r_y = \frac{\left[1 + \frac{dh}{dy}\right]^{\frac{3}{2}}}{\frac{d^2h}{dy^2}} \tag{5}$$

The two values were averaged to estimate asperity's radius of curvature r. This parameter was calculated for every asperity, and then it was averaged to find R for all the asperities. The asperity density η was found by dividing the number of asperities counted by the area scanned. The original G–W model assumes elastic Hertz contact and a constant value of R. Different equations are used here because this work assumes the asperities are sinusoidal in nature and the loads are large enough for yielding to occur. The following relations were used to convert the asperity radius and density to frequency and amplitude:

$$f = \sqrt{\frac{\eta}{2}} \tag{6}$$

$$\Delta = \frac{1}{4R(f\pi)^2} \tag{7}$$

For a single $\lambda \times \lambda$ ($\lambda = 1/f$) wavy asperity area to reach complete elastic contact, the pressure required is given by [46]

$$p^* = \sqrt{2}\pi E' f \Delta \tag{8}$$

where the effective elastic modulus to account for the deformations of both surfaces is given by

$$\frac{1}{E'} = \frac{1-\nu_1^2}{E_1} + \frac{1-\nu_2^2}{E_2} \tag{9}$$

assuming plane strain. Assuming an average contact pressure of $\overline{p}$, a required pressure p^* for complete contact, and letting

$$P_e = \frac{\overline{p}}{p^*} \tag{10}$$

The following asymptotic solutions were found by Johnson, Greenwood, and Higginson [46]:

$$\left(\overline{A}_{JGH}\right)_1 = \pi\lambda^2\left[\frac{3P_e}{8\pi}\right]^{\frac{2}{3}} \tag{11}$$

for small values of P_e and

$$\left(\overline{A}_{JGH}\right)_2 = \frac{\lambda^2}{2}\left(1 - \frac{3}{2\pi}[1 - P_e]\right) \tag{12}$$

when P_e approached unity.

Jackson and Streator [47] fitted a polynomial combining these equations using experimental data from Johnson et al. [46]:

$$\widetilde{A} = \begin{cases} \left(\overline{A}_{JGH}\right)_1\left(1 - {P_e}^{0.51}\right) + \left(\overline{A}_{JGH}\right)_2 {P_e}^{1.04}, P_e < 0.8 \\ \left(\overline{A}_{JGH}\right)_2, P_e \geq 0.8 \end{cases} \tag{13}$$

These equations neglect asperity yielding, so the elastic–plastic model developed by Krithivasan and Jackson is used instead. They derived an expression for the contact area above which elastic–plastic contact occurs. This was derived from spherical contact, so this work uses a model developed by Jackson et al. [48] that computes the critical interference above which elastic–plastic relations are used. That expression from Ghaednia et al. [49] is

$$\Delta_c = \frac{\sqrt{2}S_y}{E' f\pi\left[3e^{-\frac{2}{3}(\nu+1)} + 2\left(\frac{1-2\nu}{1-\nu}\right)\right]} \tag{14}$$

Using this value of critical interference, the following equation was fitted to the FEM data of Krithivasan and Jackson [24] that links the pressures required for complete contact under elastic and elastic–plastic loading:

$$\frac{{P^*}_{ep}}{p^*} = 0.992^{\left[\left\{\frac{\Delta}{\Delta_c}\right\}^{\frac{10}{3}\left(\frac{\Delta}{\Delta_c}\right)^{-0.39} + \frac{9}{4}\nu^4 + 0.64} - 1\right]} \tag{15}$$

The contact area for low loads is found using

$$A_p = 2\left(\frac{A_c}{2}\right)^{\frac{1}{1+d}}\left(\frac{3\overline{p}}{4Cf^2S_y}\right)^{\frac{d}{1+d}} \tag{16}$$

where

$$d = 3.8\left(\frac{E' f\Delta}{S_y}\right)^{0.11} \tag{17}$$

and

$$A_c = \frac{2}{\pi}\left(\frac{CS_y}{8E' f^2\Delta}\right)^2 \tag{18}$$

is the critical contact area at which elastic–plastic contact begins.

The equation that links the contact area for low and high loads is

$$\overline{A} = A_p\left(1 - {P_{ep}}^{1.51}\right) + \left(\overline{A}_{JGH}\right)_2 {P_{ep}}^{1.04} \tag{19}$$

In this equation, $\left(\overline{A}_{JGH}\right)_2$ is calculated by replacing P_e with P_{ep} in Equations (12) and (16).

To apply sinusoidal asperities to the GW model, the surface separation needs to be calculated. Rostami and Jackson [50] derived expressions by averaging the surface separation from a finite element model. Their fitted equations for the dimensionless surface separation G are

$$G = \left(1 - \sqrt{P_e}\right)^{2.5} \tag{20}$$

for elastic contact and

$$G = \left(1 - {P_{ep}}^{A_1 P_{ep} + A_2}\right)^{2.5} \tag{21}$$

for elastic–plastic contact. In these equations,

$$G = \frac{g}{\Delta}, \tag{22}$$

$$A_1 = -0.08 \ln B^* \tag{23}$$

$$A_2 = \frac{1}{15}(B^* - 1)^{0.44} + 0.99^{0.41\{B^*-1\}} - 0.5 \tag{24}$$

and

$$B^* = \frac{\Delta}{\Delta_c} \tag{25}$$

The integrals in Equations (2) and (3) were numerically evaluated for specified values of surface separation. To find the corresponding load, Equation (19) for elastic contact or Equation (20) for elastic–plastic contact was solved numerically.

2.2. Fluid Lubrication Model

The piston ring–cylinder wall interface is not exclusively boundary lubrication; fluid film lubrication plays an integral role in the overall system behavior. To calculate the hydrodynamic lift, the modified Reynolds equation that considers roughness effects via flow factors is employed. For rough surfaces in the mixed lubrication regime, the modified Reynolds equation is given as Equation (1).

Locker et al. [45] used deterministic modeling of flow around the measured rough surfaces of a ring a cylinder to find the averaged flow factors over the entire surface and fitted empirical equations to the predicted flow factors:

$$\varphi_x = 2.48\left(\frac{h}{\sigma}\right)^{-1.777} + 1 \tag{26}$$

$$\varphi_z = 1 - 0.4824e^{-0.2477\left(\frac{h}{\sigma}\right)} \tag{27}$$

For the surfaces being studied, φ_x and φ_z are related through the film thickness, roughness, and surface anisotropy index as

$$\varphi_x\left(\frac{h}{\sigma}, \gamma\right) = \varphi_z\left(\frac{h}{\sigma}, \frac{1}{\gamma}\right) \tag{28}$$

Because the model predicted mostly hydrodynamic lubrication at very low sliding speeds, a shear thinning model was introduced. A version of the Carreau model adapted from Jang, Khonsari, and Bair [51] was introduced as follows:

$$\mu_{eff} = \mu_2 + (\mu_1 - \mu_2)\left(1 + \left(\frac{\mu_1 U}{hG}\right)^2\right)^{\frac{n-1}{2}} \tag{29}$$

μ_2 was set to zero, and a fit was generated at a temperature of 60 °C using the data in Table 1 below. Figure 9 shows the fit to the shear thinning data, and Table 2 shows the values of G and n. Although there is a limited amount of data available, the usage of a well-accepted phenomenological equation should limit the uncertainty.

Table 1. Lubricant Viscosity Dependence on Shear Rate (the first two measurements were conducted with a Cone and Plate Viscometer, while the other values were measured by a Tapered Bearing Simulator Viscometer and modified relative to ASTM D6616).

Shear Rate (1/s)	μ (Pa·s)
10^3	0.02
10^4	0.01921
10^4	0.01971
10^5	0.01777
10^6	0.01601
3.5×10^6	0.01537

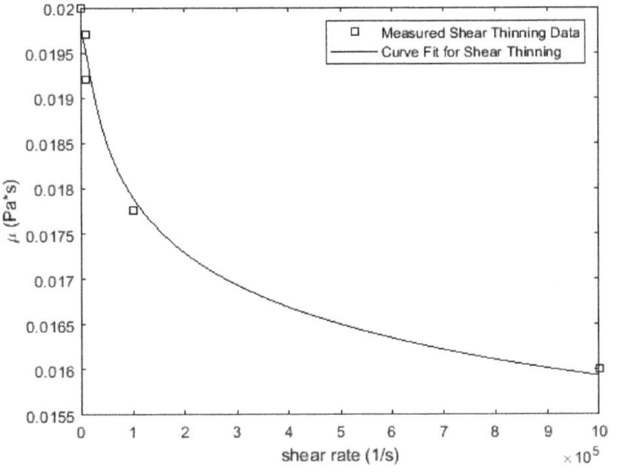

Figure 9. Shear Thinning Viscosity Model Fit.

Table 2. Coefficients Used in Equation (29).

G (Pa)	299
n	−0.0254

It was presumed that the coefficients remained valid for different temperatures. A shear factor that reflects the surface roughness was added later in the model's development when predicting the shear stress and friction from viscous shearing:

$$\tau = \mu \frac{U_2 - U_1}{h}\left(\phi_f \pm \phi_{fs}\right) \pm \frac{h}{2}\frac{\partial p}{\partial x} \tag{30}$$

As an improved comparison to experimental data was sought, pressure and temperature viscosity effects were incorporated. This model uses the Barus equation to account for increased viscosity under loading.

$$\mu = \mu_0 e^{\xi P} \tag{31}$$

Table 3 lists the values used for the pressure viscosity coefficient. There are concerns in the literature about properly accounting for pressure viscosity and shear rate to viscosity relations, but this is more important for contacts where higher pressures are likely, such as in rolling element bearings. The issues are most with Roelands Equation for piezo-viscosity [52], which we are not using it for. The equations employed are well recognized in the literature and probably adequate for the limited ranges of temperature and pressure that are expected in this model.

Table 3. Pressure Viscosity Coefficient Values for the Lubricant.

Temperature (°C)	ξ (Pa·s)
25	2.05×10^{-8}
100	1.248×10^{-8}

With only two values provided, the coefficient was assumed to vary linearly. If the temperature was beyond the range of the provided values, the closer value was used without modification.

The Roelands equation was used to adjust the viscosity due to changes in temperature.

$$\frac{\mu}{\mu_0} = \frac{\mu_\infty}{\mu_0} 10^{G_0 (1 + \frac{t_m}{135})^{-S_0}} \tag{32}$$

Figure 10 illustrates the viscosity fit as a function of temperature.

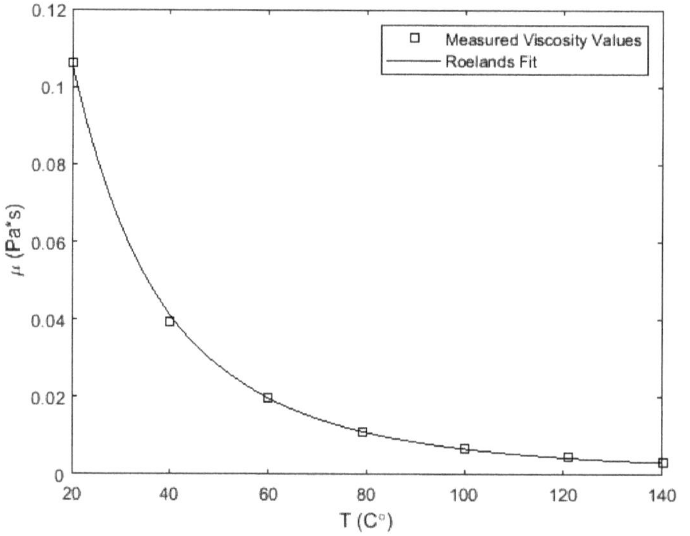

Figure 10. Temperature-Dependent Viscosity Variation.

2.3. Frictional Force Calculation

To determine the frictional force from rough surface contact, Amonton's Law of Friction is used.

$$F_f = F_{contact} \mu_k \tag{33}$$

To determine the value of the friction coefficient, the following equation derived from a curve fit of the experimental data is used. Note that this description possesses a decrease in friction with load which is also predicted and observed by previous works [53–56].

$$\mu_k = 0.1565 - 0.2 F_n^* \tag{34}$$

The frictional force from viscous shear, F_v, is calculated by multiplying the shear stress calculated from Equation (30) by the area on which it is applied. This was performed on the ring surface. The total frictional force can be calculated as follows:

$$F_t = F_v + F_f \tag{35}$$

2.4. Temperature Adjustments Due to Frictional Heating

Due to the high friction coefficient in boundary lubrication, a large amount of heat can be generated, especially when the load is large. On the local scale, this is known as flash temperature. The following equation can be used to calculate the generated heat.

$$Q = \mu_k PV \tag{36}$$

The heat depends on the total frictional force and the velocity–higher speeds and/or loads will result in a greater amount of heat generated. That manifests itself in a temperature increase in the surfaces as given by Equation (37) for a moving surface and 38 for a stationary surface [57]. In the following equations, k is the thermal conductivity, r_c is the applicable contact area, and κ is the thermal diffusivity.

$$T - T_0 = \frac{Q}{4.56 r_c k \sqrt{0.66 + Pe}} \tag{37}$$

$$T - T_0 = \frac{Q}{4 r_c k} \tag{38}$$

$$Pe = \frac{V r_c}{2\kappa} \tag{39}$$

Based off an analogy with electrical current, Equations (37) and (38) can be combined for the case in which neither surface is adiabatic [30]:

$$Q = 4 r_c k_1 (T - T_0) + 4 r_c k_2 (T - T_0) \tag{40}$$

Equation (38) for heat conduction away from the surfaces assumes a low sliding speed for both of them but can be easily adapted if either surface is sliding rapidly. Solving it for the temperature change results in

$$T - T_0 = \frac{Q}{4.56 r_c k \sqrt{0.66 + Pe} + 4 r_c k} \tag{41}$$

There are several ways that the temperature will directly influence the model. First, the temperature will change the viscosity, as described by Equation (31). Second, the temperature could influence the strength of the material locally. The model incorporates the reduction in metals' yield strength as the temperature increases modeled by Johnson and Cook [58]. This would result in the surfaces becoming more liquid-like and less resistant to flow, meaning that the predicted dry friction coefficient would decrease. This assumes the adhesive friction model mentioned previously [52,53]. Following how shear strength is lowered by Johnson and Cook, then the friction can be approximately modified by the following:

$$\mu_k = \mu_b (1 - T^*) \tag{42}$$

$$T^* = \frac{T - T_{room}}{T_m - T_{room}} \tag{43}$$

2.5. Finite Element Model

The deformation of the surfaces is divided into two scales, the asperity and macro scale. The asperity-scale deformations of both surfaces are considered by the statistical rough surface contact model by using the effective elastic modulus E' (Section 2.1), while the macro-scale deformation is considered by a finite element model. The combined surface contact pressure and oil hydrodynamic pressure deforms the ring, so the model is combined with a finite element analysis performed in Abaqus™. The abaqus2matlab toolbox is used to allow the MATLAB program to read the displacements from the finite element analysis to determine the new ring profile for the next iteration [59]. Figure 11 shows the axisymmetric mesh of the piston ring with 1726 linear CAX4R elements. In the finite element analysis, a known load is applied to the inside (left side of the figure) of the ring, while the balancing loads due to rough surface contact and hydrodynamic lift are applied to the outside (right side). Also applied are frictional forces that would induce ring tilt; the center of the ring is held stationary in the z-direction. As stated earlier, a three-dimensional model was considered to be too computationally expensive but could be implemented in future versions of the model. Due to the large variation in deformations observed at high loads, the relaxation factor was set to 0.2 to assist with convergence.

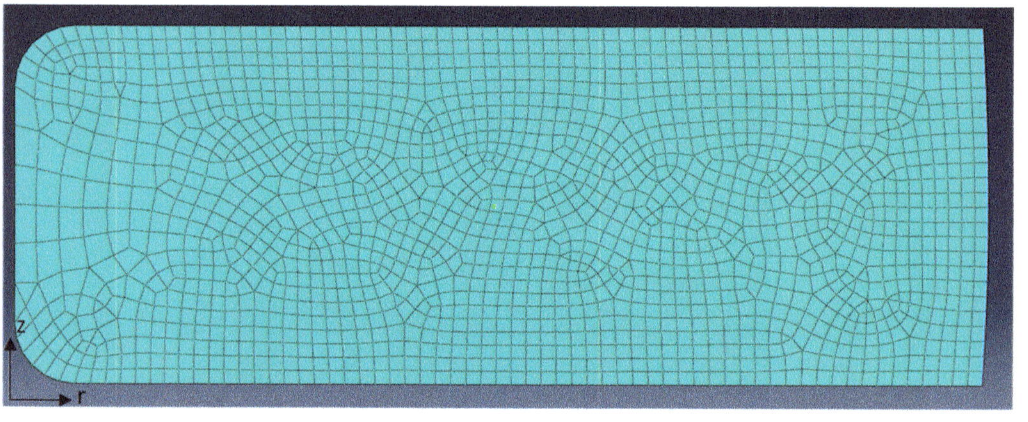

Figure 11. Mesh of Piston Ring.

3. Results

This work assesses three slightly different models that predict the friction coefficient of the piston ring–cylinder liner system. The first model neglects the effects of thermal heating and does not incorporate Equations (36)–(43) at all. The second model includes thermal heating and Equations (39) and (41) to calculate the temperature increase. The third model reflects metals' decreased yield strength by adding Equations (42) and (43). To determine the validity of the models, they are compared to experimental measurements of a reciprocating piston ring sliding against a cylinder liner in a Phoenix Tribology TE77 High Frequency Friction Machine [60,61] (see Figures 3 and 12). The same ring, liner, and lubricant considered in the model is used in the experiment. The experiments featured loads of 50 N, 100 N, or 150 N that simulates actual forces applied to the piston ring, average sliding speeds of 0.3 or 0.6 m/s by running the machine with a stroke length of 0.15 m and a frequency of 5 or 10 Hz, and base oil temperatures of 30 °C, 50 °C, 80 °C, or 120 °C, that represent typical operating conditions of a combustion engine. This results in 24 different cases being experimentally measured and then predicted by the described model. Figure 12 shows the setup.

Figure 12. Experimental Setup.

To compare to the experiments, the model was run over a range of sliding speeds as shown in Figures 13 and 14. These represent the different sliding speeds of the piston during one-reciprocating stroke. Then, the average friction coefficients over the stroke are calculated. As shown, the model is weakest at predicting transient effects such as when the sliding velocity changes direction at 0 s. In these transition regions, the contacts might actually stick briefly due to mechanical play or the ring rotating. This behavior is ignored in the model. Then, the average friction coefficients for average piston speeds of 0.3 m/s and 0.6 m/s are calculated from the model for comparison to the experiments, as will be discussed next.

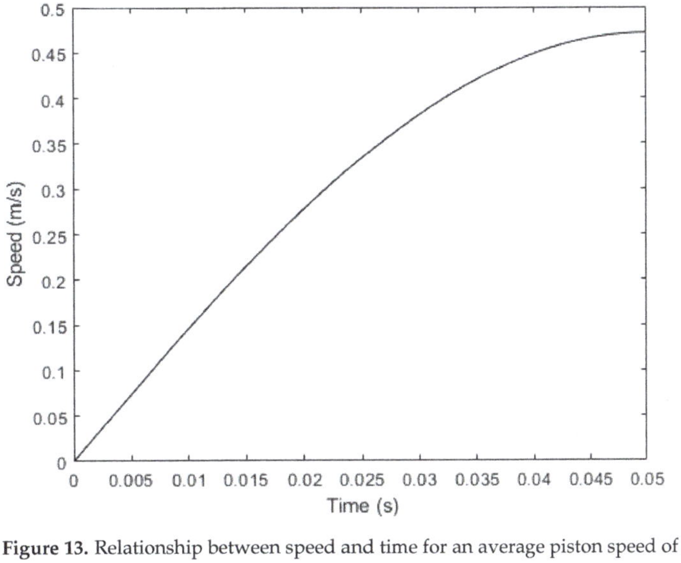

Figure 13. Relationship between speed and time for an average piston speed of 0.3 m/s.

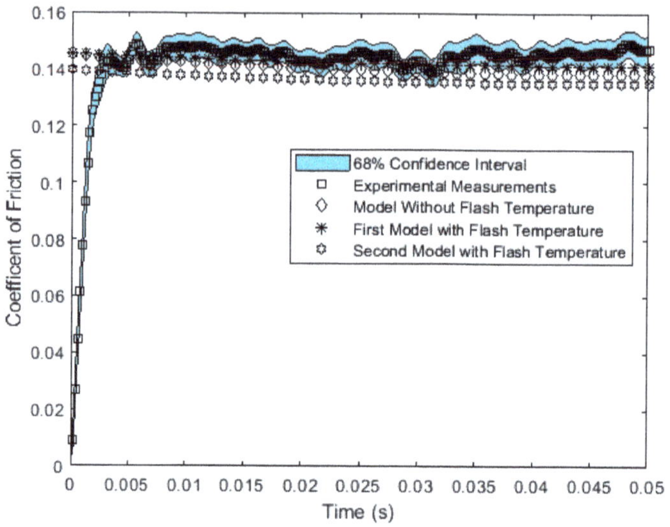

Figure 14. Comparison of model and experiment over varying velocity for base temperature of 80 °C, a 100 N load, and an average sliding speed of 0.3 m/s.

Figures 15 and 16 show the overall comparisons. Overall, there is good agreement between the models and the experiments. The largest differences are at the extremities of load and speed. Tables 4–7 compare the average means and rms error between the model versions and the experiments. These values were calculated by averaging every point in time during the repeating tests whose friction coefficient value was at least 10% of the maximum observed during its test. Based on these results, the adjustments for frictional heating, in particular metal softening with increasing temperature in the second flash temperature model, do not improve the model prediction. For a baseline temperature of 120 °C, the second flash temperature model predicts a much lower friction coefficient than the model that does not adjust for frictional heating at high temperatures. However, it predicts the highest friction coefficient at the lowest loads and temperatures. This is likely because the lower dry friction coefficient due to metal softening does not counteract the decreased load-carrying capacity of the lubricant due to the increased temperature and decreased viscosity.

Table 4. Average Friction Coefficient for a Mean Piston Speed of 0.3 m/s.

Data Set	Average	Standard Deviation
Experimental Measurements	0.1275	0.0143
No Adjustments for Frictional Heating	0.1280	0.0031
First Flash Temperature Model (Equation (41))	0.1302	0.0017
Second Flash Temperature Model (Equations (41)–(43))	0.1262	0.0016

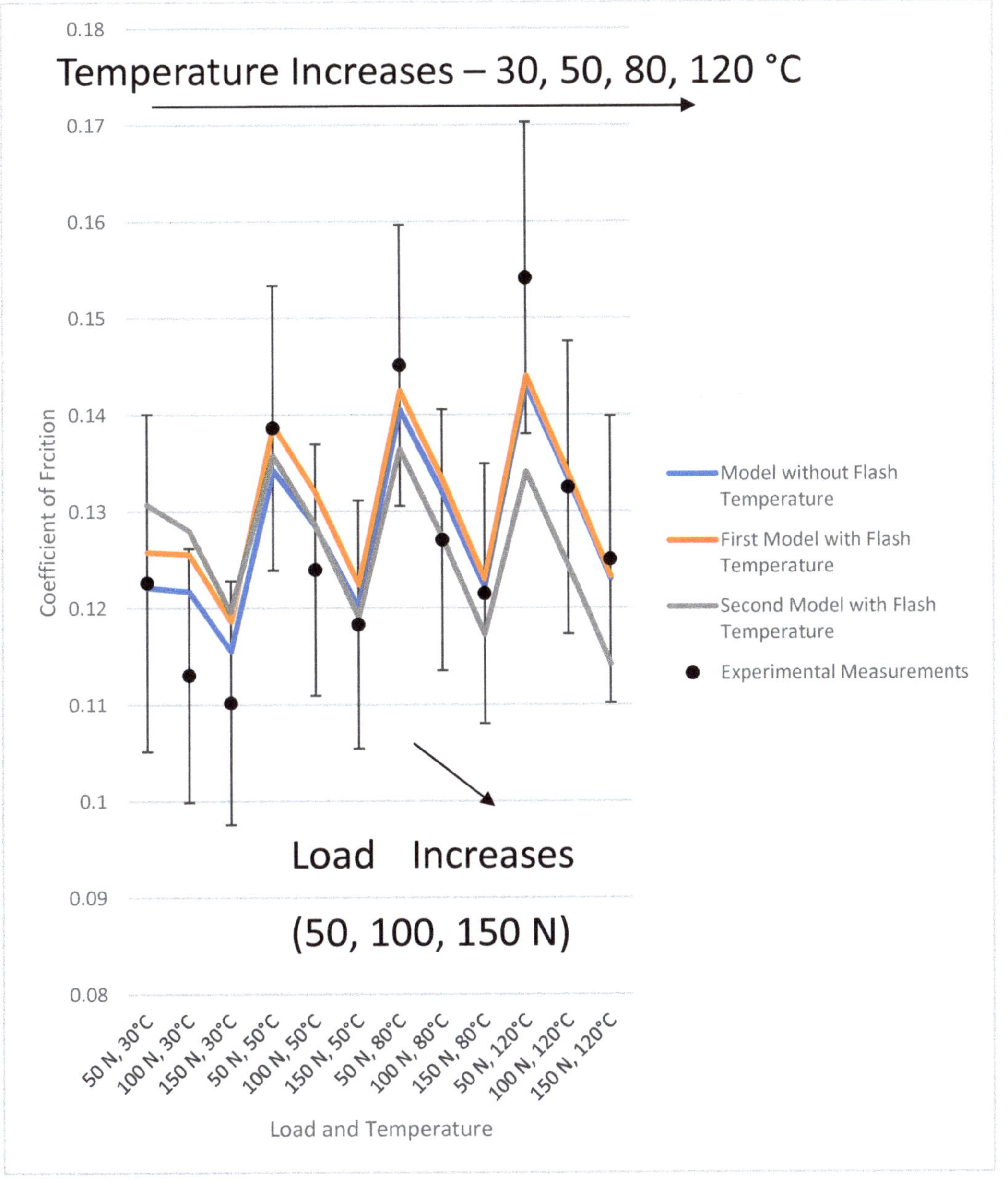

Figure 15. Overall Model Comparisons to Experiments, Average Speed 0.3 m/s.

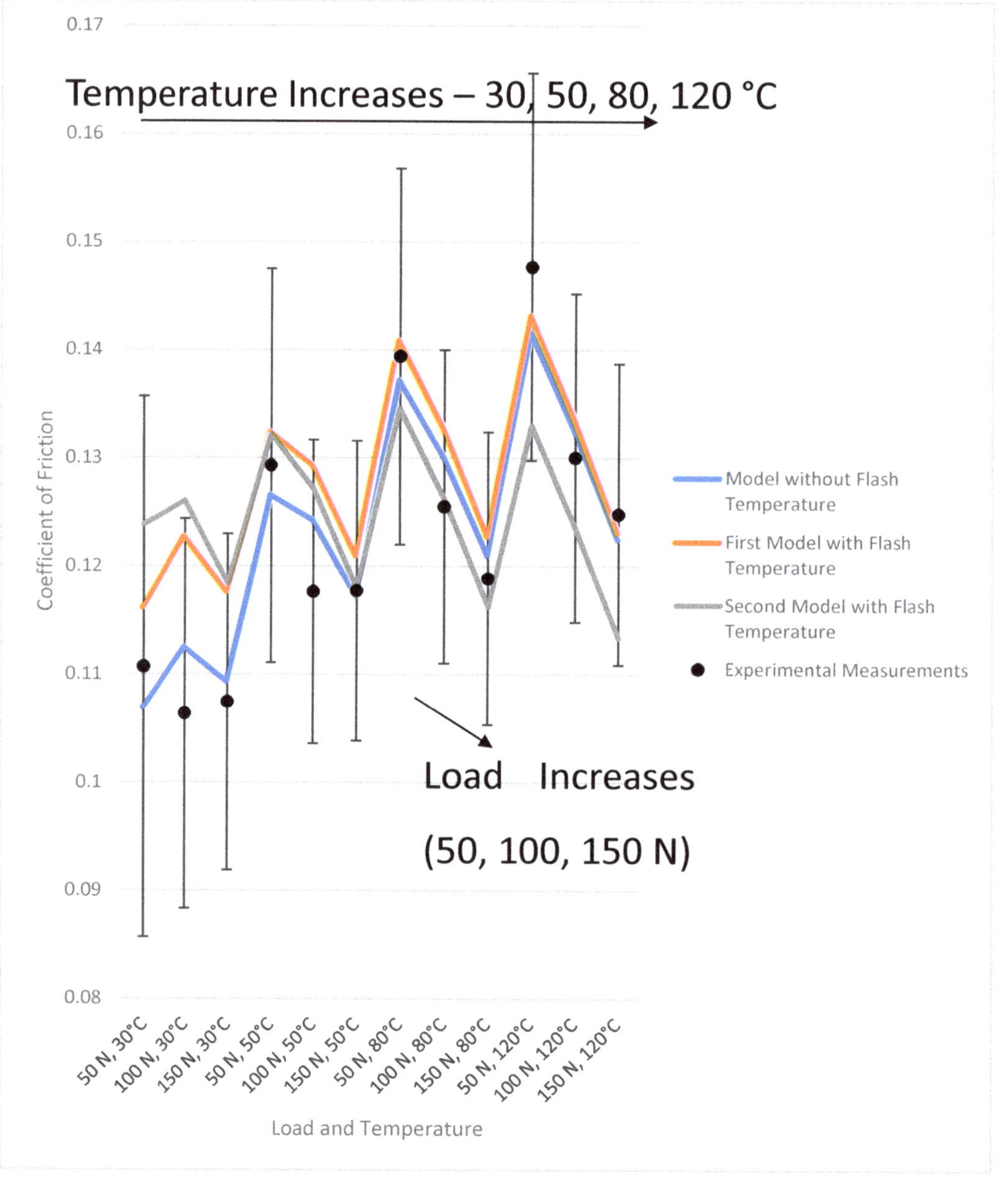

Figure 16. Overall Model Comparisons to Experiments, Average Speed 0.6 m/s.

Table 5. Average Friction Coefficient for a Mean Piston Speed of 0.6 m/s.

Model	Average	Standard Deviation
Experimental Measurements	0.1237	0.0164
No Adjustments for Frictional Heating	0.1235	0.0053
First Flash Temperature Model (Equation (41))	0.1280	0.0027
Second Flash Temperature Model (Equations (41)–(43))	0.1244	0.0023

Table 6. Average RMS Error Relative to the Experiments for a Mean Piston Speed of 0.3 m/s.

Model	RMS Error	Number of Cases For Which Model Was Closest to Experimental Average
No Adjustments for Frictional Heating	0.0014866	6
First Flash Temperature Model (Equation (41))	0.0018159	4
Second Flash Temperature Model (Equations (41)–(43))	0.0027476	2

Table 7. Average RMS Error Relative to the Experiments for a Mean Piston Speed of 0.6 m/s.

Model	RMS Error	Number of Cases For Which Model Was Closest to Experimental Average
No Adjustments for Frictional Heating	0.0011368	7
First Flash Temperature Model (Equation (41))	0.0021501	3
Second Flash Temperature Model (Equations (41)–(43))	0.0028841	2

4. Conclusions

This work presents three slightly different mixed lubrication and solid contact models of a piston ring–cylinder liner interface. The first model does not account for temperature changes at all due to frictional heating. The other two models increase the temperature due to the frictional force and account for that in different ways. One of those models decreases the solid friction coefficient due to metal softening with increasing temperature using Equations (41) and (42). This effect becomes more pronounced as the base temperature increases because the heating pushes the system towards the melting temperature. The models overall display good agreement with experimental measurements performed over a wide range of operating conditions, but larger discrepancies exist for low or high temperatures and high loads and speeds. This can partially be attributed to the shear thinning fit not working as well farther from the temperature from which it was found. The models predict an increase in friction coefficient as temperature increases or load or speed decreases. However, they do not predict as large an increase at high temperature and low load, the condition under which the largest friction coefficients were measured. Overall, these models can be used to design or evaluate friction-reduction technologies such as better lubricants or surface textures.

Author Contributions: Conceptualization, R.L.J. and H.G.; Methodology, N.R.C. and H.G.; Formal analysis, N.R.C. and R.L.J.; Investigation, A.G.; Resources, H.G. and A.G.; Writing—original draft, N.R.C.; Writing—review & editing, R.L.J.; Visualization, N.R.C.; Supervision, H.G. and A.G.; Funding acquisition, H.G. All authors have read and agreed to the published version of the manuscript.

Funding: Ford University Research Program (URP).

Data Availability Statement: On request.

Conflicts of Interest: The authors declare no conflict of interest.

Nomenclature

P	Total load
A_n	Nominal contact area
A_r	Real contact area
$\overline{A}$	Single asperity contact area
ω	Interference between surfaces
ω_c	Critical interference
S_y	Yield strength
E	Elastic modulus
ν	Poisson's ratio
E'	Effective elastic modulus
$\overline{F}$	Single asperity contact force
η	Asperity density
R	Asperity radius
σ	Composite root mean square (RMS) roughness of surfaces
σ_s	Root mean square (RMS) asperity height
f	Spatial frequency
λ	Wavelength of sinusoidal surface ($1/f$)
Δ	Amplitude of sinusoidal surface
$\overline{p}$	Average pressure over surface
p^*	Average pressure for complete elastic contact
P_e	Ratio of surface pressure to pressure needed for complete elastic contact
p^*_{ep}	Average pressure for complete elastic–plastic contact
A_c	Contact area above which elastic–plastic contact occurs
G	Dimensionless surface separation
F^*_n	Local dimensionless load
h	Surface separation or film thickness
F_v	Frictional Force due to viscous shear
F_f	Frictional force due to rough surface contact
ρ	Density of lubricant
μ	Dynamic viscosity of lubricant
p	Hydrodynamic pressure
q_x, q_z	Flow rate in axial and transverse directions
U_1, U_2	Velocity of surfaces in sliding direction
γ	Peklenik number in the principal direction
φ_s	Combined shear flow factor
φ_x, φ_z	Pressure flow factors
μ_k	Sliding friction coefficient due to rough surface contact
k	Thermal conductivity of components
r_c	Contact area radius
Pe	Peclet number
κ	Thermal diffusivity of components
T_0	Baseline surface temperature
T_{room}	Room temperature (presumed to be 20 °C)
T_m	Melting point of cylinder liner/piston ring

References

1. IEA. Electric Vehicles, IEA, Paris. 2022. Available online: https://www.iea.org/reports/electric-vehicles (accessed on 1 December 2022).
2. Kah, M.; Wong, H.X.; Chiu, J.; Lang, S. *Forecasts of Electric Vehicle Penetration and Its Impact on Global Oil Demand*; Columbia Center on Global Energy Policy: New York, NY, USA, 2022. Available online: https://www.energypolicy.columbia.edu/research/report/forecasts-electric-vehicle-penetration-and-its-impact-global-oil-demand (accessed on 1 December 2022).

3. Holmberg, K.; Andersson, P.; Erdemir, A. Global energy consumption due to friction in passenger cars. *Tribol. Int.* **2012**, *47*, 221–234. [CrossRef]
4. Furuhama, S.; Sumi, T. A Dynamic Theory of Piston-Ring Lubrication: 3rd Report, Measurement of Oil Film Thickness. *Bull. JSME* **1961**, *4*, 744–752. [CrossRef]
5. Ma, M.-T.; Sherrington, I.; Smith, E.H. Analysis of lubrication and friction for a complete piston-ring pack with an improved oil availability model: Part 1: Circumferentially uniform film. *Proc. Inst. Mech. Eng. Part J J. Eng. Tribol.* **1997**, *211*, 1–15. [CrossRef]
6. Alakin, O.; Newaz, G.M. Piston Ring-Cylinder Bore Friction Modeling in Mixed Lubrication Regime: Part I—Analytical Results. *J. Trib.* **2001**, *123*, 211–218.
7. Jeng, Y.-R. Theoretical Analysis of Piston-Ring Lubrication Part II—Starved Lubrication and Its Application to a Complete Ring Pack. *Tribol. Trans.* **1992**, *35*, 707–714. [CrossRef]
8. Furuhama, S.; Sasaki, S. New Device for the Measurement of Piston Frictional Forces in Small Engines. *SAE Trans.* **1983**, *92*, 781–792. [CrossRef]
9. Tian, T. Dynamic behaviours of piston rings and their practical impact. Part 2: Oil transport, friction and wear of ring/liner interface and the effects of piston and ring dynamics. *Proc. Inst. Mech. Eng. Part J J. Eng. Tribol.* **2002**, *216*, 229–248. [CrossRef]
10. Harigaya, Y.; Suzuki, M.; Takiguchi, M. Analysis of oil film thickness on a piston ring of diesel engine: Effect of oil film temperature. *J. Eng. Gas Turbines Power* **2003**, *125*, 596–603. [CrossRef]
11. Rahmani, R.; Rahnejat, H.; Fitzsimons, B.; Dowson, D. The effect of cylinder liner operating temperature on frictional loss and engine emissions in piston ring conjunction. *Appl. Energy* **2017**, *191*, 568–581. [CrossRef]
12. Morris, N.; Mohammadpour, M.; Rahmani, R.; Rahnejat, H. Optimisation of the piston compression ring for improved energy efficiency of high performance race engines. *Proc. Inst. Mech. Eng. Part D J. Automob. Eng.* **2017**, *231*, 1806–1817. [CrossRef]
13. Bewsher, S.R.; Leighton, M.; Mohammadpour, M.; Rahnejat, H.; Offner, G.; Knaus, O. Atomic force microscopic measurement of a used cylinder liner for prediction of boundary friction. *Proc. Inst. Mech. Eng. Part D J. Automob. Eng.* **2019**, *233*, 1879–1889. [CrossRef]
14. Howell-Smith, S.; Rahnejad, H.; King, P.D.; Dowson, D. Reducing In-Cylinder Parasitic Losses through Surface Modification and Coating. *Proc. Inst. Mech. Eng. Part D J. Automob. Eng.* **2014**, *228*, 391–402. [CrossRef]
15. Li, C.-D.; Jin, M.; Du, F.-M.; Wang, W.-W.; Shen, Y.; Xu, J.-J. Wear behavior of Al-Si alloy cylinder liner prepared by laser finishing. *Proc. Inst. Mech. Eng. Part D J. Automob. Eng.* **2018**, *232*, 1944–1949. [CrossRef]
16. Senatore, A.; Risitano, G.; Scappaticci, L.; D'Andrea, D. Investigation of the Tribological Properties of Different Textured Lead Bronze Coatings under Severe Load Conditions. *Lubricants* **2021**, *9*, 34. [CrossRef]
17. Wang, X.; Liu, W.; Zhou, F.; Zhu, D. Preliminary investigation of the effect of dimple size on friction in line contacts. *Tribol. Int.* **2009**, *42*, 1118–1123. [CrossRef]
18. Kligerman, Y.; Etsion, I.; Shinkarenko, A. Improving Tribological Performance of Piston Rings by Partial Surface Texturing. *J. Tribol.* **2005**, *127*, 632–638. [CrossRef]
19. Spencer, A. Optimizing Surface Texture for Combustion Engine Cylinder Liners. Ph.D. Thesis, Luleåtekniska Universitet, Luleå, Sweden, 2010.
20. Lu, P.; Wood, R.J.K. Tribological performance of surface texturing in mechanical applications—A review. *Surf. Topogr. Metrol. Prop.* **2020**, *8*, 043001. [CrossRef]
21. Abril, S.O.; Fonseca-Vigoya, M.D.S.; Pabón-León, J. CFD Analysis of the Effect of Dimples and Cylinder Liner Honing Groove on the Tribological Characteristics of a Low Displacement Engine. *Lubricants* **2022**, *10*, 61. [CrossRef]
22. Greenwood, J.A.; Williamson, J.B.P. Contact of nominally flat surfaces. *Proc. R. Soc. London Ser. A Math. Phys. Sci.* **1966**, *295*, 300–319. [CrossRef]
23. Jackson, R.L.; Green, I. A Statistical model of Elastic-Plastic Asperity Contact between Rough Surfaces. *Trib. Int.* **2006**, *39*, 906–914. [CrossRef]
24. Krithivasan, V.; Jackson, R.L. An analysis of three-dimensional elasto-plastic sinusoidal contact. *Tribol. Lett.* **2007**, *27*, 31–43. [CrossRef]
25. Bush, A.; Gibson, R.; Thomas, T. The elastic contact of a rough surface. *Wear* **1975**, *35*, 87–111. [CrossRef]
26. Ciavarella, M.; Greenwood, J.; Paggi, M. Inclusion of "interaction" in the Greenwood and Williamson contact theory. *Wear* **2008**, *265*, 729–734. [CrossRef]
27. Afferrante, L.; Carbone, G.; Demelio, G. Interacting and coalescing Hertzian asperities: A new multiasperity contact model. *Wear* **2012**, *278–279*, 28–33. [CrossRef]
28. Vakis, A.I. Asperity Interaction and Substrate Deformation in Statistical Summation Models of Contact Between Rough Surfaces. *J. Appl. Mech.* **2013**, *81*, 041012. [CrossRef]
29. Chu, N.; Jackson, R.; Wang, X.; Gangopadhyay, A.; Ghaednia, H. Evaluating Elastic-Plastic Wavy and Spherical Asperity-Based Statistical and Multi-Scale Rough Surface Contact Models with Deterministic Results. *Materials* **2021**, *14*, 3864. [CrossRef]
30. Khonsari, M.M.; Booser, E.R. *Applied Tribology: Bearing Design and Lubrication*; John Wiley & Sons Ltd.: Chichester, UK, 2008; p. 566.
31. Jackson, R.L. The Wear and Thermo-Elastohydrodynamic Behavior of Thrust Washer Bearings under Non-Axisymmetric Loads. Ph.D. Thesis, Woodruff School of Mechanical Engineering, Georgia Institute of Technology, Atlanta, GA, USA, 2004.

32. Patir, N.; Cheng, H.S. An Average Flow Model for Determining Effects of Three-Dimensional Roughness on Partial Hydrodynamic Lubrication. *J. Lubr. Technol.* **1978**, *100*, 12–17. [CrossRef]
33. Peklenik, J. Paper 24: New Developments in Surface Characterization and Measurements by Means of Random Process Analysis. *Proc. Inst. Mech. Eng. Conf. Proc.* **1967**, *182*, 108–126. [CrossRef]
34. Wilson, W.R.D.; Marsault, N. Partial Hydrodynamic Lubrication With Large Fractional Contact Areas. *J. Tribol.* **1998**, *120*, 16–20. [CrossRef]
35. Peeken, H.J.; Knoll, G.; Rienäcker, A.; Lang, J.; Schönen, R. On the Numerical Determination of Flow Factors. *J. Tribol.* **1997**, *119*, 259–264. [CrossRef]
36. Hu, Y.; Zheng, L. Some Aspects of Determining the Flow Factors. *J. Tribol.* **1989**, *111*, 525–531. [CrossRef]
37. Morales-Espejel, E.G. Flow factors for non-Gaussian roughness in hydrodynamic lubrication: An analytical interpolation. *Proc. Inst. Mech. Eng. Part C J. Mech. Eng. Sci.* **2009**, *223*, 1433–1441. [CrossRef]
38. Sahlin, F.; Larsson, R.; Almqvist, A.; Lugt, P.; Marklund, P. A mixed lubrication model incorporating measured surface topography. Part 1: Theory of flow factors. *Proc. Inst. Mech. Eng. Part J J. Eng. Tribol.* **2010**, *224*, 335–351. [CrossRef]
39. Tripp, J. Surface Roughness Effects in Hydrodynamic Lubrication: The Flow Factor Method. *J. Lubr. Technol.* **1983**, *105*, 458–463. [CrossRef]
40. Elrod, H.G. A General Theory for Laminar Lubrication With Reynolds Roughness. *J. Lubr. Technol.* **1979**, *101*, 8–14. [CrossRef]
41. Lunde, L.; Tonder, K. Pressure and Shear Flow in a Rough Hydrodynamic Bearing, Flow Factor Calculation. *J. Tribol.* **1997**, *119*, 549–555. [CrossRef]
42. Leighton, M.; Rahmani, R.; Rahnejat, H. Surface-specific flow factors for prediction of friction of cross-hatched surfaces. *Surf. Topogr. Metrol. Prop.* **2016**, *4*, 025002. [CrossRef]
43. Tzeng, S.T.; Saibel, E. Surface Roughness Effect on Slider Bearing Lubrication. *A S L E Trans.* **1967**, *10*, 334–348. [CrossRef]
44. Patir, N.; Cheng, H.S. Application of Average Flow Model to Lubrication Between Rough Sliding Surfaces. *J. Lubr. Technol.* **1979**, *101*, 220–229. [CrossRef]
45. Locker, A.J.; Jackson, R.L.; Ghaednia, H.; Gangopadhyay, A. Flow factor modeling of combustion engine ring and cylinder components in mixed hydrodynamic lubrication. *Proc. Inst. Mech. Eng. Part J J. Eng. Tribol.* **2022**, *237*, 210–221. [CrossRef]
46. Johnson, K.; Greenwood, J.; Higginson, J. The contact of elastic regular wavy surfaces. *Int. J. Mech. Sci.* **1985**, *27*, 383–396. [CrossRef]
47. Jackson, R.L.; Streator, J.L. A multi-scale model for contact between rough surfaces. *Wear* **2006**, *261*, 1337–1347. [CrossRef]
48. Jackson, R.L.; Saha, S.; Xu, Y. The Influence of Single Asperity Models on Predicting Contact Between Elastic Rough Surfaces Using Statistical Methods. In Proceedings of the STLE Tribology Frontiers Conference, Denver, CO, USA, 25–27 October 2015.
49. Ghaednia, H.; Wang, X.; Saha, S.; Xu, Y.; Sharma, A.; Jackson, R.L. A Review of Elastic–Plastic Contact Mechanics. *Appl. Mech. Rev.* **2017**, *69*, 060804. [CrossRef]
50. Rostami, A.; Jackson, R.L. Predictions of the average surface separation and stiffness between contacting elastic and elastic–plastic sinusoidal surfaces. *Proc. Inst. Mech. Eng. Part J J. Eng. Tribol.* **2013**, *227*, 1376–1385. [CrossRef]
51. Jang, J.Y.; Khonsari, M.M.; Bair, S. Correction Factor Formula to Predict the Central and Minimum Film Thickness for Shear-Thinning Fluids in EHL. *J. Tribol.* **2008**, *130*, 024501. [CrossRef]
52. Bair, S. Roelands' missing data. *Proc. Inst. Mech. Eng. Part J J. Eng. Tribol.* **2004**, *218*, 57–60. [CrossRef]
53. Cohen, D.; Kligerman, Y.; Etsion, I. A Model for Contact and Static Friction of Nominally Flat Rough Surfaces Under Full Stick Contact Condition. *J. Tribol.* **2008**, *130*, 031401. [CrossRef]
54. Li, L.; Etsion, I.; Talke, F.E. Contact Area and Static Friciton of Rough Surfaces with High Plasticity Index. *J. Tribol.* **2010**, *132*, 031401. [CrossRef]
55. Wang, X.; Xu, Y.; Jackson, R.L. Theoretical and Finite Element Analysis of Static Friction Between Multi-Scale Rough Surfaces. *Tribol. Lett.* **2018**, *66*, 146. [CrossRef]
56. Dickey, I.R.; Jackson, R.L.; Flowers, G.T. Measurement of Static Friction Coefficients and Comparison to Theoretical Models. *J. Tribol. Trans. ASME* **2011**, *133*, 031408. [CrossRef]
57. Tian, X.; Kennedy, F.E., Jr. Maximum and Average Flash Temperatures in Sliding Contacts. *J. Tribol.* **1994**, *116*, 167–174. [CrossRef]
58. Johnson, G.R.; Cook, W.H. Fracture characteristics of three metals subjected to various strains, strain rates, temperatures and pressures. *Eng. Fract. Mech.* **1985**, *21*, 31–48. [CrossRef]
59. Papazafeiropoulos, G.; Muñiz-Calvente, M.; Martínez-Pañeda, E. Abaqus2Matlab: A suitable tool for finite element post-processing. *Adv. Eng. Softw.* **2017**, *105*, 9–16. [CrossRef]
60. Kamps, T.; Walker, J.; Wood, R.; Lee, P.; Plint, A. Reproducing automotive engine scuffing using a lubricated reciprocating contact. *Wear* **2015**, *332–333*, 1193–1199. [CrossRef]
61. *ASTM-G0133-22*; Standard Test Method for Linearly Reciprocating Ball-on-Flat Sliding Wear, G-133. ASTM International: West Conshohocken, PA, USA, 2016.

lubricants

Article

Film Thickness Decay and Wear Behavior of Grease-Lubricated Point Contact under Cyclic Variable Loads

Yiming Han [1,*], Jing Wang [2], Hengrui Du [2], Weimin Li [1,3,4,*], Jingxin Zhao [1], Zongyi Bai [1], Meng Hu [1] and Haichao Liu [1,4]

1 State Key Laboratory of Solid Lubrication, Lanzhou Institute of Chemical Physics, Chinese Academy of Sciences, Lanzhou 730000, China
2 College of Mechanical Engineering, Donghua University, Shanghai 201620, China; jingwang@dhu.edu.cn (J.W.)
3 Zibo Innovation Center of High-End Synthetic Lubricating Materials, Zibo 255000, China
4 Qingdao Center of Resource Chemistry and New Materials, Qingdao 266000, China
* Correspondence: yiminghan@licp.cas.cn (Y.H.); liwm@licp.cas.cn (W.L.)

Abstract: For wind turbine applications, there is a cyclic load-varying process between rolling elements and raceways in pitch bearings. This kind of motion can also lead to radial fretting. However, this is seldom addressed under grease-lubricated conditions in the literature. In this study, grease-lubricated point contact problems have been investigated experimentally under cyclic load-varying conditions. The findings revealed that as the load-varying range diminishes, the variation in grease film distribution becomes more subtle and the rate of discharge of thickener fiber clusters in the stick zone decelerates. This is due to the fact that the rate of change in the Hertz contact radius is reduced and the migration of grease is weakened during the unloading process. Due to the large apparent viscosity of grease with a high soap content, entrapped grease is not easily discharged during loading, and the thickness of the film in the stick zone progressively increases as the soap content of the grease is augmented. This also causes the variable load zone to wear out more easily. As the grease is subjected to repeated loading and unloading, there is a gradual reduction in film thickness, and larger thickener fiber clusters tear, resulting in a flattened form and shear thinning. Grease containing sulphur–phosphorus additives demonstrates a superior effect on reducing fretting wear within the large variable load range but generally proves effective for smaller load-varying ranges. This study may offer insights into the degradation of grease under variable load motion and methods to prevent radial fretting wear.

Keywords: grease lubrication; dynamic load; film thickness; radial fretting; non-steady state; anti-wear additive

Citation: Han, Y.; Wang, J.; Du, H.; Li, W.; Zhao, J.; Bai, Z.; Hu, M.; Liu, H. Film Thickness Decay and Wear Behavior of Grease-Lubricated Point Contact under Cyclic Variable Loads. *Lubricants* **2024**, *12*, 42. https://doi.org/10.3390/lubricants12020042

Received: 15 December 2023
Revised: 31 January 2024
Accepted: 31 January 2024
Published: 2 February 2024

1. Introduction

Tribological interfaces in machine elements are frequently subjected to vibration or load-varying conditions, such as those encountered in the pitch bearings of wind turbines during the braking phase in the case of strong wind. Despite the absence of macroscopic motion between the rolling elements and the raceway, the contacting interfaces may suffer from dynamic loads to resist the upsetting moments induced by the blade–wind interactions. It is challenging to form hydrodynamic lubricating films for such load-varying conditions as a result of the low entrainment velocity, and mixed/boundary lubrication is expected to predominate. Surface damage, including false brinelling and surface wear [1,2], is prevalent as the interfaces undergo long periods of variable loads. These failures can exacerbate vibration and reduce the reliability and longevity of wind turbines and/or similar mechanical systems [3].

The lubrication state is considered transient when parameters such as load, type of motion, entrainment velocity, contact geometry, and lubricant supply vary over time [4], according to the (elasto)-hydrodynamic lubrication (EHL) theory. Notably, motions that typically lead to transient lubrication problems include reciprocating motion, start/stop, acceleration/deceleration, impact, and variable load motion. These transient lubrication problems exhibit unique characteristics and may not align with the classical lubrication theory of steady state. For instance, in reciprocating motion, commonly observed in pitch bearings, both the magnitude and direction of velocity vary with time. The minimum film thickness appears after but not right at the reciprocating ending points attributed to the combined effects of hydrodynamics and squeeze [5,6]. This type of oscillatory motion in wind turbines has been extensively studied using ball-on-disc model test rigs and down-sized bearing tests [7–9]. The decay of the grease-lubricating film over time has been well documented, with starvation identified as the primary cause of wear damage to the surface. However, most of the experiments have been performed with model grease, either without additives or with commercial grease whose additive packages are unknown. It is widely recognized that active sulphur and phosphorus elements present in additives contribute to the formation of tribo-films under appropriate mechano-chemical conditions, thereby enhancing anti-wear properties. The decomposition of organic molybdenum on the surface of a highly loaded friction sub-surface produces MoS_2, phosphides, and sulphides that are deposited and adsorbed in the contact area, providing friction reduction and anti-wear [10,11]. Apart from the starvation lubrication mechanism, the role of anti-wear additives and tribo-chemistry remains largely unexplored. From a practical point of view, it is interesting to know whether appropriate anti-wear additives could mitigate surface damage during reciprocating or other types of transient motion. In addition, the behavior of grease under load-varying lubrication conditions has not yet been investigated. As the applied load varies, the size of the contact zone changes, potentially causing micro-slip at the edges of the contact zone. The grease in the variable load zone experiences variable loads, leading to film thickness decay over time and an increased risk of failure. Therefore, it is necessary to study the lubrication behavior of grease under load-varying conditions by examining the film thickness variation and evaluating the effectiveness of anti-wear additives.

Grease, a non-Newtonian fluid in a semi-solid state, serves as the primary lubricant for bearings in wind turbines, as well as for other applications. It is composed of a thickener, base oil, and various additives. The presence of a thickener allows grease to form a thicker deposit and lubricating film under specific operating conditions, such as at heavy loads and/or low speeds [12–14]. Over long-term operation, the lubrication performance of grease is related to replenishment [15,16], its rheological property [17–19], and degradation [20,21]. A single dimensionless parameter, based on replenishment local to the contact, has been established between the operating parameters and the transition from the fully flooded to the starved regime by Cann [22,23]. Li et al. analyzed the formation of equilibrium oil films determined by the balance of lubricant loss and replenishment under the long-term rolling condition, as well as under the sliding–rolling condition [24,25]. This work identifies the main mechanisms that dominate oil film formation in different lubrication contacts. Many studies have explored the lubrication characteristics of grease under non-steady states, including reciprocating pure sliding [26], reciprocating pure rolling [27], cyclic impact [28–30], impact-sliding composite motion [31,32], and vibratory motion [33]. It is important to note that the influence of time-varying factors on the lubrication effect of grease is very complex. Whether unsteady motion enhances or hinders grease lubrication depends on the specific form of the motion. For instance, one study indicated that loading/unloading favors grease replenishment [34], while changes in speed direction during reciprocating motion aggravate starvation [27].

Motions with load-varying conditions have been studied extensively from the perspective of wear. Radial fretting, one of the four modes of fretting, is mainly induced by varying normal load. The contact pair undergoes a long period of radial variable load motion, either dry contact or lubricated contact, leading to radial fretting occurring on the material surface due to small-scale changes in the contact zone [35–38]. Recent investigation by the authors into the distribution of the grease film thickness during the transition from reciprocating motion to micro-oscillation revealed correlations between the lubricating film formation and surface wear [39]. Notably, the locations of reduced film thickness or rupture were found to align closely with the locations of severe surface wear. However, the lubricating film behavior in the load-varying regions remains unclear for fresh grease.

The influence of the chemical compositions of grease on the lubricating film behavior under load-varying conditions has been less well understood. In particular, studies examining the effectiveness of anti-wear additives on surface wear under such conditions are scant. It should be highlighted that this paper assumes an initial clearance (gap height) of 0 with an initially applied non-zero load, a scenario distinct from studies on impact EHL which involves an initial clearance. This study focused on the grease-lubricating film formation, decay, and surface wear in terms of grease soap content, anti-wear additives, and dynamic load variation ranges. The decay in the film thickness and changes in microstructure of the greases were observed over cycles. This study aims to gain insights into the failure mechanism of grease under radial variable load conditions from the perspective of the EHL and boundary lubrication.

2. Materials and Methods

Lubricating film thickness is an important parameter for evaluating the state of lubrication. In 1965, Cameron et al. [40,41] proposed the ball-on-disc experimental method, using optical interference techniques to observe the oil film shape in contact. Through the continuous improvement of the technology of optical interferometry and related hardware, it is possible to detect dynamic film thickness in a timely manner.

The experiment was carried out on a customized ball-on-disc test rig, as depicted schematically in Figure 1. The load variation function was facilitated by a servomotor that propels the steel ball (one-inch diameter) up and down onto a disc. The disc can be either glass, suitable for optical measurement of the lubricating film, or steel and test the effectiveness of anti-wear additives. The glass disc is 15 mm thick, and the working side in contact with the ball is coated with a thin semi-reflective Cr film to enhance the quality of interference images. The optical interference measurements utilized a red-green laser light source. The method used for film thickness measurements in this study is the dichromatic interference intensity modulation technique (DIIM) developed by Liu et al. [42]. The physical properties of the steel ball, glass disc, and steel disc are listed in Table 1. It should be noted that the steel discs are custom polished to a roughness of around 20 nm to 30 nm.

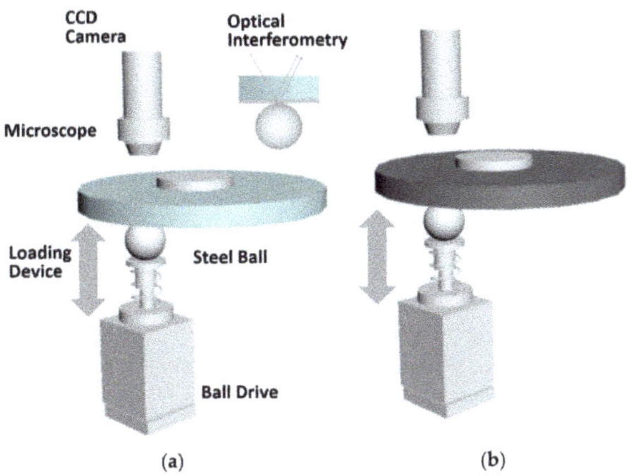

Figure 1. Schematic diagram of the experimental setup with cyclic load variation function ((**a**): ball-on-glass disc for optical film measurement, (**b**): ball-on-steel disc for evaluating the effectiveness of anti-wear additives).

Table 1. Physical properties of the specimens.

Property	Steel Ball	Glass Disk	Steel Disc
Young's modulus (GPa)	210	81	210
Poisson's ratio	0.30	0.21	0.30
Density (kg/m^3)	7850	2510	7860
Thermal conductivity (W/m·K)	23	1.11	23
Specific heat (J/kg·K)	470	840	470
Roughness (Ra, nm)	14	20	20~30

Apparent viscosity data for greases with three soap contents at 25 °C are given in Figure 2. The Anton Paar MCR302 rheometer (Anton Paar Gmbh, Graz, Austria) was applied. The apparent viscosity was tested in rotation mode with shear rates from 0.1 s^{-1} to 100 s^{-1}. The apparent viscosities of lithium greases with 20–30% soap content are not given in the figure because the grease is in too hard a state, which makes rheological testing difficult.

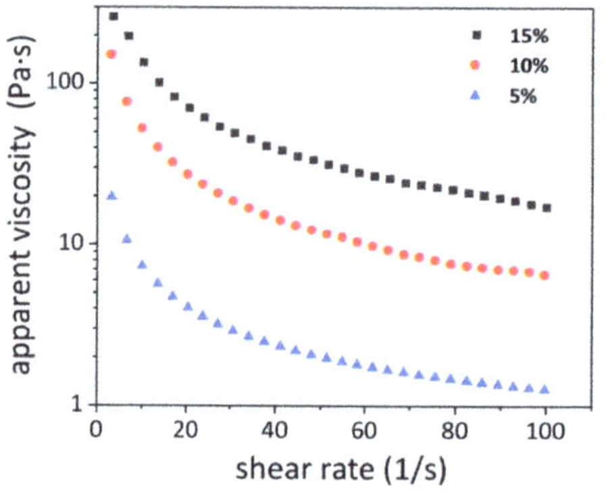

Figure 2. Apparent viscosity of lithium grease with a soap content of 5%, 10%, and 15%.

To investigate the effect of additives on fretting corrosion prevention in grease-lubricated contacts, tribological tests were first carried out with the configuration of the steel ball in contact with the steel disc. After the tribological tests, the steel disc was replaced with a glass disc to facilitate the acquisition of an optical interference image with the ball and to assess the wear and tribo-film formation on the surface of the steel ball. This is illustrated in Figure 1b.

The grease used in the experiments is laboratory-produced lithium grease. The base oil is PAO8, and the soap content varies from 5% to 30% resulting in different NLGI classes. The specific parameters of the greases are detailed in Table 2. All anti-wear additives used are commercially available. Specifically, additives 349, 353, and 232 additives are from BASF's IRGALUBE series (Ludwigshafen, Germany). The additive Mol is sourced from Vanderbilt Chemicals, LLC (Norwalk, America), and Additive 306 is from Tanee Chemical Ltd. (Beijing, China). Notably, Additives 349 and 306 exclusively contain phosphorus, whereas the additives Mol, 232, and 353 contain both sulphur and phosphorus. Each additive was added to the grease at a concentration of 2% by weight, followed by three rounds of grinding of the grease using a grinder to ensure uniform distribution. Experiments were conducted at room temperature: 25 °C.

Table 2. Properties of the prepared lithium grease samples.

Property	Grease
Thickener	Lithium
Base oil	PAO8

Table 2. *Cont.*

Property			Grease			
Base oil viscosity (40 °C mm^2/s)			73			
Base oil viscosity (100 °C mm^2/s)			8.6			
Soap content	5%	10%	15%	20%	25%	30%
Cone penetration (25 °C 0.1 mm)	395–408	270–272	194–204	160–175	116–135	92–101
NLGI	0	2	4	4	5	6

Figure 3 illustrates the motion patterns for variable loads, which are categorized into three distinct load ranges (green line:5 N–50 N, blue line:25 N–50 N, and red line:40 N–50 N). Table 3 gives all the parameters for the seven sets of experiments, including and combining tests of three load-varying ranges, six soap contents, and with/without five anti-wear additives. After the tribological and optical tests, the grease present on the rubbing track of the steel disc was re-examined to check the change in the micro-structure of the thickener. This was performed by mechanically cutting the steel disc with the tested grease intact, followed by a 72 h immersion in petroleum ether to dissolve the base oil and isolate the thickener. Subsequently, the micro-structure of the grease around the variable load zone (see Figure 3b) was characterized using field emission scanning electron microscopy (FE-SEM, JEOL Ltd., Musashino City, Japan).

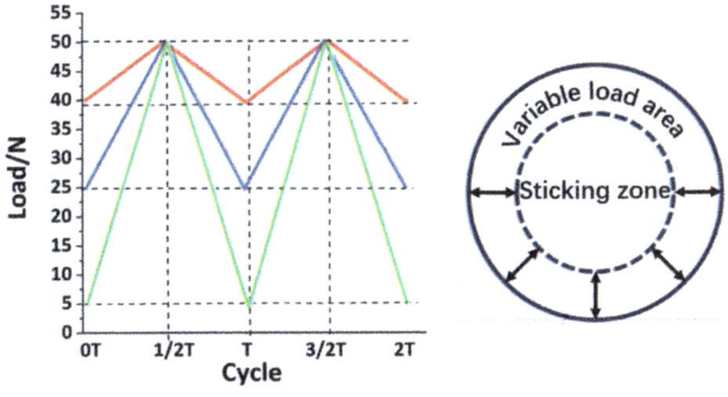

Figure 3. Schematic of the cyclic load-varying process ((**a**): three loading spectra, (**b**): possible two zones).

Table 3. Lists of tests performed and the corresponding operating conditions and grease samples.

Test No.	Grease	Soap Content	Load Variation Range (N)	Duration (min)
1	Base grease	10%	5 N–50 N, 25 N–50 N, 40 N–50 N	20
2	Base grease	5%, 15%, 20%, 25%, 30%	5 N–50 N	20
3	Grease with Mol	10%	5 N–50 N, 25 N–50 N	60
4	Grease with 232	10%	5 N–50 N, 25 N–50 N	60
5	Grease with 349	10%	5 N–50 N, 25 N–50 N	60
6	Grease with 306	10%	5 N–50 N, 25 N–50 N	60
7	Grease with 353	10%	5 N–50 N, 25 N–50 N	60

3. Results

3.1. Effect of Load-Varying Range

Figure 4 shows the optical interferometric images of grease Li-10% (10% indicates thickener contents of the base grease) during loading/unloading at various cycles, namely, the 1st, 20th, 50th, 200th, 500th, 1000th, and 3000th.

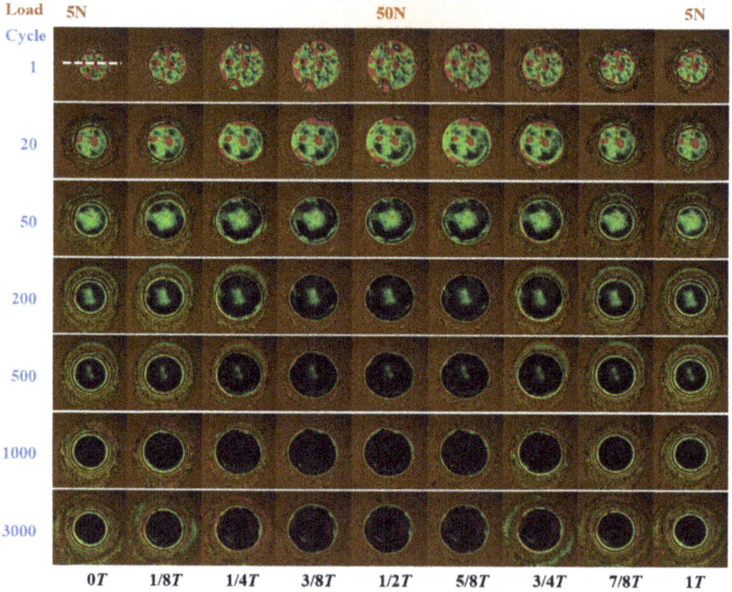

Figure 4. Optical interference image of lithium grease during cyclic load-varying motion with loads from 5 N to 50 N. Row: different instants in one cycle; Column: number of cycles (Li-10% base grease, soap content: 10%, T: 0.4 s).

Within each cycle, the load first increases from 5 N to 50 N and then decreases to 5 N, with a cycle duration of 0.4 s. Images of representative instants such as 0 T, 1/8 T, 1/4 T, 3/8 T, 1/2 T, 5/8 T, 3/4 T, 7/8 T, and 1 T are shown from left to right in a row. The initial motion reveals a contact zone abundant with thickener fibers/clusters. As the loading proceeds, the contact zone expands progressively, trapping new thickener to enter from the periphery of the contact zone, thereby increasing the total amount of thickener within the contact zone. Note that this is solely caused by the load-varying only, as there is no macroscopic rolling or sliding of the surface. Furthermore, the fiber/clusters present at the initial 0 T instant persist in the contact zone without any noticeable morphological changes. The film thickness distributions at the symmetry instants during one loading/unloading cycle exhibit high similarity, e.g., at the 1/8 T and 7/8 T instants.

With an increase in the number of cycles, the thickener clusters in the contact area become squeezed and slightly thinner by the 20th cycle, yet two distinct aggregated clusters remain at the center of the contact area, as indicated by the red segment in the interferometric image. These clusters have been present since the beginning of the motion, as evidenced by the image of the 1st cycle. As the load-varying motion continues, the fiber clusters in the contact zone decrease gradually, even though throughout the initial 500 cycles these fiber clusters consistently remain at the center of the contact zone. By the 1000th cycle, the fiber clusters are nearly invisible in the contact zone, with only a few arc-shaped clusters present during the 1/4 T–3/4 T intervals. The curved shape of film distribution bears a resemblance to the edge of the contact zone at the 0 T instant, i.e., the 5 N condition. This observation suggests that the fiber clusters which entered the contact zone at the 5 N moment affect the grease film thickness distribution for subsequent cycles.

Figure 5 shows the film thickness profile of the middle section across the contact zone corresponding to Figure 4, with measurements at the position of the dotted line marked in Figure 4 during the 1st cycle at 0 T. The figure reveals that during the first cycle, the film thickness in the contact zone exhibits a more pronounced fluctuation, with the maximum film thickness reaching approximately 200 nm. By the 50th cycle, the whole contact area gives a pattern of high film thickness in the center and low film thickness at the edges. The film thickness in the variable load zone is as low as about 50 nm. By the 1000th cycle, the central film thickness (the bump) is no longer evident, with a minimum film thickness of merely 10 nm. Upon reaching the 3000th cycle, the film thickness distribution is similar to the 1000th cycle.

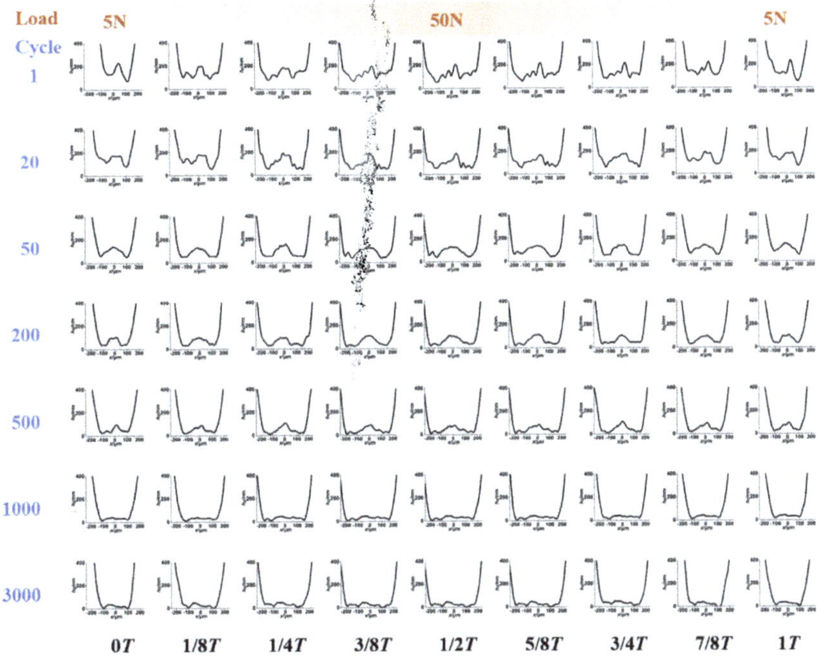

Figure 5. Mid-section film thickness distribution corresponding to Figure 4.

Figure 6 shows the optical interference images during the load-varying cyclic process with a reduced range of load variation from 25 N to 50 N. Since the film morphology of the unloading process closely mirrors that of the loading process, only the image during the loading process is given here. The phenomenon at the 1st cycle is consistent with that in Figure 4, where an increase in load results in a larger amount of thickener fiber in the contact zone. As the number of cycles increases, the amount of thickener in the contact zone diminishes. By the 50th cycle, a ring-shaped region of low film thickness appears in the variable load zone, caused by the continuous discharge of fiber clusters outside the contact zone throughout the cyclic loading process. The grease in the variable load zone is discharged quickly, contrasting with the slower grease discharge observed in the stick zone. In contrast to the results of the 5 N–50 N range of load variation in Figure 4, at the 1000th cycle, a distinct presence of thickener fiber remains in the center of the contact zone. Until the 3000th cycle, the thickener fiber clusters gradually vanish.

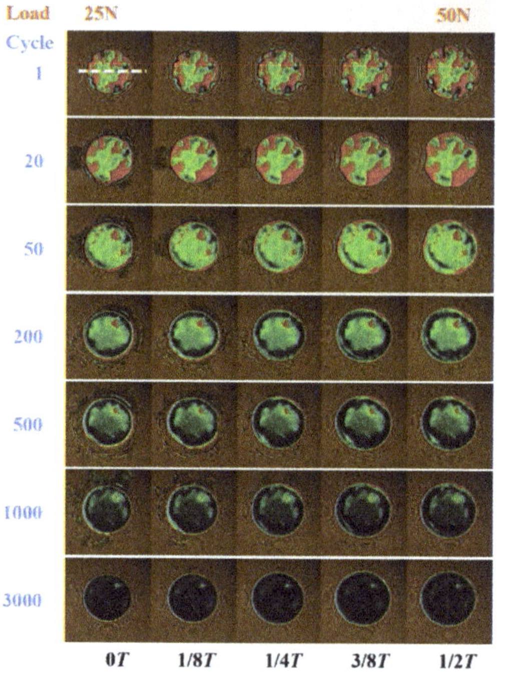

Figure 6. Optical interference image of lithium-based grease during cyclic load-varying motion with loads from 25 N to 50 N. Row: different instants in one cycle; Column: number of cycles (Li-10% base grease, soap content: 10%, T: 0.4 s).

The corresponding mid-section film thickness curves for the aforementioned working conditions are given in Figure 7. The film thickness was measured at the position of the dotted white line in Figure 6. The trend of variation in film thickness is similar to that observed in Figure 5. During the initial stages of the load-varying motion, the film thickness is notably high, approximately 150 nm. From the 50th cycle to the 1000th cycle, the film thickness profile in the contact zone shows a high film thickness in the middle and a low film thickness around. This is consistent with the information from the optical interferogram, which indicates a reduced amount of fiber clusters in the variable load zone and an increased presence in the stick zone. By the 3000th cycle, the film thickness across the entire contact zone is approximated to be between 10 nm and 20 nm.

Figure 8 shows the optical interference images for a more limited range of load variations, specifically from 40 N to 50 N. The images demonstrate significant differences when compared to the previous results. Throughout the motion which lasted up to 3000 cycles, the thickener fiber in the contact zone remained largely unchanged. Only a minor discharge of thickener fiber was observed near the edge of the contact zone. This can be attributed to the small load variations, resulting in a small variable load zone. The distribution of grease thickener is dynamic; the thickener discharge and entry into the contact zone depends on the grease flow caused by the surface movement. Under this condition, the magnitude of surface movement is small, and therefore the grease flow effect is small.

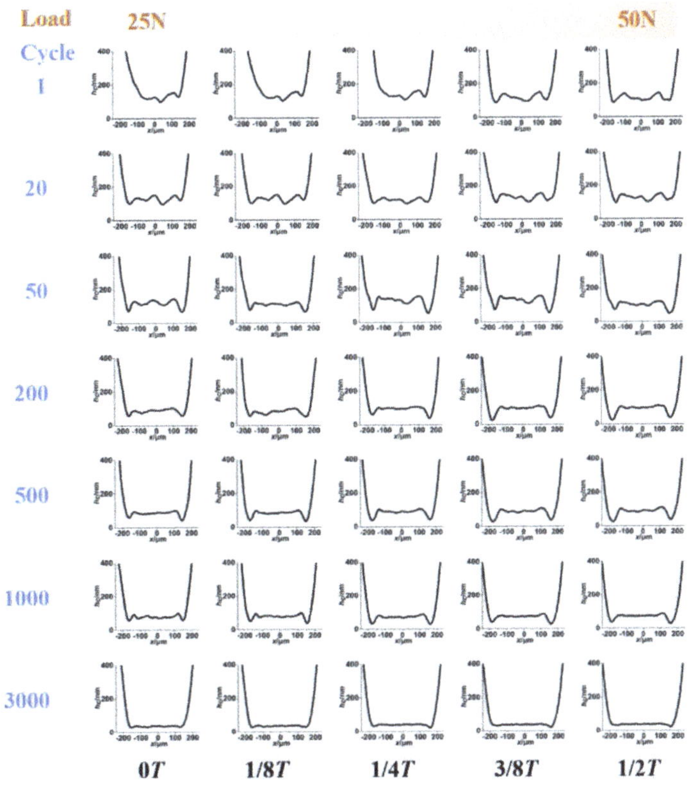

Figure 7. Mid-section film thickness distribution corresponding to Figure 6.

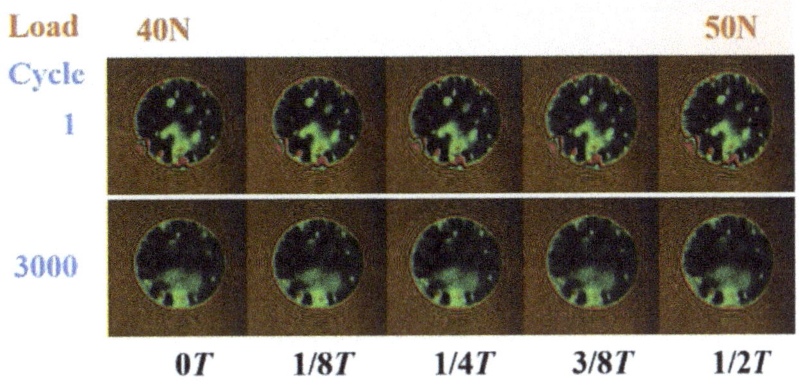

Figure 8. Optical interference image of lithium-based grease during cyclic load-varying motion with loads from 40 N to 50 N. Row: different instants in one cycle; Column: number of cycles (Li-10% base grease, soap content: 10%, T: 0.4 s).

3.2. Effect of Soap Content

The optical interferograms of grease with a 5% soap content (Li-5% base grease) under varying load conditions are given in Figure 9 for selected cycles: 1st, 20th, 50th, 200th,

500th, 1000th, and 3000th. In the first cycle, an arcuate distribution of grease thickener fiber clusters appeared between 1/4 T and 1/2 T in the contact zone. By the 20th cycle, there was a notable decrease in these clusters, and by the 1000th cycle, they were almost invisible. These clusters vanished from both the slip and stick zones. It is noteworthy that at the 3000th cycle, few curved-distribution fiber clusters appeared, potentially due to the migration of grease. Compared to the results with the soap content of 10% in Figure 4, the discharge rate of thickener fiber from Li-5% is faster. For the same number of cycles, the low soap content grease had fewer thickener fibers in the contact zone. In addition, ring-shaped thickener fibers were already present in the contact zone at 3/8 T and 1/2 T of the 20th cycle. Figure 10 shows the corresponding middle-section film thickness profile. The film thickness was measured at the position of the dotted white line in Figure 9. The minimum film thickness is approximated to be 10 nm at the 20th cycle, which is less pronounced than that in the Li-10% grease. From the 20th to the 500th cycle, the distribution diameter of the central residual thickener fiber clusters decreases from 180 nm to 100 nm, and the size of the bump film thickness also decreases. By the 1000th cycle, the residual thickener fiber clusters disappear, giving a minimum film thickness of about 10 nm in the contact zone.

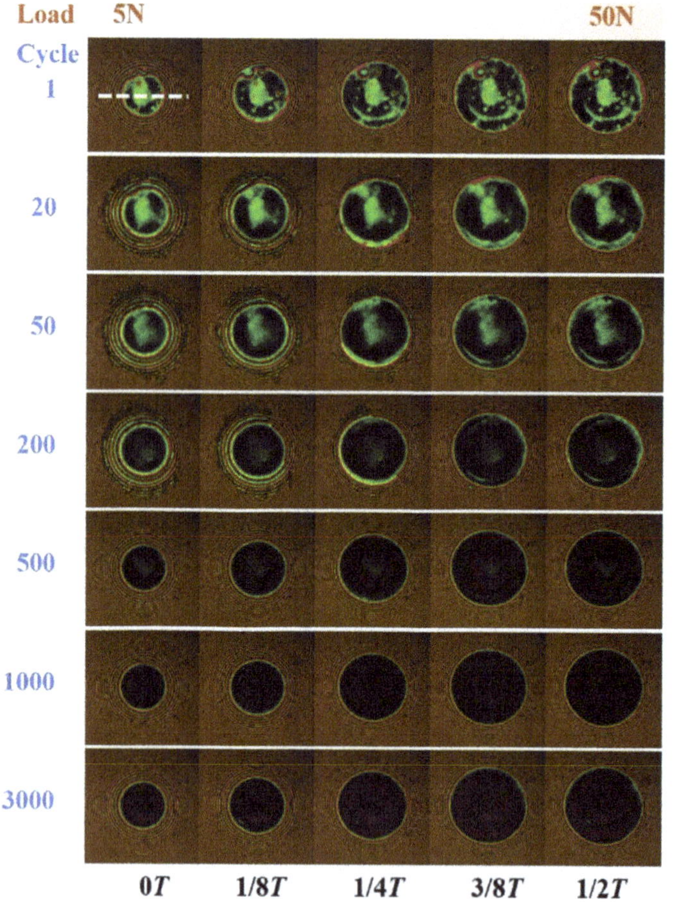

Figure 9. Optical interference image of lithium-based grease during cyclic load-varying motion with a soap content of 5%. Row: different instants in one cycle; Column: number of cycles (Li-5% base grease, load: 5 N–50 N, T: 0.4 s).

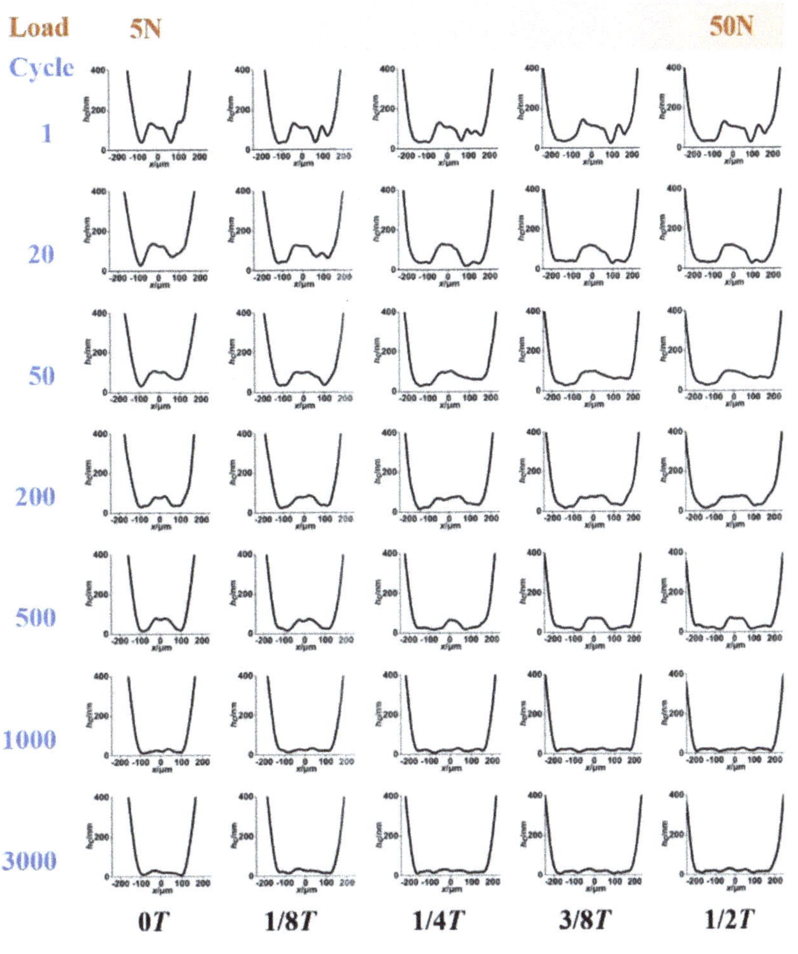

Figure 10. Mid-section film thickness distribution corresponding to Figure 9.

Figure 11 gives the results for the grease Li-30% with a high soap content of 30%. It can be observed that an increase in soap content corresponds to an increase in the amount of thickener fiber in the contact zone. The amount of the thickener fiber remaining in the stick zone gradually decreases as the number of cycles increases. This trend persists even up to 3000 cycles, where a significant amount of thickener fiber remains in the contact. However, it should be noted that at the 3000th cycle, a distinct scratch appeared on the right side of the contact surface. The underlying causes are discussed later. Analyzing the middle-section film thickness (as shown in Figure 12, the film thickness was measured at the position of the dotted white line in Figure 11.)reveals that the residual grease thickener fiber clusters maintain a maximum thickness of 100 nm at the 3000th cycle, while at the 50th and 500th cycles the thickness is 180 nm and 130 nm, respectively. By the 20th cycle, the central film thickness is about 180 nm and does not show a central bump until the 50th cycle. At the 3000th cycle, the minimum film thickness was recorded as 0.

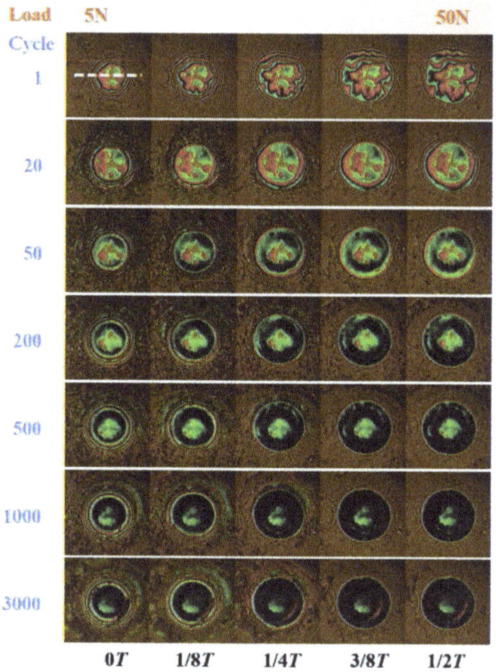

Figure 11. Optical interference image of lithium-based grease during cyclic load-varying motion with a soap content of 30%. Row: different instants in one cycle; Column: number of cycles (Li-30% base grease, load: 5 N–50 N, T: 0.4 s).

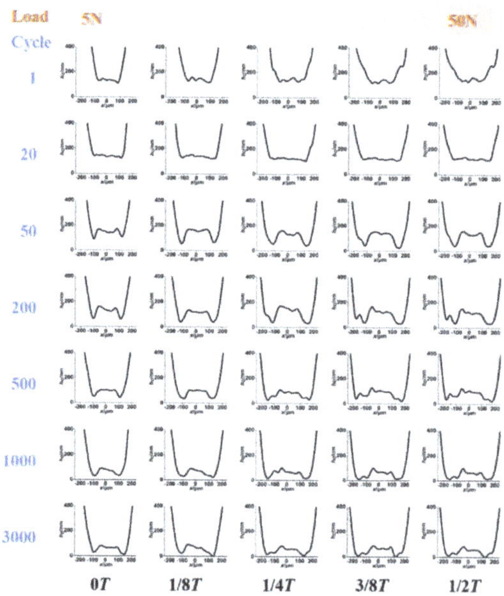

Figure 12. Mid-section film thickness distribution corresponding to Figure 11.

Figure 13 compares the variation in the minimum and central film thickness with the number of cycles for greases with two soap contents. As depicted, both greases show film decay as the number of cycles increases. The minimum film thickness of the Li-5% grease decreases from 30 nm to 10 nm, while the central film thickness reduces from 120 nm to 20 nm. For the Li-30% grease, the minimum film thickness experiences a significant decrease at the 20th cycle, dropping from 120 nm to 0 and approaching dry contact. Concurrently, the central film thickness decreases from 140 nm to 60 nm. It is noteworthy that prior to 1000 cycles, the minimum film thickness of the Li-30% grease is larger than that of the Li-5%. After reaching 1000 cycles, this trend reverses, with the minimum film thickness of Li-30% becoming very low, nearly equivalent to dry contact. This observation aligns with the damage on the surface observed at the 3000th cycle as shown in Figure 11. In terms of the central film thickness in Figure 13b, Li-30% grease has a higher film thickness compared to Li-5%, which can be attributed to the amount of residual thickener.

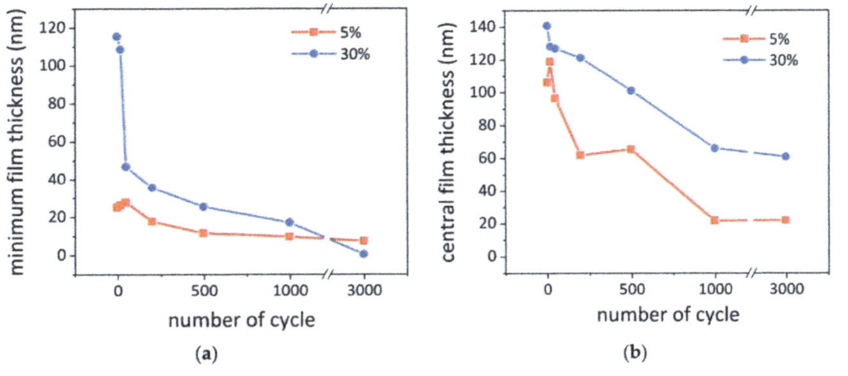

Figure 13. Variation in film thickness with the number of cycles for greases with different soap contents ((**a**): minimum film thickness, (**b**): central film thickness).

Figure 14 further shows the results of base greases produced at varying soap contents, specifically 5%, 10%, 15%, 20%, 25%, and 30%. The number of load-varying cycles was set to 500. As shown, the thickener fiber clusters remaining in the central area of the contact zone become larger and thicker as the soap content increases. The thickener fiber in the central area of the contact zone becomes visible when the soap content is greater than 15%. Note that the area occupied by the thickener fiber is smaller than the actual stick zone. All tested soap contents resulted in an arc-shaped thickener distribution. Figure 15 gives the corresponding film thickness profile along the middle section. The film thickness was measured at the position of the dotted white line in Figure 14. It can be seen that the central film thickness increases with increasing soap content. The average film thickness in the contact zone is relatively lower when the soap content is 5%. A slight increase in the film thickness in the stick zone was observed when the soap content was increased to 10%. When the soap content was 15%, the film thickness profile was flatter due to the absence of large fiber clusters along the center line.

In continuation of the 500 cycles described above, the experimental results after the 3000th cycle for greases with different soap contents are given in Figure 16. The grease with a 30% soap content has the highest number of thickener fiber clusters remaining in the contact zone. As the soap content exceeds 10%, surface damage begins to appear inside the contact area. The corresponding film thickness distribution along the middle section is depicted in Figure 17. The film thickness was measured at the position of the dotted white line in Figure 16. For greases with soap contents of 25% and 30%, the film thickness distribution in the contact area becomes non-uniform and exhibits significant fluctuations. When the soap content is greater than 15%, the contact area appears to undergo dry contact.

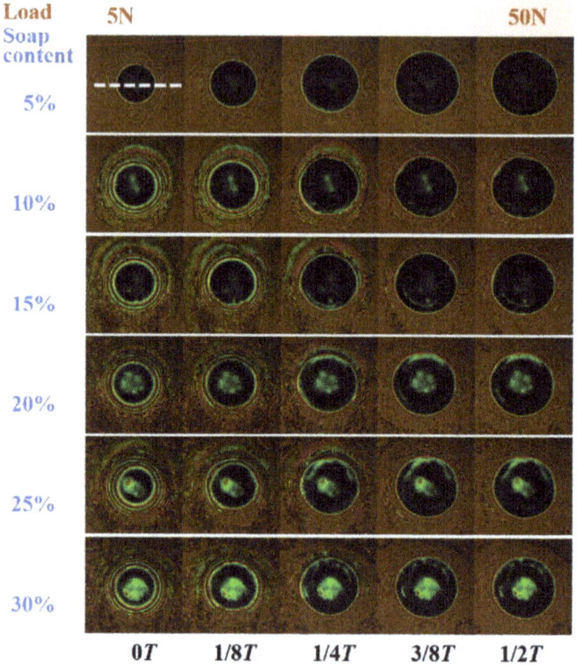

Figure 14. Optical interference images at the 500th cycle for greases with different soap contents (Li base grease, load: 5 N–50 N, T: 0.4 s).

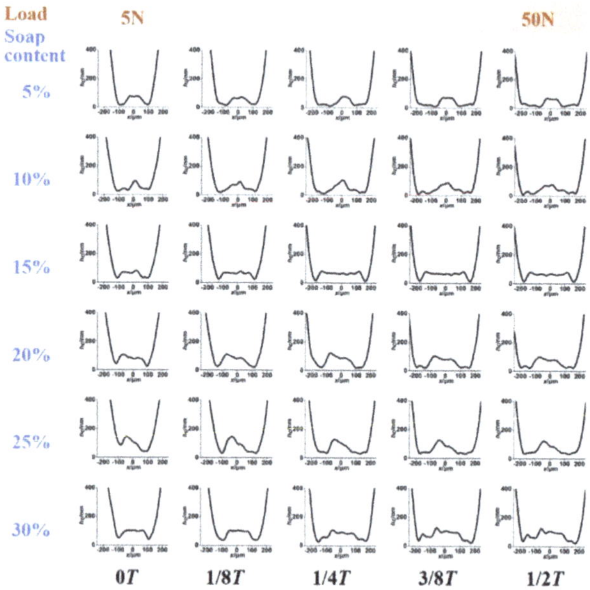

Figure 15. Mid-section film thickness distribution corresponding to Figure 14.

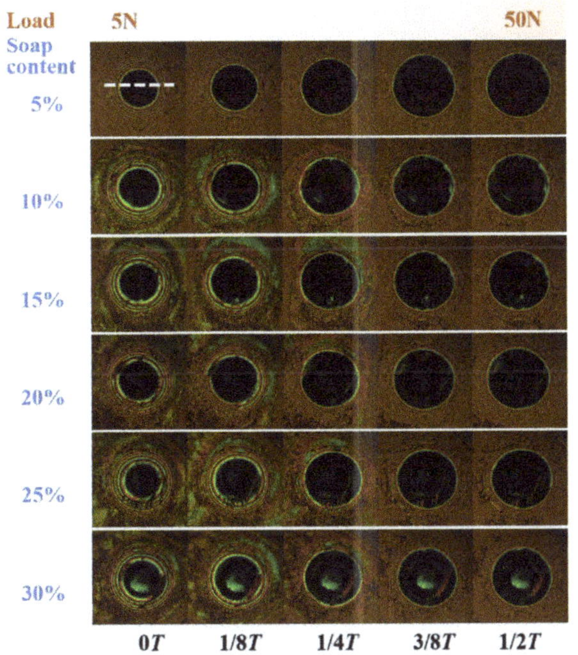

Figure 16. Optical interference images at the 3000th cycle for greases with different soap contents (Li base grease, load: 5 N–50 N, T: 0.4 s).

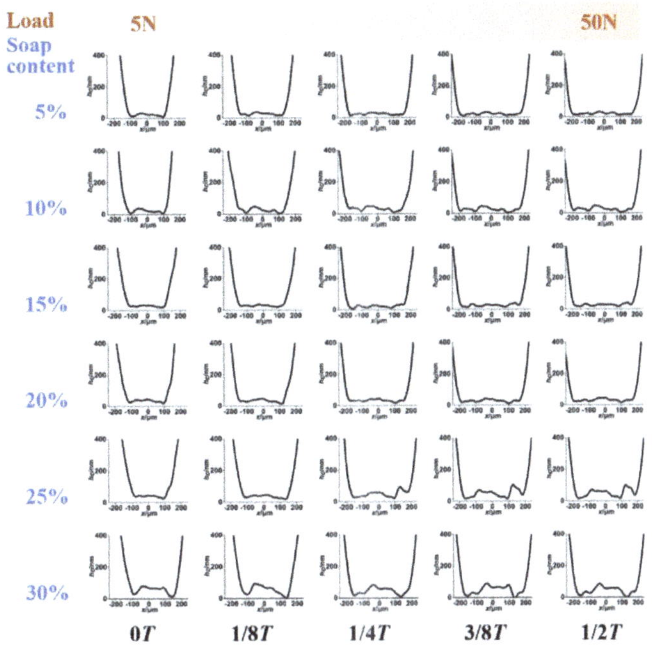

Figure 17. Mid-section film thickness distribution corresponding to Figure 16.

3.3. *Effect of Anti-Wear Additive*

Figure 18 shows the surface wear of base grease and grease with additives under cyclic load-varying conditions. The test lasted for 1 h and experienced 9000 cycles. Initially, a steel disc was used in contact with a steel ball for the experiment. Following this, the steel disc was replaced with a glass disc, allowing for the observation of wear marks and potential tribo-films on the steel ball through light interferometry at 60 N. The results are presented for two load ranges, i.e., 5 N–50 N and 25 N–50 N.

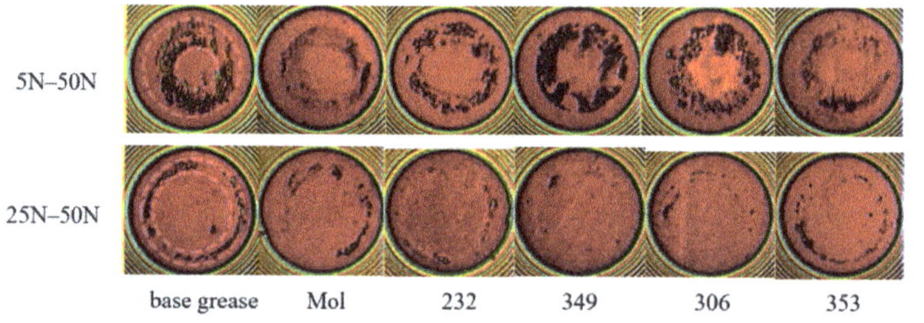

Figure 18. Surface wear of base grease and grease with additives under loaded/unloaded conditions (Load: 5 N–50 N and 25 N–50 N, T: 0.4 s, duration: 1 h).

For the base grease, severe wear was observed on the ball surface in the variable load region from 5 N to 50 N, while no significant wear was found in the stick zone due to the presence of thickener fiber clusters. All five anti-wear additives demonstrated effective resistance to surface wear. Notably, the Mol additive and 353 additive were the most effective against variable load wear in this study. The dark color around the variable load zone may indicate the formation of non-homogeneous tribo-films rather than wear. However, it is not clear through the interferograms. Additional physical and chemical analysis will be carried out in future work. For the small load variation range from 25 N to 50 N, all additives showed good anti-wear effects compared to the base grease. For Additives 232, 349, and 306, there was no wear or tribo-film generation in this operating condition. Overall, the small variable load range was less worn, which is consistent with the results demonstrated by the optical interferograms in Figure 6.

4. Discussion

Greases are distinguished by their thickener composition, which can entrain the contact zone and yield a thicker film than neat base oil at low-speed operating conditions [12–14]. This phenomenon is also observed during the load-varying process in this study. Figure 19 gives the optical interferometric images during the cyclic load-varying process for the base oil PAO-8 with a load variation range of 5 N–50 N and a variation cycle time of 0.4 s. It can be clearly seen that the shape and distribution of the film differ significantly from that of the grease in Figure 4. Compared to grease, lubricated oil cannot provide a comparable film thickness to grease during the load-varying process, particularly in the stick zone. During variable loads, the film thickness of an oil-lubricated contact does not change noticeably as the interference color does not vary. Mainly, the contact area expands as a result of elastic deformation. When loading and unloading were performed for 3000 cycles, the optical interference image remained unchanged. There was no wear of the surface and no cavitation around the contact zone. This is similar to the image of grease lubrication with 5% soap content after 1000 cycles.

Figure 20 compares the optical interference image for greases with 5% and 30% soap contents after 3000 cycles. The Li-5% grease has only a small amount of thickener fiber clusters throughout the contact zone, while the Li-30% grease retains a significant amount of thickener fibers in the stick zone. It was thought that the grease with a higher soap content enhanced lubrication at low speeds. However, the low consistency and the poor fluidity as a result of the high soap content limit the replenishment effect of grease from both outside of the contact zone and inside of the stick zone. This results in starved lubrication in the variable load (slip) zone, leading to wear on its surface. Therefore, an optimal soap content is necessary to ensure effective lubrication for load-varying running conditions. Zhu et al. [2,36,43] pointed this out in their experiments on radial fretting from the wear point of view. In all radial fretting tests, the minimum value of the cyclic imposed normal loads must be positive (above 0 N) in order to avoid an impact effect. This results in a center sticking zone, and the relative sliding of surfaces during variable loads occurs only at the edge area, forming an annular shape. Similarly, in grease lubrication, both the stick zone and the variable load zone exist in the contact area. Over time, the relative sliding of surfaces during variable loads causes the film thickness to decrease gradually to nearly zero, resulting in radial fretting.

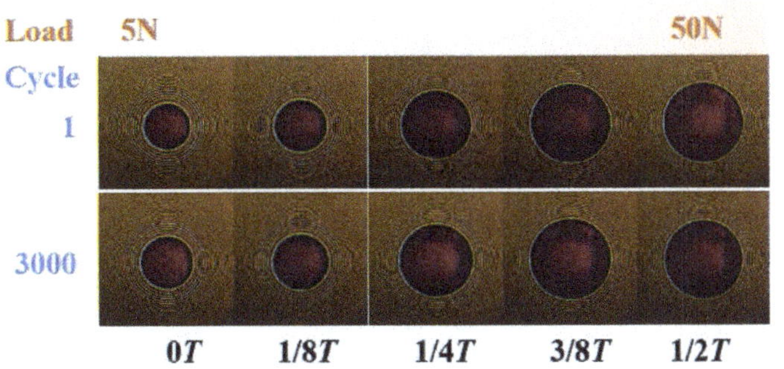

Figure 19. Optical interference image during cyclic load-varying process lubricated with PAO-8 oil (Load: 5 N–50 N, T: 0.4 s, with operating conditions identical to those in Figure 3).

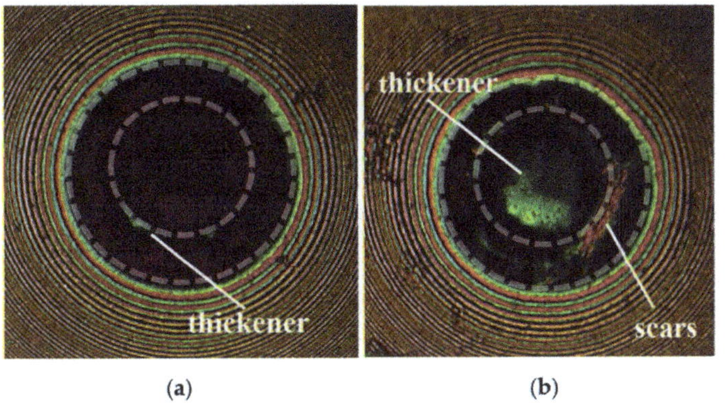

Figure 20. Wear of the disc surface lubricated with greases of different soap contents after 3000 cycles of the cyclic load-varying process.(The dotted line represent the variable load area, (**a**): grease with 5% soap content, (**b**): grease with 30% soap content).

Figure 21 gives the variation in the cavitation zone for a grease-lubricated contact with greases of different soap contents and number of cycles. The dotted white lines clearly show the boundaries of the cavitation. The shape and size of cavitation may indicate the degree of grease fluidity under cyclic motion. The interferograms were collected at the instant of 1/8 T over cycles. For the Li-5% (soap content) grease, the cavitation disappeared after 500 cycles because of oil replenishment. As the soap content increases, the grease becomes less fluid, and the size of the cavitation gradually increases with the increase in cycles. The discharge and entry of grease in the variable load zone is a dynamic process throughout the experiments. Apart from the variable load zone, there is also the discharge of grease fiber clusters at the periphery of the stick zone. This explains why the area occupied by the deposited thickener fiber clusters is always smaller than that of the stick zone. According to the Hertz contact equation, $a = (3Rw/2E')^{1/3}$, where a is the Hertzian contact radius and w is the load, it can be seen that during unloading, as the load decreases, the gradient of decrease in Hertzian contact radius a increases. This results in an increase in the rate at which the grease returns to the contact zone. At the end of the unloading process, the size of cavitation is reduced.

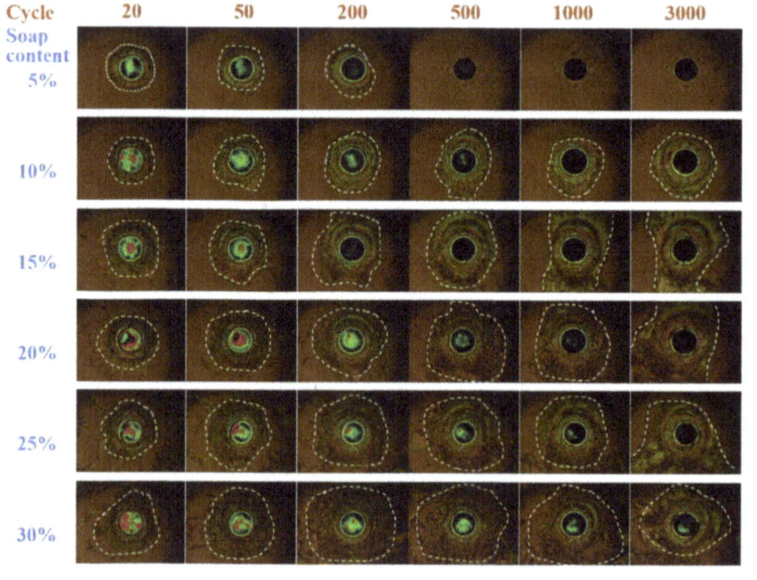

Figure 21. Variation in cavitation under grease lubrication at 1/8 T over cycles (Column: soap content of 5%, 10%, 15%, 20%, 25%, and 30%; Row: number of cycles of 20, 50, 200, 500, 1000, and 3000, The dotted white lines show the boundaries of the cavitation).

As shown in Figure 2, the grease samples simultaneously exhibit a shear thinning phenomenon in which the apparent viscosity decreases with the increasing shear rate. As the shear rate increases, the grease thickener fibers break, and the viscosity value appears to drop. It can be seen that as the soap content increases, the apparent viscosity of the grease increases and the flow performance is poorer. The migration of grease from the external to the internal is controlled by the unloading process, during which a certain amount of grease is drawn into the contact zone. During loading, more grease is entrapped in the contact zone. The whole process is governed by the dynamic flow of grease. Greases with a high soap content are less fluid and the amount of grease thickener entrapped in the contact zone is always at a high level. The grease with a low soap content has a small apparent viscosity and good flow performance. In the unloading process, the migration phenomenon of grease in the variable load area is relatively obvious. In addition, the production of bleed oil after grease degradation promotes the flow phenomenon, so the film thickness of the contact area of grease with a low soap content is lower.

For another rheological property of grease, the higher the soap content, the worse the thixotropic properties of the grease, i.e., the worse the recovery. The recovery process during thixotropic change in lithium grease is much slower than the destruction process, and the initial recovery process may take several minutes or even longer. For the experimental cycle time of 0.4 s, the grease entered the next cycle without having time to exhibit thixotropy. The range of load variations for the experimental conditions is not very large, nor is the range of Hertzian contact radius variations, and for the magnitude of the force that the grease is subjected to during variable loading, the yield stress of the grease has not yet been reached. It can also be seen by the results of optical interference images over one cycle that the images of the loading and unloading processes are essentially similar and do not reflect the thixotropic nature of the grease.

Figure 22 shows the SEM image of the grease sample, Li-10%, after the tribological experiments. Images were captured from the load-varying region of the steel disc, as well as at the periphery of the contact zone, as indicated by the white box in the figure. Magnifications of 100×, 600×, 5000×, and 20,000× were chosen for analysis. Upon examining the 5000× magnified image, a significant tear was observed in the compacted thickener fiber mass in the variable load region. Furthermore, it was noted that the grease thickener and base oil were not completely separated. The 20,000× image revealed sporadic grease thickener fibers, indicating an incomplete micro-structure of the grease, i.e., grease degradation. Figure 23 presents the SEM images of the micro-structures of both fresh grease and grease after the cyclic load-varying test. The thickener in fresh Li-grease is composed of interconnected fibers forming a 3D structure. This structure is characterized by a distinct fiber architecture and overlapping fibers. However, following cyclic loading, base oil is released from the grease thickener, leading to the destruction of its micro-structure. Consequently, most of the fibers are flattened, leaving only a small number of thickener fibers exposed on the surface.

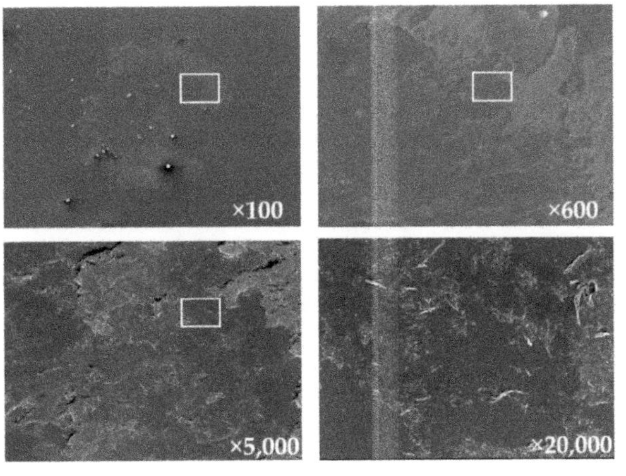

Figure 22. SEM image of grease thickener after cyclic load-varying process. (The white box represents the magnified position).

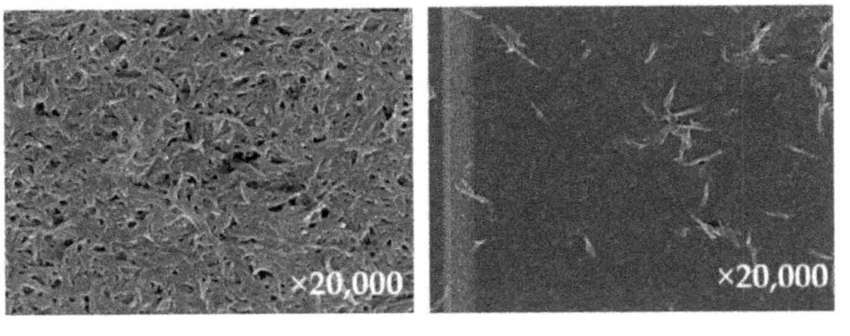

(**a**) (**b**)

Figure 23. SEM image of grease thickener microstructure.((**a**): before cyclic load-varying process, (**b**): after cyclic load-varying process).

For the effectiveness of anti-wear additives under cyclic load-varying conditions, all selected anti-wear additives demonstrated an improvement in the resistance to radial fretting wear compared to base grease without additives. This suggests that the additives are at least partly activated and contribute to the formation of a thin tribo-chemical film on the variable load zone. However, in the current study it is challenging to distinguish whether the color change in the interferograms in Figure 18 is due to tribo-film formation or surface wear, or a combination of both. In addition, anti-wear additives containing the elements sulphur and phosphorus appeared to be more effective in resisting radial fretting than the phosphorus-containing additive at larger load variations.

5. Conclusions

In this study, the decay in lubricating film thickness and surface wear of lithium grease were investigated during a cyclic load-varying process. The influences of load-varying ranges, number of variable load cycles, grease soap content, and anti-wear additives were examined. The degradation of grease was analyzed by capturing the micro-structure of the grease. The main conclusions are presented below.

1. The thickener content of grease significantly influences the initial and subsequential film thickness in the stick zone during prolonged cyclic load-varying experiments. However, an excessively high content impedes lubricant replenishment and renders the variable load zone susceptible to radial fretting wear. An optimal soap content is identified that facilitates the most effective lubrication by maintaining a thick film thickness and avoiding surface wear;
2. During the cyclic load-varying process, there is a gradual reduction in both minimum and central film thickness in the grease-lubricated contact. Micro-structure degradation of the grease occurs in this process, resulting in large clusters of thickened fibers that are torn and flattened in contact. It is observed that as the variable load range decreases and the rate of film decay and grease degradation slows down. The apparent viscosity of the grease also determines this process;
3. Anti-wear additives in grease have proven effective in mitigating wear during prolonged cyclic load-varying tests. Notably, these additives, when composed of different mechano-chemically active elements, exhibit distinct responses to wear across diverse load variation ranges. Tribofilm might be generated during such conditions.

Author Contributions: Conceptualization, Y.H. and H.L.; methodology, Y.H.; software, Y.H. and H.L.; validation, H.D. and J.W.; formal analysis, W.L.; investigation, J.Z.; resources, Z.B.; data curation, M.H.; writing—original draft preparation, Y.H.; writing—review and editing, H.L.; visualization, Y.H.; supervision, W.L.; project administration, W.L. All authors have read and agreed to the published version of the manuscript.

Funding: This research was funded by National Natural Science Foundation of China (Grant Nos. 51975560 and 52305227); Key Projects of Lanzhou Institute of Chemical Physics, Chinese Academy of Sciences (Grant No. KJZLZD-1); Taishan Scholars Project; Key R&D Program of Shandong Province (Grant No. 2020CXGC11002); and Gansu Postdoctoral Science Foundation (E202C6SK).

Data Availability Statement: Data are contained within the article.

Acknowledgments: The authors are grateful to Li Zhang for support in the performance analysis of grease.

Conflicts of Interest: The authors declare no conflicts of interest.

References

1. Waterhouse, R.B. Fretting fatigue. *Int. Mater. Rev.* **1992**, *37*, 77–98. [CrossRef]
2. Zhu, M.; Zhou, Z. An experimental study on radial fretting behaviour. *Tribol. Int.* **2001**, *34*, 321–326. [CrossRef]
3. Godfrey, D. Fretting corrosion or false brinelling? *Tribol. Lubr. Technol.* **2003**, *59*, 28–31.
4. Sugimura, J.; Jones, W.; Spikes, H.A. EHD film thickness in non-steady state contacts. *J. Tribol.-Trans. ASME.* **1998**, *120*, 442–452. [CrossRef]

5. Wang, J.; Kaneta, M.; Yang, P. Numerical analysis of TEHL line contact problem under reciprocating motion. *Tribol. Int.* **2005**, *38*, 165–178. [CrossRef]
6. Glovnea, R.; Spikes, H.A. Behavior of EHD Films during Reversal of Entrainment in Cyclically Accelerated/Decelerated Motion. *Tribol. Trans.* **2002**, *45*, 177–184. [CrossRef]
7. Wandel, S.; Bader, N.; Glodowski, J.; Lehnhardt, B.; Leckner, J.; Schwack, F.; Poll, G. Starvation and re-lubrication in oscillating bearings: Influence of grease parameters. *Tribol. Lett.* **2022**, *70*, 114. [CrossRef]
8. Wandel, S.; Bader, N.; Schwack, F.; Glodowski, J.; Lehnhardt, B.; Poll, G. Starvation and relubrication mechanisms in grease lubricated oscillating bearings. *Tribol. Int.* **2022**, *165*, 107276. [CrossRef]
9. Schwack, F.; Schneider, V.; Wandel, S.; De La Presilla, R.J.; Poll, G.; Glavatskih, S. On the critical amplitude in oscillating rolling element bearings. *Tribol. Int.* **2021**, *163*, 107154. [CrossRef]
10. Shchegolkov, A.V.; Albairmani, A.; Al-Zahiwat, M.M.; Zemtsova, N.; Shchegolkov, A. Mathematical model of the mechanoactivation process of modibden disulfide and carbon nanotubes. In Proceedings of the 2022 4th International Conference on Control Systems, Mathematical Modeling, Automation and Energy Efficiency (SUMMA), Lipetsk, Russia, 9–11 November 2022; pp. 122–124.
11. Bojarska, Z.; Kopytowski, J.; Mazurkiewicz-Pawlicka, M.; Bazarnik, P.; Gierlotka, S.; Rożeń, A.; Makowski, Ł. Molybdenum Disulfide-Based Hybrid Materials as New Types of Oil Additives with Enhanced Tribological and Rheological Properties. *Tribol. Int.* **2021**, *160*, 106999. [CrossRef]
12. Cen, H.; Lugt, P.M.; Morales-Espejel, G.E. Film thickness of mechanically worked lubricating grease at very low speeds. *Tribol. Trans.* **2014**, *57*, 1064–1071. [CrossRef]
13. De Laurentis, N.; Kadiric, A.; Lugt, P.M.; Cann, P. The influence of bearing grease composition on friction in rolling/sliding concentrated contacts. *Tribol. Int.* **2016**, *94*, 624–632. [CrossRef]
14. Kochi, T.; Sakai, M.; Nogi, T.; Dong, D.; Kimura, Y. Experimental study on the physics of thick EHL film formation with grease at low speeds. *Tribol. Lett.* **2019**, *67*, 55. [CrossRef]
15. Huang, L.; Guo, D.; Wen, S.; Wan, G.T.Y. Effects of slide/roll ratio on the behaviours of grease reservoir and film thickness of point contact. *Tribol. Lett.* **2014**, *54*, 263–271. [CrossRef]
16. Huang, L.; Guo, D.; Wen, S. Film thickness decay and replenishment in point contact lubricated with different greases: A study into oil bleeding and the evolution of lubricant reservoir. *Tribol. Int.* **2016**, *93*, 620–627. [CrossRef]
17. Tiejun, S.; Wang, D.; Yun, J.; Qinglian, L.; Liu, X.; Peng, Z. Mechanical Stability and Rheology of Lithium–Calcium-Based Grease Containing ZDDP. *RSC Adv.* **2016**, *6*, 11637–11647.
18. Delgado, M.; Valencia, C.; Sánchez, M.M.; Franco, J.M.; Gallegos, C. Thermorheological Behaviour of a Lithium Lubricating Grease. *Tribol. Lett.* **2006**, *23*, 47–54. [CrossRef]
19. Rezasoltani, A.; Khonsari, M.M. On the Correlation Between Mechanical Degradation of Lubricating Grease and Entropy. *Tribol. Lett.* **2014**, *56*, 197–204. [CrossRef]
20. Zhou, Y.; Lugt, P.M. On the application of the mechanical aging master curve for lubricating greases to rolling bearings. *Tribol. Int.* **2020**, *141*, 105918. [CrossRef]
21. Zhou, Y.; Bosman, R.; Lugt, P.M. An experimental study on film thickness in a rolling bearing for fresh and mechanically aged lubricating greases. *Tribol. Trans.* **2019**, *62*, 557–566. [CrossRef]
22. Cann, P. Starvation and reflow in a grease-lubricated elastohydrodynamic contact. *Tribol. Trans.* **1996**, *39*, 698–704. [CrossRef]
23. Cann, P.; Damiens, B.; Lubrecht, A. The transition between fully flooded and starved regimes in EHL. *Tribol. Int.* **2004**, *37*, 859–864. [CrossRef]
24. Li, X.; Guo, F.; Poll, G.; Yang, F.; Yang, P. Grease film evolution in rolling elastohydrodynamic lubrication contacts. *Friction* **2020**, *9*, 179–190. [CrossRef]
25. Jin, X.; Li, X.; Chen, Q.; Yang, P.; Guo, F.; Jiang, N. Observation of grease film behavior in sliding-rolling concentrated contacts. *J. Tribol.* **2021**, *144*, 011602. [CrossRef]
26. Wang, J.; Meng, X.-H.; Wang, S.; Zou, Q. Grease film variation in reciprocating sliding motion. *Tribol. Int.* **2017**, *114*, 373–388. [CrossRef]
27. Han, Y.; Wang, J.; Wang, S.; Zou, Q.; Barber, G.C. Response of grease film at low speeds under pure rolling reciprocating motion. *Friction* **2020**, *8*, 115–135. [CrossRef]
28. Nishikawa, H.; Miyazaki, H.; Kaneta, M.; Guo, F. Effects of two-stage impact load on point contact elastohydrodynamic lubrication films. *Proc. Inst. Mech. Eng. Part J-J. Eng. Tribol.* **2008**, *222*, 807–814. [CrossRef]
29. Kaneta, M.; Wang, J.; Guo, F.; Křupka, I.; Hartl, M. Effects of loading process and contact shape on point impact elastohydrodynamics. *Tribol. Trans.* **2012**, *55*, 772–781. [CrossRef]
30. Frýza, J.; Šperka, P.; Kaneta, M.; Křupka, I.; Hartl, M. Effects of lubricant rheology and impact speed on EHL film thickness at pure squeeze action. *Tribol. Int.* **2017**, *106*, 1–9. [CrossRef]
31. Lv, Z.; Han, Y.; Zhang, R.; Wang, J. Occurrence of Grease Lubricated Impact-Sliding Composite Wear. *Lubricants* **2022**, *10*, 284. [CrossRef]
32. Zhang, X.; Glovnea, R. An experimental investigation of grease lubricated EHD contact subjected to normal sinusoidally variable loading. *Tribol. Int.* **2020**, *147*, 106272. [CrossRef]

33. Frýza, J.; Šperka, P.; Křupka, I.; Hartl, M. Effects of lateral harmonic vibrations on film thickness in EHL point contacts. *Tribol. Int.* **2018**, *117*, 236–249. [CrossRef]
34. Cann, P.; Lubrecht, A. The effect of transient loading on contact replenishment with lubricating greases. In *Tribology Series*; Elsevier: Amsterdam, The Netherlands, 2003; Volume 43, pp. 745–750.
35. Zhu, M.; Yu, H.; Cai, Z.; Zhou, Z. Radial fretting behaviours of dental feldspathic ceramics against different counterbodies. *Wear* **2005**, *259*, 996–1004. [CrossRef]
36. Zhu, M.; Zhou, Z. On the mechanisms of various fretting wear modes. *Tribol. Int.* **2011**, *44*, 1378–1388. [CrossRef]
37. Cubillas, D.; Olave, M.; Llavori, I.; Ulacia, I.; Larrañaga, J.; Zurutuza, A.; Lopez, A. A novel formulation for radial fretting wear: Application to false brinelling in thrust bearings. *Wear* **2022**, *488*, 204078. [CrossRef]
38. Cubillas, D.; Olave, M.; Llavori, I.; Ulacia, I.; Larrañaga, J.; Zurutuza, A.; Lopez, A. Semi-analytical methodology to predict fretting damage areas in angular contact ball bearing raceways under variable loading. *Wear* **2022**, *508*, 204477. [CrossRef]
39. Han, Y.; Wang, J.; Liu, W.; Ma, R.; Jin, X. Oil film variation and surface damage in the process of reciprocation-oscillation transformation. *Tribol. Int.* **2019**, *140*, 105828. [CrossRef]
40. Cameron, A.; Gohar, R. Theoretical and experimental studies of the oil film in lubricated point contact. *Proc. R. Soc. London Ser. A* **1966**, *291*, 520–536.
41. Foord, C.A.; Wedeven, L.D.; Westlake, F.J.; Cameron, A. Optical elastohydrodynamics. *Proc. Inst. Mech. Eng.* **1969**, *184*, 487–505. [CrossRef]
42. Liu, H.C.; Guo, F.; Guo, L.; Wong, P.L. A dichromatic interference intensity modulation approach to measurement of lubricating film thickness. *Tribol. Lett.* **2015**, *58*, 15. [CrossRef]
43. Zhu, M.; Yu, H.; Zhou, Z. Radial fretting behaviours of dental ceramics. *Tribol. Int.* **2006**, *39*, 1255–1261. [CrossRef]

lubricants

Article

Effect of Laminar, Turbulent and Slip Conditions on the Dynamic Coefficients of a Dry Gas Seal

Youngjun Park, Mibbeum Hahn and Gunhee Jang *

Department of Mechanical Convergence Engineering, Hanyang University, Seoul 04763, Republic of Korea
* Correspondence: ghjang@hanyang.ac.kr

Abstract: The dynamic coefficients of a dry gas seal affect the dynamic characteristics of rotor-seal systems. Fluid films in a dry gas seal can be laminar, turbulent or with slip conditions, according to various operating conditions and design parameters. They can be defined as laminar or turbulent, depending on the Reynolds number, and as slip or non-slip, depending on the Knudsen number. However, previous research did not consider the effect of laminar, turbulent and slip conditions on the dynamic coefficients of a dry gas seal. We proposed a mathematical perturbation method to calculate the dynamic coefficients of the dry gas seal according to laminar, turbulent, and slip effects. We derived the perturbed equations of the modified Reynolds equation, which includes the effects of laminar, turbulent and slip conditions. The pressure of the modified Reynolds equation was solved using the finite element method and the Newton–Raphson method, and the perturbed pressures with respect to three degrees of freedom were calculated by substituting the calculated pressure into the perturbed equations. We verified the proposed method by comparing the simulated results with prior studies. The dynamic coefficients of a T-grooved dry gas seal were investigated according to laminar, turbulent, and slip conditions in a fluid film with different clearances.

Keywords: dry gas seal; dynamic coefficient; mathematical perturbation; numerical analysis; slip boundary condition; turbulent flow

Citation: Park, Y.; Hahn, M.; Jang, G. Effect of Laminar, Turbulent and Slip Conditions on the Dynamic Coefficients of a Dry Gas Seal. *Lubricants* **2023**, *11*, 98. https://doi.org/10.3390/lubricants11030098

Received: 31 January 2023
Revised: 21 February 2023
Accepted: 23 February 2023
Published: 24 February 2023

1. Introduction

A dry gas seal is a non-contacting and dry-running mechanical face seal that prevents leakage of gas in various high-speed machines using gas as a working fluid. The dynamic coefficients of the dry gas seal, which are stiffness and damping coefficients, are important parameters to determine the vibration and the stability of a rotor–seal system during operation. Figure 1 shows the mechanical structure of a dry gas seal. It is composed of a rotating seal and a stationary seal. Pressure is developed in a small fluid gap between the rotating and stationary seals in such a way to balance the closing force and the spring force. The closing force is generated by the outer pressure of the working fluid. As Han et al. showed, the fluid in the gap can be laminar, turbulent or in slip conditions depending on the operating condition and design variables [1]. The behavior of a fluid film can be defined as laminar or turbulent, depending on the Reynolds number ($Re = \rho U h/\mu$), and as slip or non-slip, depending on the Knudsen number ($Kn = \mu\sqrt{0.5\pi RT}/hp$), where μ, ρ, U, h, R, T, and p represent the viscosity coefficient, density, fluid velocity, fluid film thickness, gas constant, temperature, and pressure of the fluid film, respectively. A high rotating velocity and large clearance of the fluid in the grooved area (generally located at the outer part of the rotating seal) increase the Reynolds number. As such, the fluid state may change from laminar to turbulent. Additionally, slip may occur between the fluid and the solid because low pressure and small clearance of the area in the inner part of the seal increase the Knudsen number. Han et al. investigated the pressure and the leakage of dry gas seals for laminar, turbulent, and slip conditions, but they did not investigate the dynamic coefficient of the dry gas seal for these conditions.

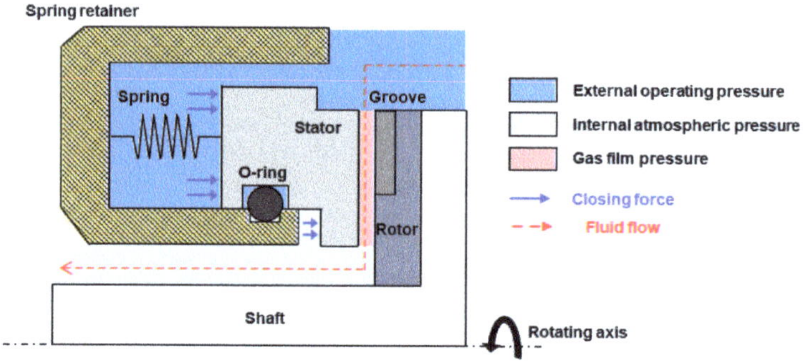

Figure 1. Mechanical structure of a dry gas seal.

Several researchers analyzed the dynamic coefficients of dry gas seals using the physical perturbation. The dynamic coefficients in the physical perturbation were determined by numerically differentiating the opening force of the seal with respect to finite displacements and velocities at steady-state clearance. Chen et al. analyzed the stiffness, opening force, and leakage rate of a compliant foil face gas seal (CFFGS) with different design parameters using the Reynolds equation. They showed that CFFGS has stable and good leakage performance [2]. Lu et al. calculated the gas film stiffness of dry gas seal by considering the slip effect of fluid films and optimized the groove shape to improve the gas film stiffness [3]. Other researchers also investigated dry gas seals using the physical perturbation method which extracts dynamic coefficients [4,5]. However, the physical perturbation method requires more calculation time than the mathematical perturbation method, and its accuracy is dependent on the finite displacements and velocities. Moreover, prior researchers used the Reynolds equation considering the laminar flow or slip condition of the fluid film to calculate the dynamic coefficients but did not consider the turbulent flow.

Some researchers have derived the perturbed Reynolds equation mathematically to study the dynamic coefficients of dry gas seals, which extracts dynamic coefficients from the perturbed pressure and film thickness in the quasi-static equilibrium state of the dry gas seal [6–16]. Chen et al. analyzed the dynamic behavior of the dry gas seal using the mathematically derived perturbed Reynolds equation [6]. The simulated leakage rate was compared with the experimental one. Ruan et al. calculated the stiffness and damping coefficients of a dry gas seal using the perturbed Reynolds equation. Additionally, they analyzed the stability of the seal by constructing a spring–damper–mass system [7]. Liu et al. analyzed the dynamic coefficients of the dry gas seal and showed that the interactions between angular and axial perturbation were negligible [8]. Other researchers also investigated dry gas seals using the mathematical perturbation of the Reynolds equation [9–15]. However, they did not consider turbulent flow and slip conditions of the fluid film.

In this paper, we proposed a mathematical perturbation method to calculate the dynamic coefficients of the dry gas seal according to laminar, turbulent, and slip effects. We derived the perturbed equation of the modified Reynolds equation, which includes the effects of laminar, turbulent, and slip conditions. The pressure of the modified Reynolds equation was solved using the finite element method and the Newton–Raphson method. Perturbed pressures with respect to three degrees of freedom were calculated by substituting the calculated pressure into the perturbed equations. We verified the proposed method by comparing the simulated results with prior studies. Finally, the dynamic coefficients of a T-grooved dry gas seal were investigated according to laminar, turbulent, and slip conditions in a fluid film with different clearances.

2. Method of Analysis

The conventional compressible Reynolds equation can be derived from the Navier–Stokes equation, assuming Newtonian laminar flow, ideal gas conditions, no body force, no inertial force, and no slippage, and ignoring the pressure gradient along the film thickness. On the other hand, Ng-Pan proposed a turbulent compressible Reynolds equation that includes turbulent effect in the compressible Reynolds equation [16,17], and Fukui–Kaneko proposed a modified Reynolds equation that includes the slip effect [18,19]. Recently, Han et al. [1] defined fluid state coefficients according to fluid flow to consider laminar, turbulent, and slip conditions, and proposed a modified Reynolds equation considering laminar, turbulent, and slip conditions as follows:

$$C_r \frac{h^3}{\mu}\frac{\partial}{\partial r}\left(p\frac{\partial p}{\partial r}\right) + C_\theta \frac{h^3}{\mu}\frac{\partial}{r\partial\theta}\left(p\frac{\partial p}{r\partial\theta}\right) = \frac{U_0}{2}\frac{\partial(hp)}{r\partial\theta} + \frac{\partial(hp)}{\partial t} \tag{1}$$

where h, μ, p, U, C_r and C_θ are the clearance, viscosity, pressure, fluid velocity, and fluid state coefficients of radial and circumferential directions considering laminar, turbulent, and slip conditions of the fluid film, respectively. Table 1 defines the fluid state coefficients according to laminar, turbulent, and slip conditions.

Table 1. Fluid state coefficients according to laminar, turbulent, and slip conditions [1].

Fluid Condition	C_r	C_θ
Laminar	$1/12$	$1/12$
Turbulent	$1/G_r$	$1/G_\theta$
Slip	$q_p/12$	$q_p/12$

In Table 1, G_r, G_θ, and q_p are radial turbulent coefficient, circumferential turbulent coefficient, and slip coefficient defined in Equations (2)–(4), and Table 2 [1,16–24]:

$$G_r = 12 + 0.0043Re^{0.96} \tag{2}$$

$$G_\theta = 12 + 0.0136Re^{0.9} \tag{3}$$

$$q_p = c_0 + c_1(Kn) + c_2(Kn)^2 + c_3(Kn)^3 \tag{4}$$

where Re and Kn are the Reynolds number and Knudsen number, respectively.

Table 2. Slip coefficients according to $1/Kn$ [1].

Range of Inverse Kn	c_0	c_1	c_2	c_3
$5 < 1/Kn \leq 1000$	1.000	6.097	6.391	−12.812
$0.15 < 1/Kn \leq 5$	0.831	7.505	0.939	−0.058
$1/Kn \leq 0.15$	−13.375	12.640	0.099	0.0004

Figure 2 shows the perturbed film thickness h for a three-DOF (z, θ_x and θ_y) of a dry gas seal. The perturbed film thicknesses with respect to z, θ_x and θ_y are represented by Δh, $\Delta\theta_x$, and $\Delta\theta_y$. The polar coordinate (r, θ') is introduced to define the film thickness on a dry gas seal, and the fixed angular coordinate θ' is defined from the –x axis in a counterclockwise direction. The perturbed film thickness h and pressure p are defined as follows:

$$h = h_0 + \Delta z - r\sin\theta'\Delta\theta_x + r\cos\theta'\Delta\theta_y \tag{5}$$

$$\frac{\partial h}{\partial t} = \Delta\dot{z} - r\sin\theta'\Delta\dot{\theta}_x + r\cos\theta'\Delta\dot{\theta}_y \tag{6}$$

$$p = p_0 + p_z\Delta z + p_{\theta_x}\Delta\theta_x + p_{\theta_y}\Delta\theta_y + p_{\dot{z}}\Delta\dot{z} + p_{\dot{\theta}_x}\Delta\dot{\theta}_x + p_{\dot{\theta}_y}\Delta\dot{\theta}_y \tag{7}$$

$$p_z = \left(\frac{\partial P}{\partial z}\right)_0,\ p_{\theta_x} = \left(\frac{\partial P}{\partial \theta_x}\right)_0,\ p_{\theta_y} = \left(\frac{\partial P}{\partial \theta_y}\right)_0,\ p_{\dot{z}} = \left(\frac{\partial P}{\partial \dot{z}}\right)_0,\ p_{\dot{\theta}_x} = \left(\frac{\partial P}{\partial \dot{\theta}_x}\right)_0,\ p_{\dot{\theta}_y} = \left(\frac{\partial P}{\partial \dot{\theta}_y}\right)_0 \tag{8}$$

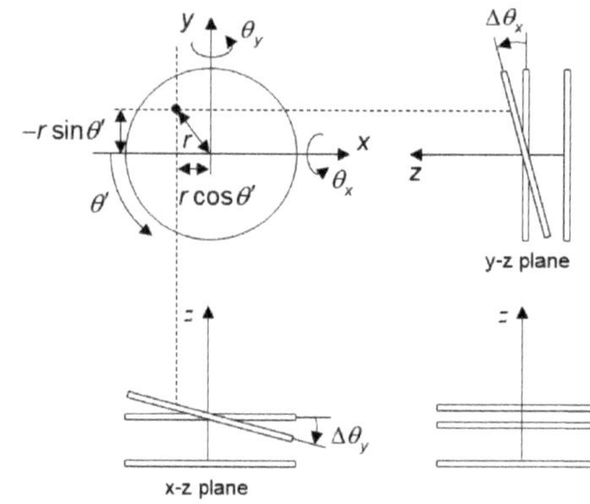

Figure 2. Relation between perturbed film thickness in a dry gas seal.

Here, h_0 is the film thickness in the equilibrium. p_z, p_{θ_x}, and p_{θ_y} are the perturbed pressure generated by infinitesimal displacement with respect to z, θ_x and θ_y, respectively. Additionally, $p_{\dot{z}}$, $p_{\dot{\theta}_x}$, and $p_{\dot{\theta}_y}$ are the perturbed pressure generated by infinitesimal velocity with respect to $\dot{z}$, $\dot{\theta}_x$, and $\dot{\theta}_y$, respectively.

Substituting Equations (5)–(7) into (1) yields the perturbed Reynolds equation as follows:

$$
\begin{aligned}
&\frac{\partial}{\partial r}\left(\begin{array}{c} C_r\frac{(h_0+\Delta z-r sin\theta'\,\Delta\theta_x+r\cos\theta'\,\Delta\theta_y)^3}{\mu}\left(\begin{array}{c} p_0+p_z\Delta z+p_{\theta_x}\Delta\theta_x+p_{\theta_y}\Delta\theta_y \\ +p_{\dot{z}}\Delta\dot{z}+p_{\dot{\theta}_x}\Delta\dot{\theta}_x+p_{\dot{\theta}_y}\Delta\dot{\theta}_y \end{array}\right) \\ \frac{\partial\left(p_0+p_z\Delta z+p_{\theta_x}\Delta\theta_x+p_{\theta_y}\Delta\theta_y+p_{\dot{z}}\Delta\dot{z}+p_{\dot{\theta}_x}\Delta\dot{\theta}_x+p_{\dot{\theta}_y}\Delta\dot{\theta}_y\right)}{\partial r} \end{array}\right) \\
&+ \\
&\frac{\partial}{r\partial\theta}\left(\begin{array}{c} C_\theta\frac{(h_0+\Delta z-r sin\theta'\,\Delta\theta_x+r\cos\theta'\,\Delta\theta_y)^3}{\mu}\left(\begin{array}{c} p_0+p_z\Delta z+p_{\theta_x}\Delta\theta_x+p_{\theta_y}\Delta\theta_y \\ +p_{\dot{z}}\Delta\dot{z}+p_{\dot{\theta}_x}\Delta\dot{\theta}_x+p_{\dot{\theta}_y}\Delta\dot{\theta}_y \end{array}\right) \\ \frac{\partial\left(p_0+p_z\Delta z+p_{\theta_x}\Delta\theta_x+p_{\theta_y}\Delta\theta_y+p_{\dot{z}}\Delta\dot{z}+p_{\dot{\theta}_x}\Delta\dot{\theta}_x+p_{\dot{\theta}_y}\Delta\dot{\theta}_y\right)}{r\partial\theta} \end{array}\right) \\
&= \\
&\frac{r\omega}{2}\frac{\partial}{r\partial\theta}\left\{(h_0+\Delta z-r sin\theta'\,\Delta\theta_x+r\cos\theta'\,\Delta\theta_y)\left(\begin{array}{c} p_0+p_z\Delta z+p_{\theta_x}\Delta\theta_x+p_{\theta_y}\Delta\theta_y \\ +p_{\dot{z}}\Delta\dot{z}+p_{\dot{\theta}_x}\Delta\dot{\theta}_x+p_{\dot{\theta}_y}\Delta\dot{\theta}_y \end{array}\right)\right\} \\
&+\frac{\partial}{\partial t}\left\{(h_0+\Delta z-r sin\theta'\,\Delta\theta_x+r\cos\theta'\,\Delta\theta_y)\left(\begin{array}{c} p_0+p_z\Delta z+p_{\theta_x}\Delta\theta_x+p_{\theta_y}\Delta\theta_y \\ +p_{\dot{z}}\Delta\dot{z}+p_{\dot{\theta}_x}\Delta\dot{\theta}_x+p_{\dot{\theta}_y}\Delta\dot{\theta}_y \end{array}\right)\right\}
\end{aligned} \quad (9)
$$

By expanding Equation (9) and assuming that higher order terms are negligible, the governing equations for each variable can be expressed as follows:

$$
\frac{\partial}{\partial r}\left(C_r\frac{h_0{}^3}{\mu}p_0\frac{\partial p_0}{\partial r}\right)+\frac{\partial}{r\partial\theta}\left(C_\theta\frac{h_0{}^3}{\mu}p_0\frac{\partial p_0}{r\partial\theta}\right)=\frac{r\omega}{2}\frac{\partial(h_0p_0)}{r\partial\theta} \quad (10)
$$

$$
\begin{aligned}
&\frac{\partial}{\partial r}\left(C_r\frac{h_0{}^3}{\mu}p_0\frac{\partial p_z}{\partial r}+C_r\frac{h_0{}^3}{\mu}p_z\frac{\partial p_0}{\partial r}\right) \\
&+\frac{\partial}{r\partial\theta}\left(C_\theta\frac{h_0{}^3}{\mu}p_z\frac{\partial p_0}{r\partial\theta}+C_\theta\frac{h_0{}^3}{\mu}p_0\frac{\partial p_z}{r\partial\theta}\right)-\frac{r\omega}{2}\frac{\partial}{r\partial\theta}(h_0p_z) \\
&=-\frac{\partial}{\partial r}\left(C_r\frac{3h_0{}^2}{\mu}p_0\frac{\partial p_0}{\partial r}\right)-\frac{\partial}{r\partial\theta}\left(C_\theta\frac{3h_0{}^2}{\mu}p_0\frac{\partial p_0}{r\partial\theta}\right) \\
&+\frac{r\omega}{2}\frac{\partial}{r\partial\theta}(p_0)+h_0p_{\dot{z}}
\end{aligned} \quad (11)
$$

$$\begin{aligned}&\frac{\partial}{\partial r}\left(C_r\frac{h_0{}^3}{\mu}p_0\frac{\partial p_{\theta_x}}{\partial r}+C_r\frac{h_0{}^3}{\mu}p_{\theta_x}\frac{\partial p_0}{\partial r}\right)\\&+\frac{\partial}{r\partial\theta}\left(C_\theta\frac{h_0{}^3}{\mu}p_{\theta_x}\frac{\partial p_0}{r\partial\theta}+C_\theta\frac{h_0{}^3}{\mu}p_0\frac{\partial p_{\theta_x}}{r\partial\theta}\right)-\frac{r\omega}{2}\frac{\partial}{r\partial\theta}(h_0p_{\theta_x})\\&=\frac{\partial}{\partial r}\left(C_r\frac{3h_0{}^2}{\mu}r\sin\theta' p_0\frac{\partial p_0}{\partial r}\right)+\frac{\partial}{r\partial\theta}\left(C_\theta\frac{3h_0{}^2}{\mu}r\sin\theta' p_0\frac{\partial p_0}{r\partial\theta}\right)\\&+\frac{r\omega}{2}\frac{\partial}{r\partial\theta}(-r\sin\theta' p_0)+h_0p_{\dot\theta_x}\end{aligned}\tag{12}$$

$$\begin{aligned}&\frac{\partial}{\partial r}\left(C_r\frac{h_0{}^3}{\mu}p_0\frac{\partial p_{\theta_y}}{\partial r}+C_r\frac{h_0{}^3}{\mu}p_{\theta_y}\frac{\partial p_0}{\partial r}\right)\\&+\frac{\partial}{r\partial\theta}\left(C_\theta\frac{h_0{}^3}{\mu}p_{\theta_y}\frac{\partial p_0}{r\partial\theta}+C_\theta\frac{h_0{}^3}{\mu}p_0\frac{\partial p_{\theta_y}}{r\partial\theta}\right)-\frac{r\omega}{2}\frac{\partial}{r\partial\theta}\left(h_0p_{\theta_y}\right)\\&=-\frac{\partial}{\partial r}\left(C_r\frac{3h_0{}^2}{\mu}r\cos\theta' p_0\frac{\partial p_0}{\partial r}\right)-\frac{\partial}{r\partial\theta}\left(C_\theta\frac{3h_0{}^2}{\mu}r\cos\theta' p_0\frac{\partial p_0}{r\partial\theta}\right)\\&+\frac{r\omega}{2}\frac{\partial}{r\partial\theta}(r\cos\theta' p_0)+h_0p_{\dot\theta_y}\end{aligned}\tag{13}$$

$$\begin{aligned}&\frac{\partial}{\partial r}\left(C_r\frac{h_0{}^3}{\mu}p_{\dot z}\frac{\partial p_0}{\partial r}+C_r\frac{h_0{}^3}{\mu}p_0\frac{\partial p_{\dot z}}{\partial r}\right)\\&+\frac{\partial}{r\partial\theta}\left(C_\theta\frac{h_0{}^3}{\mu}p_{\dot z}\frac{\partial p_0}{r\partial\theta}+C_\theta\frac{h_0{}^3}{\mu}p_0\frac{\partial p_{\dot z}}{r\partial\theta}\right)-\frac{r\omega}{2}\frac{\partial}{r\partial\theta}(h_0p_{\dot z})\\&=p_0+h_0p_z\end{aligned}\tag{14}$$

$$\begin{aligned}&\frac{\partial}{\partial r}\left(C_r\frac{h_0{}^3}{\mu}p_{\dot\theta_x}\frac{\partial p_0}{\partial r}+C_r\frac{h_0{}^3}{\mu}p_0\frac{\partial p_{\dot\theta_x}}{\partial r}\right)\\&+\frac{\partial}{r\partial\theta}\left(C_\theta\frac{h_0{}^3}{\mu}p_{\dot\theta_x}\frac{\partial p_0}{r\partial\theta}+C_\theta\frac{h_0{}^3}{\mu}p_0\frac{\partial p_{\dot\theta_x}}{r\partial\theta}\right)-\frac{r\omega}{2}\frac{\partial}{r\partial\theta}\left(h_0p_{\dot\theta_x}\right)\\&=-p_0r\sin\theta'+h_0p_{\theta_x}\end{aligned}\tag{15}$$

$$\begin{aligned}&\frac{\partial}{\partial r}\left(C_r\frac{h_0{}^3}{\mu}p_{\dot\theta_y}\frac{\partial p_0}{\partial r}+C_r\frac{h_0{}^3}{\mu}p_0\frac{\partial p_{\dot\theta_y}}{\partial r}\right)\\&+\frac{\partial}{r\partial\theta}\left(C_\theta\frac{h_0{}^3}{\mu}p_{\dot\theta_y}\frac{\partial p_0}{r\partial\theta}+C_\theta\frac{h_0{}^3}{\mu}p_0\frac{\partial p_{\dot\theta_y}}{r\partial\theta}\right)-\frac{r\omega}{2}\frac{\partial}{r\partial\theta}\left(h_0p_{\dot\theta_y}\right)\\&=p_0r\cos\theta'+h_0p_{\theta_y}\end{aligned}\tag{16}$$

Equations (10)–(16) are the governing equations of $p_0, p_z, p_{\theta_x}, \ p_{\theta_y}, \ p_{\dot z}, p_{\dot\theta_x}$, and $p_{\dot\theta_y}$. Equation (10) is, in fact, the modified Reynolds equation in Equation (1). A finite element equation of the Reynolds equation can be obtained by multiplying Equation (10) with the weighting function and integrating using Green's theorem:

$$\begin{aligned}&w\int_\Gamma\left(C_r\frac{h_0^3}{\mu}p_0\frac{\partial p_0}{\partial r}+C_\theta\frac{h_0^3}{\mu}p_0\frac{\partial p_0}{r\partial\theta}-\frac{r\omega(h_0p_0)}{2}\right)\cdot\mathbf{n}d\Gamma\\&-\int_A\left(C_r\frac{h^3}{\mu}\frac{\partial w}{\partial r}p_0\frac{\partial p_0}{\partial r}+C_\theta\frac{h^3}{\mu}\frac{\partial w}{r\partial\theta}p_0\frac{\partial p_0}{r\partial\theta}-\frac{r\omega h_0}{2}p_0\frac{\partial w}{r\partial\theta}\right)dA=0\end{aligned}\tag{17}$$

Here, $\Gamma, \ w$, and $\mathbf{n}$ are the boundary, weighting function, and normal unit vector of the boundary, respectively. The boundary integral terms of Equation (17) are zero because the weighting functions at the boundary are zero. The pressure in a four-node element can be expressed by the nodal pressure $\mathbf{P}_e$ and shape function $\mathbf{N}$ as follows:

$$p=\mathbf{N}^{\mathrm{T}}\mathbf{P}_e\tag{18}$$

Similarly, the weighting function can be expressed by an arbitrary vector $\boldsymbol{\eta}_e$ as follows:

$$w=\boldsymbol{\eta}_e^{\mathrm{T}}\mathbf{N}\tag{19}$$

Substituting Equations (18) and (19) into (17) yields the local matrix of the finite-element equation:

$$\int_A\boldsymbol{\eta}_e{}^{\mathrm{T}}\left(\begin{array}{c}C_r\frac{h^3}{\mu}\frac{\partial\mathbf{N}}{\partial r}\mathbf{N}^{\mathrm{T}}\mathbf{P}_e\frac{\partial\mathbf{N}^{\mathrm{T}}}{\partial r}\mathbf{P}_e+C_\theta\frac{h^3}{\mu}\frac{\partial\mathbf{N}}{r\partial\theta}\mathbf{N}^{\mathrm{T}}\mathbf{P}_e\frac{\partial\mathbf{N}^{\mathrm{T}}}{r\partial\theta}\mathbf{P}_e\\-\frac{r\omega h}{2}\frac{\partial\mathbf{N}}{r\partial\theta}\mathbf{N}^{\mathrm{T}}\mathbf{P}_e\end{array}\right)dA=0\tag{20}$$

Equation (20) is non-linear and can be solved by using the Newton–Raphson method as follows [1]:

$$\begin{aligned}
&\mathbf{R}^{(n)} + \alpha_{relax}\frac{\partial \mathbf{R}^{(n)}}{\partial \mathbf{P}^{(n)}}\Delta \mathbf{P}^{(n)} = 0 \\
&\mathbf{R} = \int_A \left\{ \frac{h^3}{\mu}\mathbf{N}^T\mathbf{P}\left(C_r\frac{\partial \mathbf{N}}{\partial r}\frac{\partial \mathbf{N}^T}{\partial r}\mathbf{P} + C_\theta \frac{\partial \mathbf{N}}{r\partial\theta}\frac{\partial \mathbf{N}^T}{r\partial\theta}\mathbf{P}\right) - \frac{r\omega h}{2}\frac{\partial \mathbf{N}}{r\partial\theta}\mathbf{N}^T\mathbf{P}\right\} dA = 0 \\
&\frac{\partial \mathbf{R}}{\partial \mathbf{P}} = \int_A \left\{ \begin{array}{l} \frac{h^3}{\mu}\left(\begin{array}{l} \mathbf{N}^T\mathbf{P}\left(C_r\frac{\partial \mathbf{N}}{\partial r}\frac{\partial \mathbf{N}^T}{\partial r} + C_\theta\frac{\partial \mathbf{N}}{r\partial\theta}\frac{\partial \mathbf{N}^T}{r\partial\theta}\right) \\ +\left(C_r\frac{\partial \mathbf{N}}{\partial r}\frac{\partial \mathbf{N}^T}{\partial r} + C_\theta\frac{\partial \mathbf{N}}{r\partial\theta}\frac{\partial \mathbf{N}^T}{r\partial\theta}\right)\mathbf{P}\mathbf{N}^T \\ +\frac{\partial \mathbf{N}}{\partial r}\frac{\partial \mathbf{N}^T}{\partial r}\mathbf{P}\mathbf{N}^T\mathbf{P}\frac{\partial C_r}{\partial \mathbf{P}} + \frac{\partial \mathbf{N}}{r\partial\theta}\frac{\partial \mathbf{N}^T}{r\partial\theta}\mathbf{P}\mathbf{N}^T\mathbf{P}\frac{\partial C_\theta}{\partial \mathbf{P}} \end{array} \right) \\ -\frac{r\omega h}{2}\frac{\partial \mathbf{N}}{r\partial\theta}\mathbf{N}^T \end{array} \right\} dA
\end{aligned} \tag{21}$$

where $\mathbf{R}$ and $\partial\mathbf{R}/\partial\mathbf{P}$ are Equation (20) and the pressure derivative of Equation (20), respectively, and n is the iteration number. The relaxation factor used to increase the stability of the analysis results, α_{relax}, was set to 0.5. The local matrices of $\mathbf{R}$ and the local matrix of $\partial\mathbf{R}/\partial\mathbf{P}$ were assembled to generate the global matrices of $\mathbf{R}$ and $\partial\mathbf{R}/\partial\mathbf{P}$, respectively. The analysis was repeated until the ratio of the summation of the all components of the global matrix of $\mathbf{R}$ to that of the global matrix of $\partial\mathbf{R}/\partial\mathbf{P}$ became less than 10^{-4}, and in most of the analyses in this paper, they converged within 20 iterations. Equation (22) shows the boundary condition applied to solve the modified Reynolds equation in which internal and external pressures are applied to internal and external boundaries and pressure along the circumferential direction is continuous.

$$p(r_i,\theta) = p_i, p(r_o,\theta) = p_e, p(r,\theta) = p(r,\theta+2\pi) \tag{22}$$

where p_i and p_e are internal and external pressures applied to internal and external boundaries.

The finite element equations of the perturbed modified Reynolds equation can be obtained by multiplying the weight function and integrating using Green's theorem, as in Equation (17). The finite element equations of the left-hand terms are as follows:

$$\begin{aligned}
&w\int_\Gamma \left(\begin{array}{l} C_r\frac{h_0{}^3}{\mu}p_0\frac{\partial p_i}{\partial r} + C_r\frac{h_0{}^3}{\mu}p_i\frac{\partial p_0}{\partial r} + C_\theta\frac{h_0{}^3}{\mu}p_0\frac{\partial p_i}{r\partial\theta} \\ +C_\theta\frac{h_0{}^3}{\mu}p_i\frac{\partial p_0}{r\partial\theta} - \widetilde{V}_\theta\frac{\partial(h_0p_i)}{r\partial\theta} \end{array} \right)\cdot \hat{\mathbf{n}}d\Gamma \\
&-\int_A \left(\begin{array}{l} C_r\frac{h_0{}^3}{\mu}\frac{\partial w}{\partial r}p_0\frac{\partial p_i}{\partial r} + C_r\frac{h_0{}^3}{\mu}\frac{\partial w}{\partial r}p_i\frac{\partial p_0}{\partial r} + C_\theta\frac{h_0{}^3}{\mu}\frac{\partial w}{r\partial\theta}p_0\frac{\partial p_i}{r\partial\theta} \\ +C_\theta\frac{h_0{}^3}{\mu}\frac{\partial w}{r\partial\theta}p_i\frac{\partial p_0}{r\partial\theta} - \widetilde{V}_\theta\frac{\partial w}{r\partial\theta}h_0p_i \end{array} \right) dA
\end{aligned} \tag{23}$$

where $i = z, \theta_x, \theta_y, \dot{z}, \dot{\theta}_x, \dot{\theta}_y$. The boundary integral terms of Equation (23) are zero because the weighting functions at the boundary are zero. Substituting Equations (18) and (19) into (23) yields the left-hand finite element equation of the local matrix:

$$\int_A \boldsymbol{\eta}^{\mathrm{T}} \left(\begin{array}{l} C_r\frac{h_0{}^3}{\mu}\frac{\partial \mathbf{N}}{\partial r}\mathbf{N}^{\mathrm{T}}\mathbf{P}_0\frac{\partial \mathbf{N}^{\mathrm{T}}}{\partial r}\mathbf{P}_i + C_r\frac{h_0{}^3}{\mu}\frac{\partial \mathbf{N}}{\partial r}\mathbf{N}^{\mathrm{T}}\mathbf{P}_i\frac{\partial \mathbf{N}^{\mathrm{T}}}{\partial r}\mathbf{P}_0 \\ +C_\theta\frac{h_0{}^3}{\mu}\frac{\partial \mathbf{N}}{r\partial\theta}\mathbf{N}^{\mathrm{T}}\mathbf{P}_0\frac{\partial \mathbf{N}^{\mathrm{T}}}{r\partial\theta}\mathbf{P}_i + C_\theta\frac{h_0{}^3}{\mu}\frac{\partial \mathbf{N}}{r\partial\theta}\mathbf{N}^{\mathrm{T}}\mathbf{P}_i\frac{\partial \mathbf{N}^{\mathrm{T}}}{r\partial\theta}\mathbf{P}_0 \\ -\frac{r\omega}{2}h_0\frac{\partial \mathbf{N}}{r\partial\theta}\mathbf{N}^{\mathrm{T}}\mathbf{P}_i \end{array} \right) dA \tag{24}$$

where $i = z, \theta_x, \theta_y, \dot{z}, \dot{\theta}_x, \dot{\theta}_y$.

The right-hand finite element equations of the local matrices corresponding to p_z, p_{θ_x}, p_{θ_y}, $p_{\dot{z}}$, $p_{\dot{\theta}_x}$, and $p_{\dot{\theta}_y}$ can be obtained similarly, as follows:

$$\int_A \boldsymbol{\eta}^{\mathrm{T}} \left(\begin{array}{l} -C_r\frac{3h_0{}^2}{\mu}\frac{\partial \mathbf{N}}{\partial r}\mathbf{N}^{\mathrm{T}}\mathbf{P}_0\frac{\partial \mathbf{N}^{\mathrm{T}}\mathbf{P}_0}{\partial r} - C_\theta\frac{3h_0{}^2}{\mu}\frac{\partial \mathbf{N}}{\partial r}\mathbf{N}^{\mathrm{T}}\mathbf{P}_0\frac{\partial \mathbf{N}^{\mathrm{T}}\mathbf{P}_0}{r\partial\theta} \\ +\frac{r\omega}{2}\frac{\partial \mathbf{N}}{r\partial\theta}\left(\mathbf{N}^{\mathrm{T}}\mathbf{P}_0\right) - h_0\mathbf{N}\mathbf{N}^{\mathrm{T}}\mathbf{P}_{\dot{z}} \end{array} \right) dA \tag{25}$$

$$\int_A \boldsymbol{\eta}^{\mathrm{T}} \begin{pmatrix} C_r \frac{3h_0{}^2}{\mu} r\sin\theta' \frac{\partial \mathbf{N}}{\partial r}\mathbf{N}^{\mathrm{T}}\mathbf{P}_0 \frac{\partial \mathbf{N}^{\mathrm{T}}\mathbf{P}_0}{\partial r} + C_\theta \frac{3h_0{}^2}{\mu} r\sin\theta' \frac{\partial \mathbf{N}}{r\partial\theta}\mathbf{N}^{\mathrm{T}}\mathbf{P}_0 \frac{\partial \mathbf{N}^{\mathrm{T}}\mathbf{P}_0}{r\partial\theta} \\ +\frac{r\omega}{2}\frac{\partial \mathbf{N}}{r\partial\theta}\left(-r\sin\theta'\mathbf{N}^{\mathrm{T}}\mathbf{P}_0\right) - h_0\mathbf{N}\mathbf{N}^{\mathrm{T}}\mathbf{P}_{\dot{\theta}_x} \end{pmatrix} dA \tag{26}$$

$$\int_A \boldsymbol{\eta}^{\mathrm{T}} \begin{pmatrix} -C_r \frac{3h_0{}^2}{\mu} r\cos\theta' \frac{\partial \mathbf{N}}{\partial r}\mathbf{N}^{\mathrm{T}}\mathbf{P}_0 \frac{\partial \mathbf{N}^{\mathrm{T}}\mathbf{P}_0}{\partial r} - C_\theta \frac{3h_0{}^2}{\mu} r\cos\theta' \frac{\partial \mathbf{N}}{r\partial\theta}\mathbf{N}^{\mathrm{T}}\mathbf{P}_0 \frac{\partial \mathbf{N}^{\mathrm{T}}\mathbf{P}_0}{r\partial\theta} \\ +\frac{r\omega}{2}\frac{\partial \mathbf{N}}{r\partial\theta}\left(r\cos\theta'\mathbf{N}^{\mathrm{T}}\mathbf{P}_0\right) - h_0\mathbf{N}\mathbf{N}^{\mathrm{T}}\mathbf{P}_{\dot{\theta}_y} \end{pmatrix} dA \tag{27}$$

$$-\int_A \boldsymbol{\eta}^{\mathrm{T}}\left(\mathbf{N}\mathbf{N}^{\mathrm{T}}\mathbf{P}_0 + h_0\mathbf{N}\mathbf{N}^{\mathrm{T}}\mathbf{P}_z\right)dA \tag{28}$$

$$-\int_A \boldsymbol{\eta}^{\mathrm{T}}\left(-r\sin\theta'\mathbf{N}\mathbf{N}^{\mathrm{T}}\mathbf{P}_0 + h_0\mathbf{N}\mathbf{N}^{\mathrm{T}}\mathbf{P}_{\theta_x}\right)dA \tag{29}$$

$$-\int_A \boldsymbol{\eta}^{\mathrm{T}}\left(r\cos\theta'\mathbf{N}\mathbf{N}^{\mathrm{T}}\mathbf{P}_0 + h_0\mathbf{N}\mathbf{N}^{\mathrm{T}}\mathbf{P}_{\theta_y}\right)dA \tag{30}$$

Once the calculated pressure from Equation (21) is substituted into the perturbed Equations (24)–(30), the local matrix equation of the perturbed pressure can be written as linear equations with respect to the perturbed pressure, as shown in Equations (31) and (32).

$$\begin{aligned} [k]\{p_j\} &= [F] + [A][p_k] \\ [k]\{p_k\} &= [F] + [B][p_j] \end{aligned} \tag{31}$$

$$\begin{bmatrix} [k] & -[A] \\ -[B] & [k] \end{bmatrix} \begin{Bmatrix} p_j \\ p_k \end{Bmatrix} = \begin{bmatrix} [F] \\ [F] \end{bmatrix} \tag{32}$$

where $j = z, \theta_x, \theta_y$ and $k = \dot{z}, \dot{\theta}_x, \dot{\theta}_y$. Subsequently, a global matrix equation can be assembled, and the perturbed pressure can be determined. Equation (33) shows the boundary condition applied to solve the perturbed Reynolds equations in which internal and external perturbed pressures are zero at the internal and external boundaries and perturbed pressure along the circumferential direction is continuous.

$$p_k(r_i,\theta) = 0, p_k(r_o,\theta) = 0, p_k(r,\theta) = p_k(r,\theta+2\pi) \tag{33}$$

where k is z, θ_x, θ_y, $\dot{z}$, $\dot{\theta}_x$, and $\dot{\theta}_y$, respectively.

The stiffness and damping matrices for a dry gas seal can be calculated from the calculated perturbed pressure as follows:

$$\mathbf{K} = \begin{bmatrix} \mathrm{K}_{zz} & \mathrm{K}_{z\theta_x} & \mathrm{K}_{z\theta_y} \\ \mathrm{K}_{\theta_x z} & \mathrm{K}_{\theta_x\theta_x} & \mathrm{K}_{\theta_x\theta_y} \\ \mathrm{K}_{\theta_y z} & \mathrm{K}_{\theta_y\theta_x} & \mathrm{K}_{\theta_y\theta_y} \end{bmatrix} = -\iint_A \begin{bmatrix} p_z & p_{\theta_x} & p_{\theta_y} \\ -p_z r\sin\theta & -p_{\theta_x} r\sin\theta & -p_{\theta_y} r\sin\theta \\ p_z r\cos\theta & p_{\theta_x} r\cos\theta & p_{\theta_y} r\cos\theta \end{bmatrix} rdrd\theta \tag{34}$$

$$\mathbf{C} = \begin{bmatrix} \mathrm{C}_{zz} & \mathrm{C}_{z\theta_x} & \mathrm{C}_{z\theta_y} \\ \mathrm{C}_{\theta_x z} & \mathrm{C}_{\theta_x\theta_x} & \mathrm{C}_{\theta_x\theta_y} \\ \mathrm{C}_{\theta_y z} & \mathrm{C}_{\theta_y\theta_x} & \mathrm{C}_{\theta_y\theta_y} \end{bmatrix} = -\iint_A \begin{bmatrix} p_{\dot{z}} & p_{\dot{\theta}_x} & p_{\dot{\theta}_y} \\ -p_{\dot{z}} r\sin\theta & -p_{\dot{\theta}_x} r\sin\theta & -p_{\dot{\theta}_y} r\sin\theta \\ p_{\dot{z}} r\cos\theta & p_{\dot{\theta}_x} r\cos\theta & p_{\dot{\theta}_y} r\cos\theta \end{bmatrix} rdrd\theta \tag{35}$$

3. Numerical Verification

To verify the developed equations and program for the perturbed Reynolds equation, we developed a finite element model of the dry gas seal analyzed by Faria [25]. Faria analyzed a spiral grooved dry gas seal whose external radius, groove radius, internal radius, groove depth, groove angle, and number of grooves were 88.9 mm, 76.4 mm, 71.1 mm, 2.54 μm, 20°, and 12, respectively. The rotating velocity, external pressure, internal pressure, and clearance were 15,000 rpm, 5.05 bar, 1.01 bar, and 2.54 μm, respectively. The developed finite element model consisted of 7920 quadrilateral elements, similar to Faria's model. The opening force, stiffness coefficient, and damping coefficient to axial direction were calculated and compared with the result obtained by Faria. Table 3 shows the simulated results of this study and Faria. Since Faria solved the compressible Reynolds equation including only the effect of laminar flow, we calculated the opening force, stiffness, and damping coefficients according to laminar flow. The opening force, stiffness, and damping coefficients of this study are very close to those of Faria. The opening forces for this study and that of Faria are 4840 N and 4837 N, respectively, and the difference was less than 0.1%. The stiffness coefficient along the z-direction in this study and that of Faria were 1.34×10^9 N/m and 1.32×10^9 N/m, and the differences were less than 1.5%. The damping coefficient along z-direction used of this study and that of Faria are 4.20×10^4 Ns/m and 4.23×10^4 Ns/m, and the difference was less than 1%.

Table 3. Comparison of the stiffness and damping coefficients with Faria's results.

	Opening Force [N]	Stiffness Coefficient (K_{zz}) [N/m]	Damping Coefficient (C_{zz}) [Ns/m]
Present analysis	4840	1.34×10^9	4.20×10^4
Faria's result [25]	4837	1.32×10^9	4.23×10^4

4. Results and Discussion

4.1. Perturbed Pressure of the Fluid Film in a Dry Gas Seal according to Laminar, Turbulent, and Slip Conditions in the Fluid Film

We analyzed the dynamic coefficients of the fluid film on a T-grooved dry gas seal under laminar, turbulent, and slip conditions. The analysis was performed using three different methods depending on the governing equation. First, the L_method, which can only consider laminar flow, was used to solve the laminar compressible Reynolds equation. Second, the LT_method was used to solve the modified Reynolds equation by considering the laminar and turbulent behaviors of the fluid film. Third, the LTS_method was used to solve the modified Reynolds equation by considering the laminar, turbulent and slip conditions of the fluid film. Figure 3a, shows the geometry of the T-grooved dry gas seal. The FE model of the dry gas seal was developed with 180,000 quadrilateral elements, and Figure 4b shows the FE model corresponding to the boxed area of Figure 4a which has one groove-ridge area. We checked the convergence of the stiffness and damping coefficients by increasing the number of finite elements of the FE model. It shows that 180,000 quadrilateral elements of the FE model used in this study were sufficient to guarantee the convergence of the stiffness and damping coefficients. Table 4 shows the design parameters of the T-grooved dry gas seal used in the numerical analysis. The internal and external pressures were assumed to be 1 bar and 60 bar. The clearance increased from 2.5 μm to 4 μm with increments of 0.5 μm. The rotating seal was rotated clockwise at 25,000 rpm. All analyses were performed under isothermal and isoviscous conditions. Figure 4 shows the distribution of the perturbed pressure p_z corresponding to the L_method, LT_method, and LTS_method for a T-groove when the clearance was 2.5 μm. Figure 5 shows the pressure distribution of the perturbed pressure $p_{\dot{z}}$ corresponding to the L_method, LT_method, and LTS_method for a T-groove when the clearance was 2.5 μm. Figure 6a shows the Reynolds number along the radial direction at $\theta = 0°$ and $\theta = 15°$. The fluid between the stator and the rotor is assumed to change from laminar to turbulent flow when the Reynolds number exceeds 1000 [1,16,17]. Figure 6b shows the Knudsen number along the radius at $\theta = 0°$ and θ =15. A slip between the fluid and wall is assumed to occur when the Knudsen number is greater than 0.001 [1,18,19].

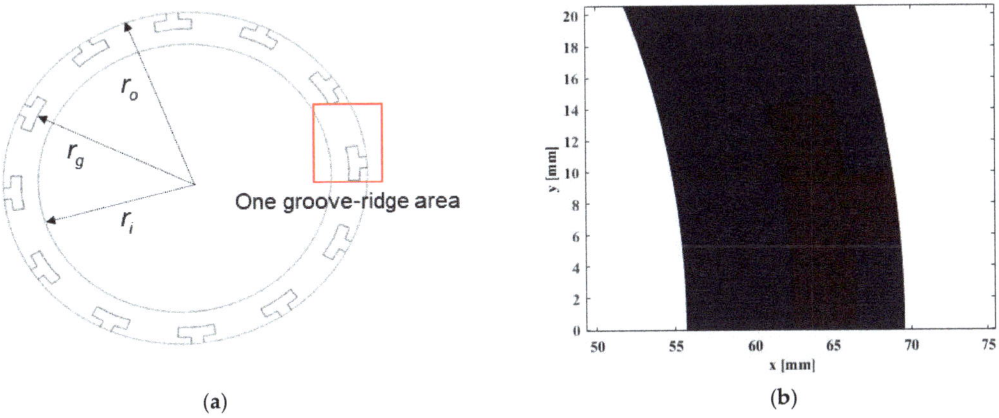

Figure 3. Geometry (**a**) and FE model (**b**) of the T-grooved dry gas seal.

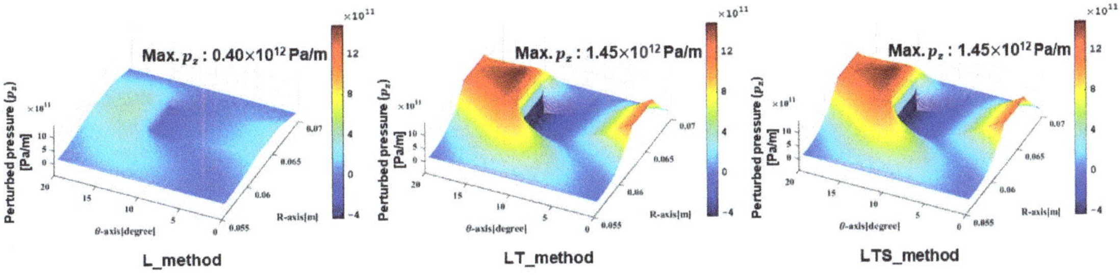

Figure 4. Perturbed pressure p_z with respect to axial displacement for laminar, turbulent, and slip conditions of the fluid film.

Table 4. Geometric parameters of the grooved seal.

Parameter	Value
Outer radius, r_o [mm]	69.5
Groove radius, r_g [mm]	62.4
Internal radius, r_i [mm]	55.7
Groove number	18
Ratio of ridge and groove	0.65
Groove depth [mm]	0.01

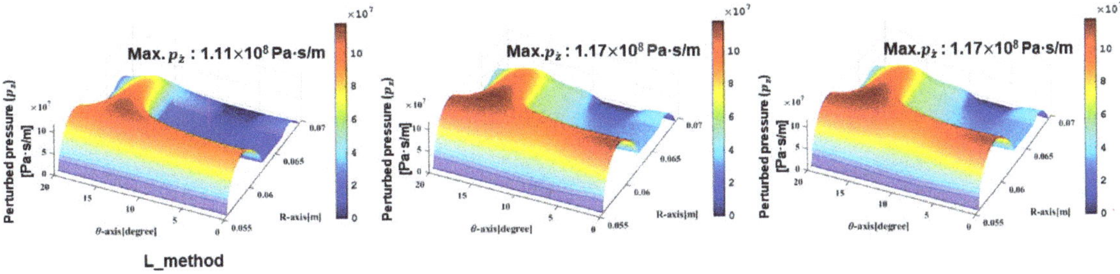

Figure 5. Perturbed pressure $p_{\dot{z}}$ with respect to axial velocity for laminar, turbulent, and slip conditions of the fluid film.

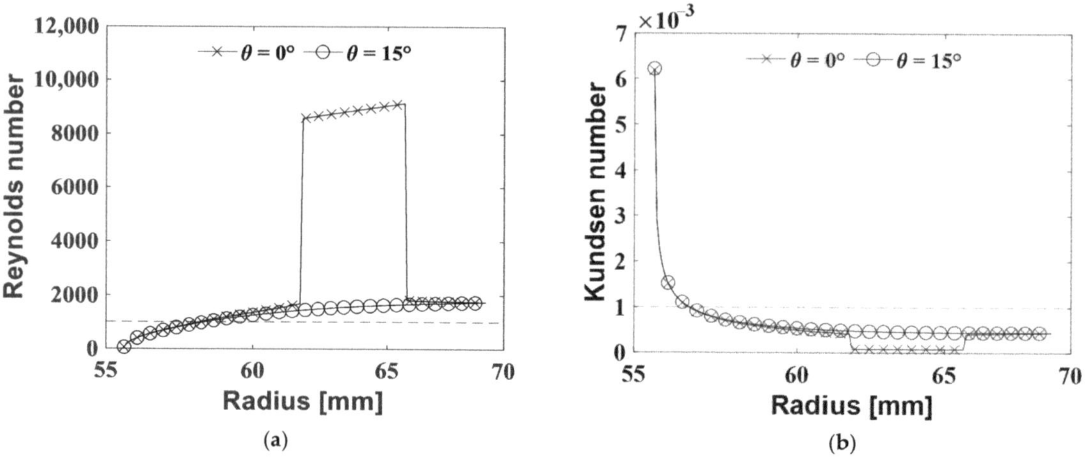

Figure 6. Reynolds number (**a**) and Knudsen number (**b**) of the fluid film at θ = 0° and θ = 15.

Figure 4 shows that the maximum perturbed pressures p_z obtained using the LT_method and LTS_method were greater than that obtained using the L_method. As shown in Figure 6a, the ridge part of the outer plane, where the maximum perturbed pressure occurred, had a large Reynolds number, indicating turbulent flow. Figure 6b also shows that the inner plane, where the perturbed pressure is small, has a large Knudsen number, indicating the occurrence of slippage. Therefore, the maximum perturbed pressure increased when the turbulent effect was considered. Furthermore, there was no difference in the maximum perturbed pressure p_z obtained using the LT_method and LTS_method because turbulent flow plays a dominant role in generating maximum perturbed pressure of the ridge part and because slippage did not occur in the groove and ridge parts of the outer plane, as shown in Figure 6b.

Figure 5 shows that the maximum perturbed pressures $p_{\dot{z}}$ obtained using the LT_method and LTS_method were greater than that obtained using the L_method. As shown in Figure 6a, the internal plane area (with a radius of around 59.8 mm) where the maximum perturbed pressure occurred had a large Reynolds number to generate turbulent flow. Therefore, the maximum perturbed pressure $p_{\dot{z}}$ increased when the turbulent effect was considered. Furthermore, there was no difference in the maximum perturbed pressure $p_{\dot{z}}$ obtained using the LT_method and LTS_method because the turbulent flow plays a dominant role in generating maximum perturbed pressure of the internal plane area and because slippage did not occur in that area as shown in Figure 6b.

4.2. Dynamic Coefficients of the Fluid Film in a Dry Gas Seal according to Clearance of the Fluid Film and External Pressure

Tables 5 and 6 show the stiffness and damping coefficients of a dry gas seal according to the clearance, as calculated by the L_method (L), LT_method (LT), and LTS_method (LTS). In all three analysis methods, the dynamic coefficients decreased as the clearance increased.

Table 5. Stiffness coefficients according to the clearance.

Clearance [μm]	Stiffness Coefficients (K)											
	K_{zz} [$\times 10^7$ N/m]			Difference [%]			$K_{\theta x \theta x}$ [$\times 10^5$ Nm/rad]			Difference [%]		
	L	LT	LTS	LT–L	LTS–L	LTS–LT	L	LT	LTS	LT–L	LTS–L	LTS–LT
2.5	74.86	249.56	248.63	233.4	232.1	−0.4	11.48	46.36	46.20	303.9	302.6	−0.3
3.0	63.17	196.07	195.42	210.4	209.4	−0.3	9.31	35.61	35.50	282.6	281.4	−0.3
3.5	55.34	155.19	154.73	180.4	179.6	−0.3	7.91	27.44	27.37	246.8	245.9	−0.3
4.0	49.92	124.35	124.03	149.1	148.5	−0.3	6.97	21.32	21.27	205.7	205.0	−0.2

Table 6. Damping coefficients according to the clearance.

Clearance [μm]	Damping Coefficients (C)											
	C_{zz} [×10⁴ Ns/m]			Difference [%]			$C_{\theta x \theta x}$ [×10 Nsm/rad]			Difference [%]		
	L	LT	LTS	LT–L	LTS–L	LTS–LT	L	LT	LTS	LT–L	LTS–L	LTS–LT
2.5	27.75	34.26	34.19	23.4	23.2	−0.2	53.34	66.57	66.44	24.8	24.6	−0.2
3.0	17.99	23.90	23.86	32.8	32.6	−0.2	34.99	47.35	47.27	35.3	35.1	−0.2
3.5	12.58	18.05	18.02	43.4	43.2	−0.2	24.69	36.30	36.25	47.1	46.8	−0.1
4.0	9.27	14.31	14.28	54.3	54.1	−0.2	18.31	29.12	29.08	59.0	58.8	−0.1

The stiffness coefficient of the LT_method was greater than that of the L_method. This can be explained by Figure 7a, which shows the difference of the perturbed pressure p_z between the LT_method and L_method at 2.5 μm clearance. Figure 7c shows the geometry corresponding to Figure 7a,b. As shown in Figure 7a, the increased area of the perturbed pressure (p_z) (red and green in the contour plot) is larger than the decreased area of the perturbed pressure (blue and sky-blue in the contour plot), and the maximum value of the increased area of the perturbed pressure (+1.12 × 10^{11} N/m) is larger than the maximum value of the decreased area of the perturbed pressure (−0.32 × 10^{11} N/m). Additionally, the damping coefficient of the LT_method was greater than that of the L_method (Table 6). This can be explained by Figure 7b, which shows the difference in the perturbed pressure $p_{\dot{z}}$ between the LT_method and L_method at 2.5 μm clearance. As shown in Figure 7b, the increased area of the perturbed pressure ($p_{\dot{z}}$) (red and green in the contour plot) was larger than the decreased area of the perturbed pressure (blue and sky-blue in the contour plot), and the maximum value of the increased perturbed area of the perturbed pressure (+3.70 × 10^7 Ns/m) was larger than the maximum value of the decreased area of the perturbed pressure (−1.18 × 10^7 Ns/m).

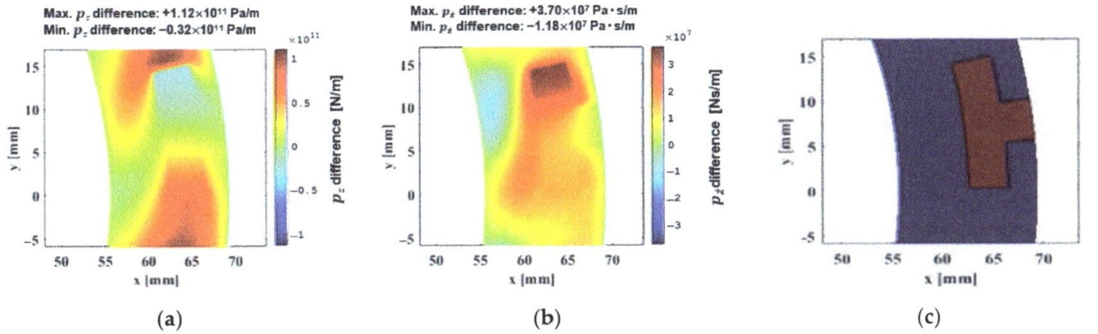

Figure 7. Perturbed pressure difference between the LT_method and L_method for p_z (**a**) and $p_{\dot{z}}$ (**b**) and the corresponding T-groove geometry (**c**).

5. Conclusions

In this study, we proposed a mathematical perturbation method of the modified Reynolds equation that includes the effects of laminar, turbulent, and slip behaviors of a fluid film. The pressure of the modified Reynolds equation was solved using the finite element method and the Newton–Raphson method, and the perturbed pressures with respect to three degrees of freedom were calculated by substituting the calculated pressure into the perturbed equations. We verified the proposed method by comparing the simulated results of the previous study. The dynamic coefficients of a T-grooved dry gas seal were investigated according to laminar, turbulent, and slip conditions in the fluid film with different clearances. This study shows that the stiffness and damping coefficients decreased as the clearance increased. The turbulent effect increased the stiffness coefficients, and it decreased as the clearance increased. Additionally, the turbulent effect increased the damping coefficients. The slip effect decreased the stiffness and damping coefficients, but the slip effect was smaller than the turbulent effect. This study shows that the laminar, turbulent, and slip conditions of a fluid film on a dry gas seal should be considered to predict the dynamic coefficients of a dry gas seal accurately. This study will make it possible to develop a dynamic model of the rotor-seal system with predicted dynamic coefficients of a dry gas seal, and it will eventually contribute to developing a robust and stable dry gas seal in various operating environments.

6. Future Work

Herein, we studied a non-contact dry-running mechanical face seal system. However, solid contact between rotating and stationary seal faces may occur due to angular misalignment, assembly tolerances, or large closing force. Friction and wear are important sources to degrade operating performance and shorten a sound life of dry gas seals [26–31]. Hong et al. proposed a numerical method to investigate the friction characteristics of oil film journal bearings considering elastomeric lubrication and contact force [32]. We hope to extend this study by considering elastomeric lubrication and the contact mechanism.

Author Contributions: Conceptualization, G.J.; methodology, Y.P. and M.H.; software, Y.P. and M.H.; validation, Y.P.; formal analysis, Y.P. and M.H.; investigation, Y.P. and M.H.; resources, Y.P.; data curation, Y.P.; writing—original draft preparation, Y.P. and M.H.; writing—review and editing, G.J.; visualization, Y.P. and M.H.; supervision, G.J.; project administration, G.J.; funding acquisition, G.J. All authors have read and agreed to the published version of the manuscript.

Funding: This work was partially supported by a Korea Energy Technology Evaluation and Planning (KETEP) grant funded by the Ministry of Trade, Industry and Energy (MOTIE) (20222B10100010, Development of silicon carbide composite material technology with improved surface lubricity, 2022). This work was also partially supported by a Korea Research Institute for Defense Technology planning and advancement (KRIT) grant funded by Defense Acquisition Program Administration (DAPA) (No. KRIT-CT-21-008, Predicting precise leakage rates of carbon seals and technology of designing aerodynamic patterns for improving lifetime, 2022).

Data Availability Statement: Not applicable.

Acknowledgments: The authors would like to acknowledge the Korea Energy Technology Evaluation and Planning (KETEP) grant funded by the Ministry of Trade, Industry and Energy (MOTIE) (20222B10100010, Development of silicon carbide composite material technology with improved surface lubricity, 2022). The authors would also like to acknowledge the Korea Research Institute for Defense Technology planning and advancement (KRIT) grant funded by Defense Acquisition Program Administration (DAPA) (No. KRIT-CT-21-008, Predicting precise leakage rates of carbon seals and technology of designing aerodynamic patterns for improving lifetime, 2022).

Conflicts of Interest: The authors declare no conflict of interest.

Nomenclature

h	Film thickness [mm]
h_0	Film thickness in equilibrium [mm]
C_r	Fluid state coefficients for radial flow
C_θ	Fluid state coefficient for circumferential flow
Kn	Knudsen number
Re	Reynolds number
$\mathbf{N}$	Shape function vector
$\mathbf{P}_e$	Element pressure vector [Pa]
P, p	Pressure of fluid film [Pa]
$p_z, p_{\theta_x}, p_{\theta_y}$	Perturbed pressure generated by infinitesimal displacement [Pa/m]
$p_{\dot{z}}, p_{\dot{\theta}_x}, p_{\dot{\theta}_y}$	Perturbed pressure generated by infinitesimal velocity [Pa·s/m]
K	Stiffness matrix [N/m]
C	Damping matrix [Ns/m]

q_p	Slip coefficient; $c_0 + c_1(1/q_p) + c_2(1/q_p)^2 + c_3(1/q_p)^3$
R	Gas constant [J/(kg·K)] (R_{air}: 287 J/(kg·K))
T	Temperature [K]
U, u	Fluid velocity [m/s]
w	Weighting function
Greek symbols	
ρ	Density [kg/m^3]
$\boldsymbol{\eta}$	Arbitrary vector
μ	Viscosity [Pa·s]
θ'	Fixed angular coordinates
ω	Rotating velocity [rad/s]

References

1. Hahn, M.; Park, Y.; Kang, M.; Jun, S.; Jang, G. Effects of laminar, turbulent, and slip conditions in a fluid film on a dry gas seal. *Machines* **2022**, *10*, 954. [CrossRef]
2. Chen, Y.; Wang, Q.; Peng, X.; Li, Y.; Wang, B.; Jin, J. Flow field and sealing performance analysis of compliant foil face gas seal. *Adv. Mech. Eng.* **2022**, *14*, 16878132221108488. [CrossRef]
3. Lu, J. Theoretical analysis and experiment on gas film stiffness with slip flow in a spiral-grooved dry gas seal. *Ind. Lubr. Tribol.* **2021**, *73*, 1226–1236. [CrossRef]
4. Jiang, J.; Peng, X.; Zong, C.; Zhao, W.; Chen, Y.; Li, J. Enhancing film stiffness of spiral groove dry gas seal via shape modification at low speed: Numerical results and experiment. *Tribol. Trans.* **2019**, *62*, 931–942. [CrossRef]
5. Ruan, B. Finite element analysis of the spiral groove gas face seal at the slow speed and the low pressure conditions—Slip flow consideration. *Tribol. Trans.* **2000**, *43*, 411–418. [CrossRef]
6. Chen, Y.; Peng, X.; Jiang, J.; Meng, X.; Li, J. Experimental and theoretical studies of the dynamic behavior of a spiral-groove dry gas seal at high-speeds. *Tribol. Int.* **2018**, *125*, 17–26. [CrossRef]
7. Blasiak, S.; Zahorulko, A.V. A parametric and dynamic analysis of non-contacting gas face seals with modified surfaces. *Tribol. Int.* **2016**, *94*, 126–137. [CrossRef]
8. Liu, Y.; Shen, X.; Xu, W. Numerical analysis of dynamic coefficients for gas film face seals. *J. Tribol.* **2002**, *124*, 743–754. [CrossRef]
9. Teng, L.; Jiang, J.; Peng, X.; Li, J.; Zheng, S.; Zang, C.-P. Study on angular free vibration stability and parameters influence of dry gas seal based on the characteristic equation. *Shock Vib.* **2022**, *2022*, 7378999. [CrossRef]
10. Kim, H.; Jang, G.; Ha, H. A generalized Reynolds equation and its perturbation equations for fluid dynamic bearings with curved surfaces. *Tribol. Int.* **2012**, *50*, 6–15. [CrossRef]
11. Kim, H.; Jang, G.; Lee, S. Complete determination of the dynamic coefficients of coupled journal and thrust bearings considering five degrees of freedom for a general rotor-bearing system. *Microsyst. Technol.* **2011**, *17*, 749–759. [CrossRef]
12. Park, D.-J.; Kim, C.-H.; Jang, G.-H.; Lee, Y.-B. Theoretical considerations of static and dynamic characteristics of air foil thrust bearing with tilt and slip flow. *Tribol. Int.* **2008**, *41*, 282–295. [CrossRef]
13. Jang, G.; Lee, S. Determination of the dynamic coefficients of the coupled journal and thrust bearings by the perturbation method. *Tribol. Lett.* **2006**, *22*, 239–246. [CrossRef]
14. Kou, G.; Li, X.; Wang, Y.; Lin, M.; Tan, C.; Mou, M. Steady performance and dynamic characteristics of a superellipse groove dry gas seal at a high-speed condition. *Ind. Lubr. Tribol.* **2020**, *72*, 789–796. [CrossRef]
15. Rui, C.; Yong, Z.; Jiakang, Y.; Zixi, W. Research on the performance of foil thrust bearings under dynamic disturbances. *Tribol. Int.* **2022**, *174*, 107744.
16. Ng, C.-W.; Pan, C.H.T. A linearized turbulent lubrication theory. *J. Basic Eng.* **1965**, *87*, 675–682. [CrossRef]
17. Taylor, C.M.; Dowson, D. Turbulent lubrication theory—Application to Design. *J. Lubr. Technol.* **1974**, *96*, 36–46. [CrossRef]
18. Fukui, S.; Kaneko, R. Analysis of ultra-thin gas film lubrication based on linearized Boltzmann equation: First report—Derivation of a generalized lubrication equation including thermal creep flow. *J. Tribol.* **1988**, *110*, 253. [CrossRef]
19. Fukui, S.; Kaneko, R. A database for interpolation of poiseuille flow rates for high knudsen number lubrication problems. *J. Tribol.* **1990**, *112*, 78–83. [CrossRef]
20. Ng, C. Fluid dynamic foundation of turbulent lubrication theory. *Tribol. Trans.* **1964**, *7*, 311–321. [CrossRef]
21. Elrod, H.G.; Ng, C.W. A Theory for turbulent fluid films and its application to bearings. *J. Lubr. Technol.* **1967**, *89*, 346–362. [CrossRef]
22. Beskok, A.; Karniadakis, G.E.; Trimmer, W. Rarefaction and compressibility effects in gas microflows. *J. Fluids Eng.* **1996**, *118*, 448–456. [CrossRef]
23. Lu, J.; Ding, X.; Zhang, W. The influence of spiral angle on lift-off speed of face gas seal under start-up. *SN Appl. Sci.* **2019**, *1*, 1068. [CrossRef]
24. Yan, R.; Chen, H.; Zhang, W.; Hong, X.; Bao, X.; Ding, X. Calculation and verification of flow field in supercritical carbon dioxide dry gas seal based on turbulent adiabatic flow model. *Tribol. Int.* **2022**, *165*, 107275. [CrossRef]
25. Faria, M.T.C. An efficient finite element procedure for analysis of high-speed spiral groove gas face seals. *J. Tribol.* **2001**, *123*, 205–210. [CrossRef]

26. Xuexing, D.; Delin, C.; Weizheng, Z.; Shurong, Y. Experiment of frictional vibration performance of the micro-texture of DLC thin film with dry gas seal rings. *Tribol. Int.* **2020**, *147*, 106267.
27. Cao, H.; Liu, Y.; Lin, C.J. Failure analysis and transformation of hydrogen compressor dry gas seal. *Appl. Mech. Mater.* **2013**, *331*, 98–101. [CrossRef]
28. Zardynezhad, S. Achieve successful compressor startup by addressing dry gas seal failure. *Hydrocarb. Process* **2015**, *94*, 63–68.
29. Ruan, B. Numerical modeling of dynamic sealing behaviors of spiral groove gas face seals. *J. Tribol.* **2002**, *124*, 186–195. [CrossRef]
30. Miller, B.A.; Green, I. Semi-analytical dynamic analysis of spiral grooved mechanical gas face seals. *J. Tribol.* **2003**, *125*, 403–413. [CrossRef]
31. Yang, H.X. Theoretical analyses and field applications of gas film lubricated mechanical face seals with her ring bone spiral grooves. *Tribol. Trans.* **2009**, *52*, 800–806.
32. Hong, G.; Kim, K.; Park, Y.; Jang, G. Numerical determination of the frictional coefficients of a fluid film journal bearing considering the elastohydrodynamic lubrication and the asperity contact force. *Machines* **2022**, *10*, 494. [CrossRef]

lubricants

MDPI

Article

Sanded Wheel–Rail Contacts: Experiments on Sand Crushing Behaviour

Bettina Suhr [1,*], William A. Skipper [2], Roger Lewis [2] and Klaus Six [1]

[1] Virtual Vehicle Research GmbH, Inffeldgasse 21a, 8010 Graz, Austria
[2] Department of Mechanical Engineering, The University of Sheffield, Mappin Street, Sheffield S1 3JD, UK
* Correspondence: bettina.suhr@v2c2.at

Abstract: In railway operation, the sanding process is used to overcome low adhesion conditions in the wheel–rail contact. In the literature, previously conducted research has been experimental, e.g., measuring adhesion coefficients (ACs) under different contact conditions (dry, wet, ...) or applying different sands. Under dry conditions, sanding can reduce measured ACs, while under wet conditions different types of rail sand can leave ACs unchanged or increase adhesion. Despite active research, the physical mechanisms causing the change in ACs under sanded conditions are still poorly understood. A possible remedy is the development of advanced models of sanding including local effects. As a basis for such a model, this study presents experimental results concerning single grain crushing behaviour of two types of rail sand under dry and wet contact conditions. Firstly, initial breakage behaviour is investigated with focus on the particle fragments' size and spread as only fragments within the running band are available to influence the AC during roll-over. Secondly, single grain crushing tests are conducted under realistic wheel–rail load showing the formation of solidified clusters of sand fragments, as well as their size and thickness. This information is important for understanding mechanisms and for future physics-based modelling of the sanding process in wheel–rail contacts.

Keywords: wheel–rail contact; low adhesion; sanding; single grain crushing tests

Citation: Suhr, B.; Skipper, W.A.; Lewis, R.; Six, K. Sanded Wheel–Rail Contacts: Experiments on Sand Crushing Behaviour. *Lubricants* **2023**, *11*, 38. https://doi.org/10.3390/lubricants11020038

Received: 16 December 2022
Revised: 12 January 2023
Accepted: 13 January 2023
Published: 20 January 2023

1. Introduction

In railway operation, the contact between wheel and rail is determined by complex tribological processes. Extremely high contact pressures and tangential stresses, caused for example by traction or braking, result in severe plastic deformation of the near-surface layers [1–7]. This influences damage and wear behaviour and the developing wheel–rail roughness affecting the transferable tangential force between wheel and rail. The tangential force is limited by the maximal adhesion coefficient (AC), which is around 0.35 or higher for dry conditions [8,9]. Between wheel and rail, third body layers (3BLs) can be embedded, such as liquids (e.g., water), solids (e.g., wear particles), or combinations thereof. Some of these 3BLs can cause low adhesion conditions with ACs below 0.1 [10–13]. Such low adhesion conditions influence the traction and braking performance of railway vehicles in service. In the worst case, they can cause safety issues [14,15]. Typical causes for low adhesion conditions are damp (wet) contact conditions [16] ('wet rail' phenomenon) or when the rail surface is contaminated with leaves [14,17].

The application of sand has been used for many years to overcome low adhesion conditions in the wheel–rail contact. Figure 1 shows the process divided into three phases. In phase I, the particles are applied: some are expelled and some are entrained into the contact. In phase II, particles are in the wheel–rail contact and influence the adhesion. Finally, in phase III particles leave the contact and some of them remain on wheel and rail.

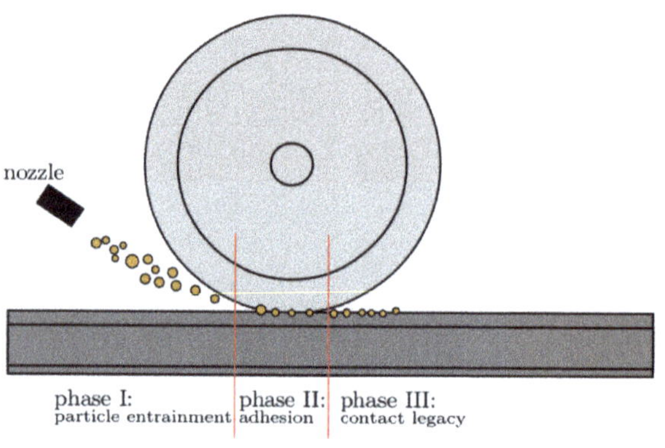

Figure 1. Phases during wheel–rail contact sanding.

Under low adhesion conditions, sanding generally increases the maximal AC between wheel and rail, but it can also lead to increased damage on both wheel and rail [18,19]. Wheel–rail sanding is a field of active research, where the adhesion increasing aspect is more extensively investigated [20] than the damage or isolation aspects [21]. Research is conducted with both field testing and lab experiments, e.g., twin-disc tests [22–25], linear full-scale rigs [26], or high pressure torsion (HPT) test rigs [27,28]. Indentations were measured on both wheel and rail as well as wheel–rail ACs under different contact conditions (dry, wet, or different kinds of contamination) but also with different types of sand [27] or different amounts of sand being applied [25]. In static loading tests [19], indentations were observed mainly on the wheel. Large indents could be caused by large particles being embedded into the surface prior to crushing or because the grain fractures into larger bits, which remain together. Smaller indents were possibly formed by crushed particles. These different mechanisms could lead to differences in the change of roughness caused by sanding. Large indents of embedded sand particles were also found in twin-disc tests under wet conditions [25] together with small amounts of spalling and pits. Under dry conditions, sanding can have little effect or even reduce measured ACs [22,27,28]. In contrast, under wet conditions different types of rail sand can leave ACs unchanged or increase adhesion, where some sands restore adhesion to nearly dry conditions [25,27,28]. The described behaviour can be seen in Figure 2, where measured ACs from HPT tests under dry and wet conditions are shown, using two different types of sand.

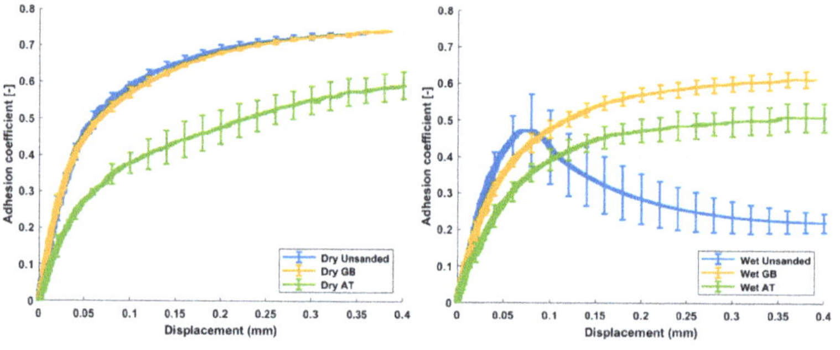

(**a**) dry conditions (**b**) wet conditions

Figure 2. Influence of sanding on adhesion coefficient under dry and wet conditions in a HPT test. Two different types of sand were applied, for more details see [28].

Despite the active research in this field, the physical mechanisms causing the change in ACs under sanded conditions are still poorly understood. Sand grains will partially crush when entering the contact, causing plastic deformations on wheel and rail surfaces (i.e., change of roughness). The amount of sand in the contact determines whether the metal surfaces are (partially) separated or not, allowing for different mechanisms of load transfer, see Figure 3. Under high loads, sand fragments solidify and could form clusters, which indent into wheel and rail surfaces (affecting roughness) and cause form closure effects. Adhesion could also be increased via form closure effects caused by sand grains directly penetrating into wheel–rail surfaces, or the sand powder solidifies and partially covers the rough wheel–rail surfaces, increasing the effective contact area between sand and steel and thereby the AC. The role of water in wet contacts is also unclear. This mentioned lack of understanding is caused by the fact that current experimental abilities do not allow for any monitoring of the aforementioned mechanisms in the contact zone during roll-over.

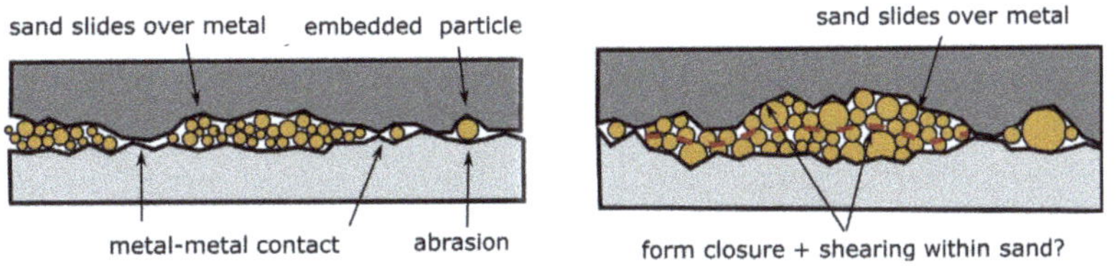

Figure 3. Possible mechanisms for increasing adhesion via sanding.

Models predicting the normal and tangential contact stresses (forces) in the wheel–rail interface are mostly based on classical theory [29–41]. Such models will not help with gaining new insights about physical phenomena occurring in the contact when sand is introduced. The only possibility to account for the influence of sand in these models is to change the coefficient of friction. Beside models based on classical theory, extended models can be found in the literature, see [29,41] for an overview. For example, in [9,10,42,43] models are presented, which account for the influence of solid 3BLs on the adhesion characteristic. However, these models assume a continuous 3BL across the contact and thus spread local effects (e.g., resulting from sand grains) within the contact region. Furthermore, the change of roughness due to local plasticity effects cannot be described. In [17], the approach from [9,10] and the model from [8] are combined. However, the model represents a semi-physical approach not explicitly accounting for the physical effects occurring in the contact zone mentioned previously. Thus, the impact of sand under low adhesion conditions cannot be described with this methodology either.

To summarise, although the positive effect of sand is well known and experimentally proven, the physics behind this improvement are not yet fully understood (effects occurring in the contact patch during roll-over of sand particles cannot be observed). Furthermore, models taking into account local effects are not available. Such models in combination with experiments are important for a better understanding of the physics occurring in sanded contacts. This could be particularly useful for aims as, e.g., reducing the amount of sand necessary to increase adhesion and thus reducing/avoiding damage due to sanding.

This study is part of a research project where sanded contacts are experimentally investigated in detail, and advanced models are to be developed to better understand the phenomena responsible for the positive effect of sand on adhesion in phase II, see Figure 3. The experimental work on single sand grains aims for a better understanding of the material behaviour and should also generate the necessary information for model development and parameterisation.

In this paper, two types of rail sand were investigated. Single grain crushing tests were conducted under dry and wet contact conditions to investigate the initial breakage behaviour. In the sanding process, the first fracture will take place several centimetres in front of the contact patch due to the narrowing gap between wheel and rail. Some of the resulting fragments are expelled from the running band (and are thus not active any more), while others stay inside and be crushed again. Therefore, focus in the conducted tests were on the particle fragments' size and spread after the first crushing. Furthermore, a Weibull statistic was fitted to the measured data.

In a second step, single grain crushing tests were conducted under a realistic wheel–rail load of 900 MPa, this included the initial crushing and subsequent further crushing of the fragments remaining in the contact area. These tests aimed to provide information about the condition of the sand after it had been fully loaded in a vertical direction: the possible formation of solidified clusters of sand fragments, as well as their size and thickness. This information is important for understanding mechanisms and for physics-based modelling of the sanding process in wheel–rail contacts. Finally, Scanning Electron Microscope (SEM) images were taken from some of the formed sand clusters after full loading to gain information about the surface structure and to see if they were composed of connected smaller fragments.

2. Single Grain Crushing: First Breakage

These tests investigate the state of sand grain fragments after the first breakage event. The analysis of fragments' size and spread shows how much of the material is available (after first breakage) in the area of an conceptual wheel–rail contact. This information is important for understanding possible mechanisms of the influence of sanding on the adhesion coefficient.

2.1. Materials and Methods

Rail sands from Great Britain (GB) and Austria (AT) were used in this study. Both sands were riffled down to 100 g to ensure a representative sample [44], each sample was then subjected to sieve analysis to determine the average particle size [45]. The average size of GB was 1.30 mm and AT was 0.98 mm. For all particle crushing tests particles of approximately average size were desired. Therefore, particles passing through a sieve aperture of 1.4 mm and settling on a sieve with an aperture of 1.18 mm were used for GB tests; for AT tests these respective sieve apertures were 1.18 mm and 0.6 mm.

Single particle crushing tests till first fracture were performed on a Bruker (Bruker, Coventry, West Midlands, UK.) universal mechanical tester (UMT) using a 1000 N load cell with a resolution of 50 mN. The set-up of this test is shown in Figure 4. The first fracture was defined by a sudden, large drop in measured load. An example of a measured path–force curve is shown in Figure 5, where the peak load before first fracture is marked. The tests were performed in displacement control with a speed of 0.05 mm/s. Both top and bottom platens had contacting faces with a 50×50 mm^2 contact area, with the bottom platen being made from EN24 grade steel and the top platen from EN5.

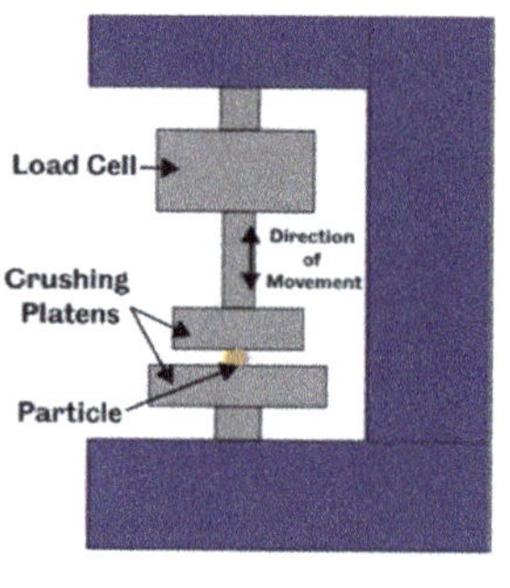

Figure 4. Test set-up for single particle crushing tests in the UMT test rig.

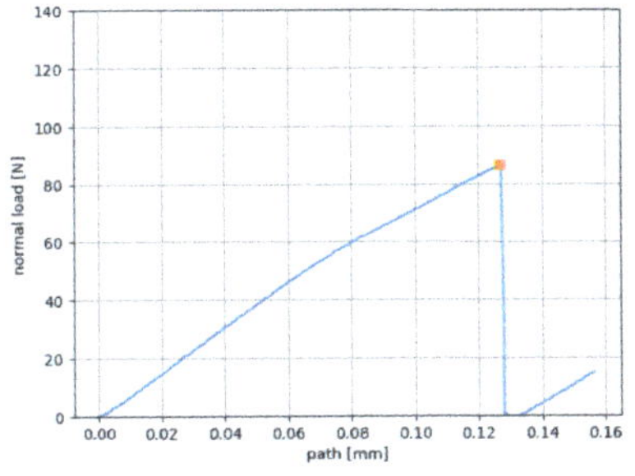

Figure 5. Examples of a measured path–force curve from a single crushing test, with the particle breakage force shown as orange square symbol.

Tests were conducted with GB sand and AT sand, respectively, in dry and wet conditions. Here, wet conditions were created by pipetting 10 or 20 μL of distilled water onto the particle. A total of 30 tests were performed for each condition. However, some tests had to be excluded from further analysis due to different problems occurring. In some tests, grains showed successively small fractures without a clear peak force, this was more often the case for AT sand. In a few tests, no clear grain breakage occurred despite a high loading path. For GB sand, there were several cases, were the path–force curve showed a yielding-like behaviour, where unusually high loading paths were reached together with very high breakage forces. Possible reasons could be an indenting of the grain in the steel plates. This behaviour was seen only for GB sand, which has a higher Young's modulus than AT sand [28]. For the fitting of the Weibull statistics, 24 tests were used for GB sand in both dry and wet conditions. For AT sand, 24 tests were used for dry conditions and 25 for wet conditions. For the analysis of the fragments' spreading a few further tests had to be excluded, where the test did not stop before a second severe breakage had taken place. For this analysis, 21 tests were used for GB sand under dry conditions and 23 tests under wet conditions. For AT sands, 23 tests each were analysed under dry and wet conditions.

Each particle was weighed before crushing and the fragments remaining on the platens were weighed after crushing to determine the amount of material expelled from the contact.

Before and after each test, photos were taken, see Figure 6. To be able to measure the initial particle's area and the fragments' area and position, these photos were segmented using the Fiji (Fiji: version 2.35.) [46] and Gimp (Gimp: version 2.10.22.) [47] software. In Fiji the photos were converted to black and white (setting 8 bit format) and then a manual threshold selection was applied. As the sand grains differed in their colour and brightness, a manual correction step was necessary using Gimp, which clearly introduced some impreciseness to the process. An example of the segmented photo after crushing can be seen in the right part of Figure 6. After the segmentation step, the Fiji software was applied using the "Analyze Particles" function to calculate the particle/fragments' area, position, diameter, and other shape descriptors. With the segmented photo after crushing, for each fragment the distance from the centre of the initial grain was calculated.

Figure 6. Examples for photos taken and post-processing.

2.2. Results: Analysis of Fragment's Size and Spread

In the following analysis, no distinction is made between tests under wet conditions using 10 µL or 20 µL of water, as for both cases the results were very similar.

In Figure 7, histograms of the fragments' distance from the centre and the fragments' area can be seen for both GB and AT sand under dry and wet conditions. For GB sand under dry conditions, the grain's fragments spread out with a slow decrease until 35 mm distance. In contrast, under wet conditions more than 96% of the GB sand's fragments have a distance smaller than 5 mm.

For AT sand under dry conditions, 68% of all fragments have a distance from the centre less than 5 mm. The probability of fragments at higher distances decays faster and is almost 0 for values higher than 25 mm. This is a clear difference to the behaviour of GB sand under dry conditions. AT sand under wet conditions shows the least spreading of all cases investigated: 98% of the fragments have a distance less than 5 mm.

In Figure 7c,d, histograms of the fragments' area are shown. Here, the results for both types of sand and for dry and wet contact conditions are similar. The vast majority of fragments are smaller than 0.1 mm (size of the histograms first bin) and only few larger particles exist.

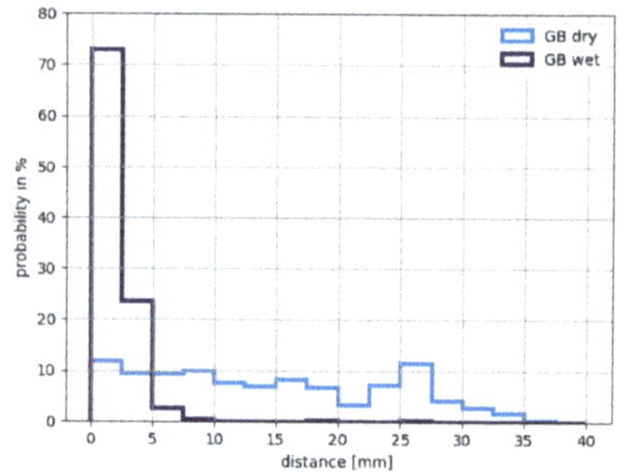

(**a**) GB sand: fragment distance from centre

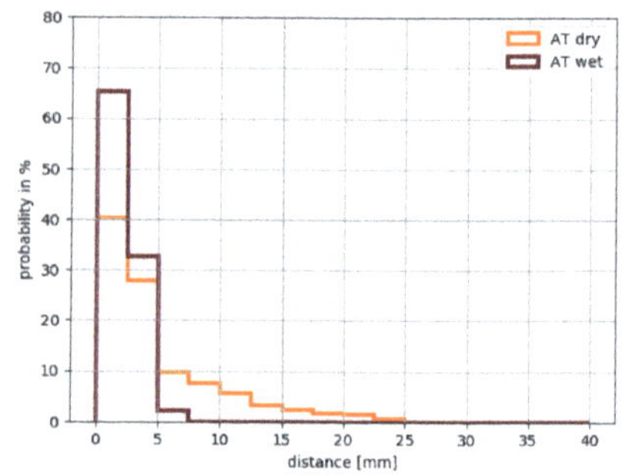

(**b**) AT sand: fragment distance from centre

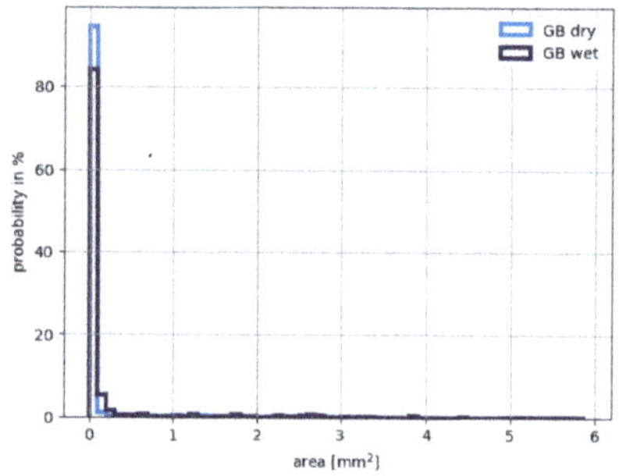

(**c**) GB sand: fragment area

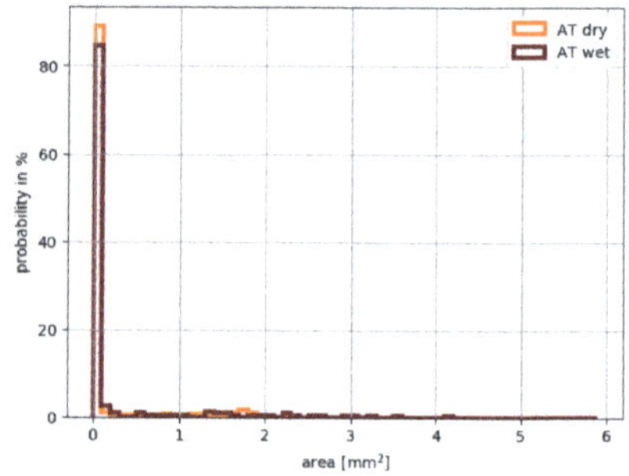

(**d**) AT sand: fragment area

Figure 7. Distance and area for GB and AT sands.

Thinking of the application of wheel–rail sanding, it is of interest to ask how much of the mass of the initial grain would stay within a radius of 5 mm: radius of a conceptual wheel–rail contact patch. This radius corresponds to a typical running band width when the tread of a wheel is in contact with the rail head. Experimental results using a full-scale test rig can be found in [48]. From the post-processing of the crushing tests, the fragments' area and distance can be related. Figure 8 shows cumulative histograms, where the fragment distance is weighted by its area divided by the sum of all fragments. The area of the conceptual wheel–rail contact is shown as grey box. For GB sand, the contact condition makes a big difference: under dry conditions 68% of the fragments' area stay within the conceptual wheel–rail contact, while it is 98% under wet conditions. This influence of the contact condition is less pronounced for AT sand: Even under dry conditions 93% of the fragments' area stays within the conceptual wheel rail contact and under wet conditions it is 100%.

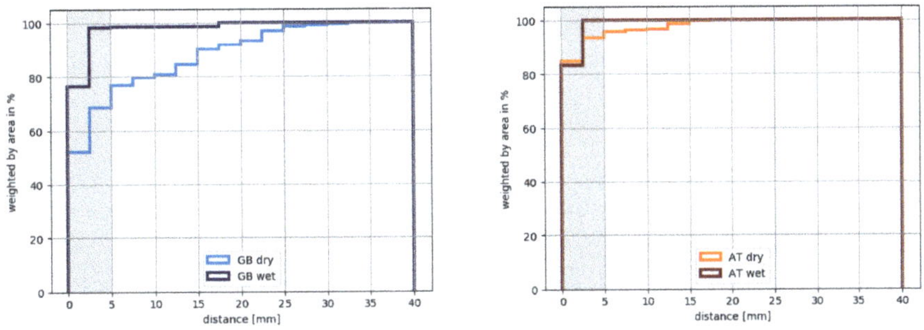

(**a**) GB sand (**b**) AT sand

Figure 8. Distance weighted by area for GB and AT sands. The area of an conceptual wheel–rail contact is shaded in grey.

The previous analysis took into account the fragments which remained on the lower plate (sized 5 cm × 5 cm) of the testing device. To check if any fragments were expelled from this lower plate, the weight of the initial grain was recorded, as well as the weight of the fragments remaining on the lower plate. In some cases, fragments stuck to the upper plate of the testing device. In the check for mass loss, these fragments' weight was added to the fragments' weight from the lower plate. Figure 9 shows box plots of the mass loss for GB and AT sands for both contact conditions. For GB sand under dry conditions, the mass loss ranged up to 10% while there were four outliers reaching up to 66%. Under wet conditions on GB sand, there were three outliers with a mass loss between 25% and 33%, while in all other cases the mass loss was below 3%. For AT sand under both dry and wet conditions the mass loss was below 7%, with the exception of four tests under dry conditions, which ranged between 15% and 42% of mass loss. It is consistent with the fragments' spread behaviour that GB sand under dry conditions has a higher mass loss than under wet conditions. The results for AT sand are similar for dry and wet conditions, which is also in accordance with their spread behaviour. Comparing mass loss results between GB and AT sand, AT sand shows higher values than expected from the spread behaviour. The observed mass loss was often close to the accuracy of the scale used for measuring and as AT sand grains are smaller than GB sand grains. This could be a possible explanation.

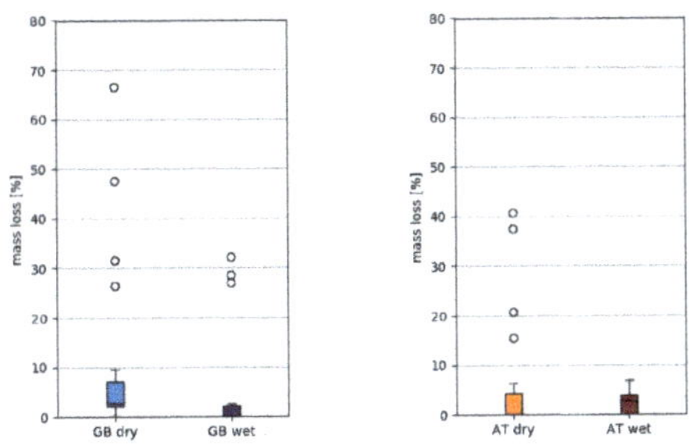

(**a**) GB sand (**b**) AT sand

Figure 9. Mass loss after grain crushing test for GB and AT sands.

2.3. Results: Fitting of Weibull Statistics

For later use in modelling of particle breakage, the Weibull statistics will be fitted to the measured breakage force/probability of survival of GB and AT sand. In this section, no distinction between dry and wet contact condition will be made, as it is assumed that the contact condition does not influence the breakage force.

The particle breakage force is defined as the highest force before a clear drop in the force can be seen, compare in Figure 5 the orange square symbol. Denoting the force at particle breakage with F_b allows the calculation of the stress at breakage [49–51]

$$\sigma_b = \frac{F_b}{d^2}, \tag{1}$$

where d denotes the grain's diameter. The survival probability of a grain of diameter d under stress σ is described using Weibull statistics [49–52]:

$$P_s(d) = \exp\left[-\left(\frac{d}{d_0}\right)^3\left(\frac{\sigma}{\sigma_0}\right)^m\right], \tag{2}$$

where d_0 is the reference diameter, m is the Weibull modulus, and σ_0 is the characteristic stress such that a particle of size d_0 has 37% survival probability. The reference diameter is the median of the used particle diameters, shown in Figure 10a. It can be seen that d_0 is clearly larger for GB sand compared to AT sand. Note that the values shown are higher than the sieve sizes mentioned in the experiment description: thin particles may pass the sieve aperture diagonally and in image analysis a larger diameter can be measured depending on the orientation of the particle. The breakage forces are shown in Figure 10b, and for GB sand they reach up to 103 N, while for AT sand they are well below 65 N. Interestingly, the breakage stresses are mostly in the same range for both types of sand, compare Figure 10c.

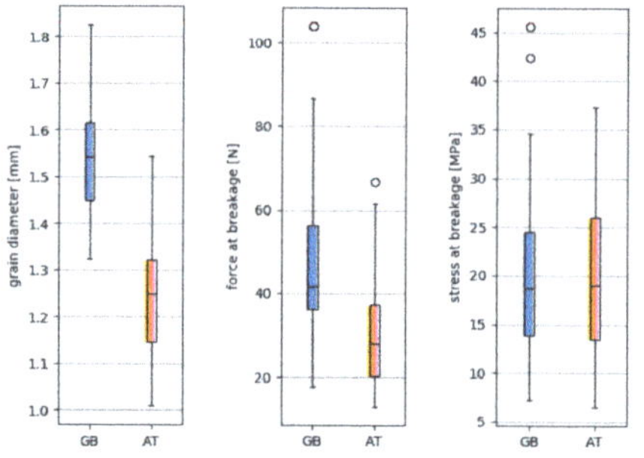

(a) (b) (c)

Figure 10. Grain diameter, force, and stress at breakage for GB and AT sands. (**a**) Diameter; (**b**) force at breakage; (**c**) stress at breakage.

Using the calculated stresses and assigning survival probabilities, the Weibull statistics is fitted to the data [50]. A comparison of measured values from the experiment and the fitted Weibull statistics is shown in Figure 11 for both types of sand. The parameters of the fitted Weibull statistics are given in Table 1. The results are surprisingly similar for both types of sand, considering the differences found in the analysis of fragment's spread.

Table 1. Parameters of the fitted Weibull statistics for both GB and AT sand.

Sand Type	d_0 (mm)	σ_0 (MPa)	m (-)
GB	1.54	22.74	2.85
AT	1.25	22.49	2.72

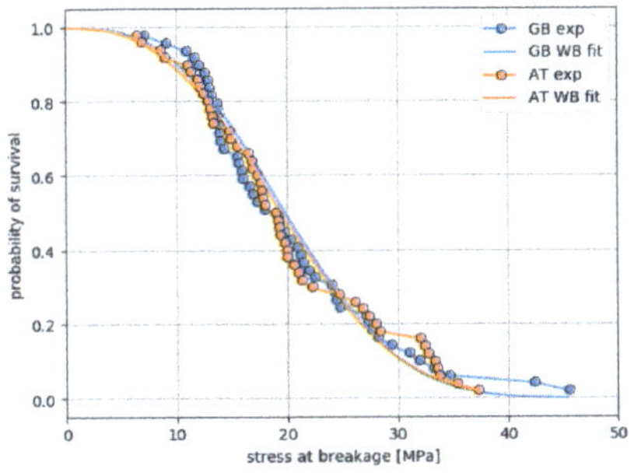

Figure 11. Fitted Weibull statistics.

3. Single Grain Crushing: High Loading

These tests aim to provide information about the condition of the sand after it has been fully loaded in the vertical direction. There is the possible formation of solidified clusters of sand fragments and their size and thickness are of interest. This information is important for understanding and physics-based modelling of the sanding process in wheel–rail contacts.

3.1. Materials and Methods

Single particle crushing tests under realistic wheel–rail contact pressures were carried out using a Denison Mayes (Denison Mayes, Leeds, West Yorkshire, UK.) hydraulic test frame. The set-up of this experiment is included in Figure 12. The bottom platen had an area of 50 × 50 mm^2 whereas the top platen had a circular area with a 11 mm diameter. This meant the applied contact pressure was 900 MPa at the load capacity of the rig (90 kN). Both platens were made of hardened O1 tool steel to minimise the amount of plastic deformation of the surfaces upon crushing the particle. For both GB sand and AT sand, 5 tests were conducted under dry and wet conditions each, totalling 20 tests. For wet conditions, 20 µL of distilled water were applied by pipette.

Mass measurements and photos were taken for each particle in the same manner as was performed for the UMT testing. Moreover, a non-contact imaging and measuring tool named Alicona (Bruker, Market Harborough, Leicestershire, UK.) InfiniteFocus SL was used to take high resolution 3D scans of the crushed grain on the lower plate after the test. These scans had a vertical resolution of 500 nm.

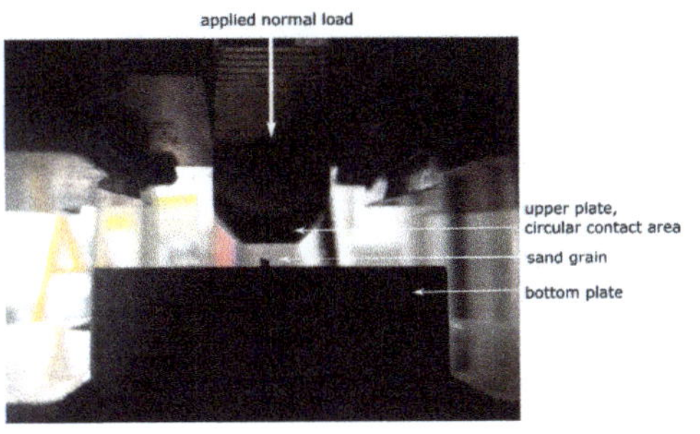

Figure 12. Test set-up for single particle crushing tests under realistic wheel–rail contact pressures.

SEM samples were prepared from the remains of the particles crushed under high load by removing the fragments from the crushing surface and mounting them onto stubs. SEM imaging was conducted using an FEI (FEI, Hillsboro, OR, USA.) Nova NanoSEM 450 with an Everhart–Thornley Detector. Imaging voltage was 5 kV with a spot size of 3 and a working distance of 2.5–3.5 mm.

3.2. Results: Fragment Spreading under High Loading

During the tests, the sand grains fractured repeatedly, with varying amounts of fragments being expelled from the contact. However, in almost all tests the formation of clusters of solidified sand fragments was seen. Their size varied for both types of sand and for dry/wet contact condition.

GB sand under dry conditions showed a high variation in the amount of fragments, which stayed within the contact area. From the five conducted tests, in three tests several small clusters formed. An example can be seen in Figure 13a, where the marked cluster has a side length of about 1 mm. The corresponding Alicona scan, Figure 13b, shows heights between 40 µm and 70 µm. In one test, nearly all fragments were expelled from the contact, while in another test a large cluster of side length 3.8 mm formed, see Figure 13c. This cluster had a height of 40 µm at the edges and 140 µm at its centre in the Alicona scan, see Figure 13d. The observed behaviour, including a high spreading of fragments, is in accordance with the results from the first breakage test.

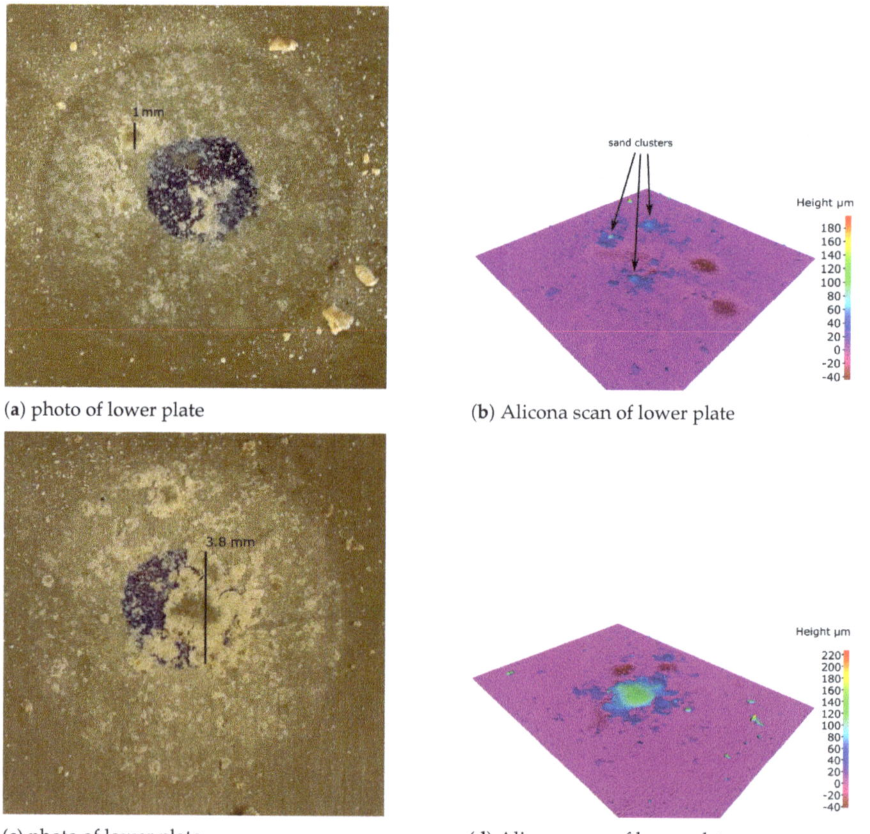

(**a**) photo of lower plate (**b**) Alicona scan of lower plate

(**c**) photo of lower plate (**d**) Alicona scan of lower plate

Figure 13. Example result for high load testing of GB sand under dry conditions.

The corresponding tests for AT sand under dry conditions showed larger sized clusters of solidified sand fragments in 4 of 5 tests. A typical result of the high load tests can be seen in Figure 14. This cluster has side length of about 3 mm and showed a height of 70 µm at the edges and 140 µm at its centre in the Alicona scan. In one test, several smaller clusters formed. These results are in accordance with the low amount of spreading of AT sand seen in the first breakage testing.

(**a**) photo of lower plate (**b**) Alicona scan of lower plate

Figure 14. Example result for high load testing of AT sand under dry conditions.

Under a wet condition, both types of sand showed typically only one big cluster of solidified sand powder, see Figures 15 and 16. This is again in agreement with the initial breakage results presented, where the least spread of fragments was seen under wet conditions for both types of sand. The cluster of GB sand shown in Figure 15 had a side length of 4.4 mm and a height between 100 µm and 220 µm. In contrast, the cluster of AT sand shown in Figure 16 was much smaller with a side length of about 3 mm and heights between 60 µm and 160 µm. For both types of sand, one test existed, where the sand fragments seemed to have drifted in the applied water drop and no cluster was formed. In this study, no detailed analysis of the formed cluster's shape is conducted. Because of the described variations in the results and because of the low number of conducted high loading tests, no reliable conclusions could be drawn from such an analysis.

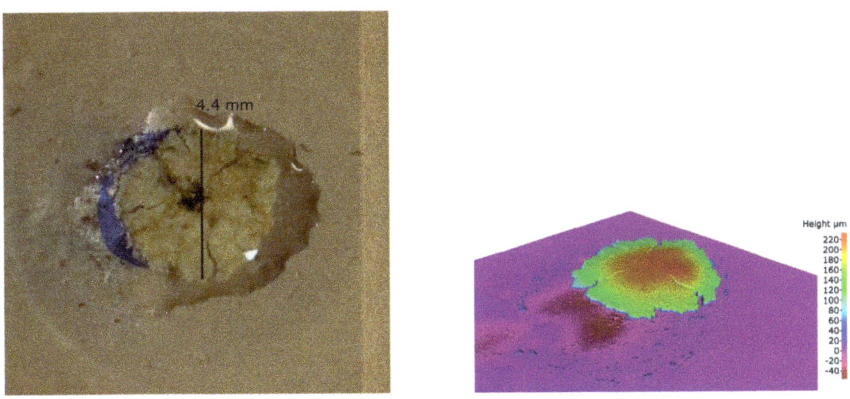

(**a**) photo of lower plate (**b**) Alicona scan of lower plate

Figure 15. Example result for high load testing of GB sand under wet conditions.

(**a**) photo of lower plate

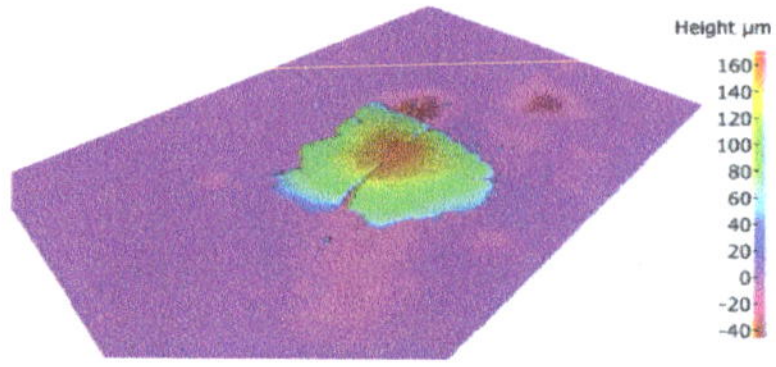

(**b**) Alicona scan of lower plate

Figure 16. Example result for high load testing of AT sand under wet conditions.

To further investigate the formed clusters' surface, SEM images were taken from solidified fragments of GB and AT sand. This was an attempt to see if the larger fragments/clusters were composed of connected smaller fragments. Figure 17 shows the obtained SEM images of one lager solidified cluster and a zoom in of both GB and AT sand. The zoom in of the GB sand cluster contains a crack, which might have occurred during the loading test or during preparation of the sample for the SEM imaging. The solidified sand looks like a solid material, no composition of smaller fragments can be seen on the surface or within the cracked area. The same holds true for the AT sand, where the zoom in shows areas of smooth surface alternating with areas of higher surface structure.

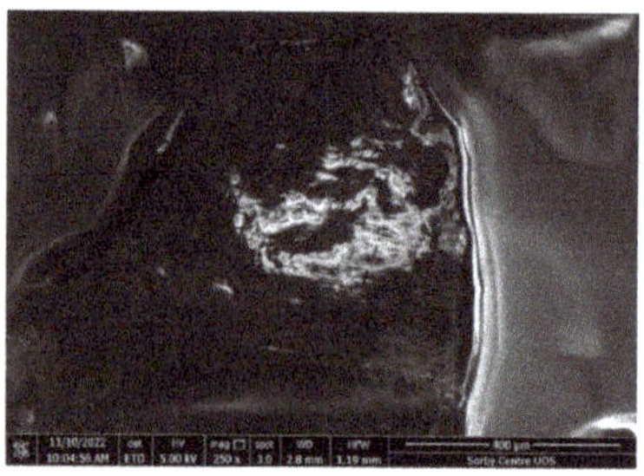

(**a**)

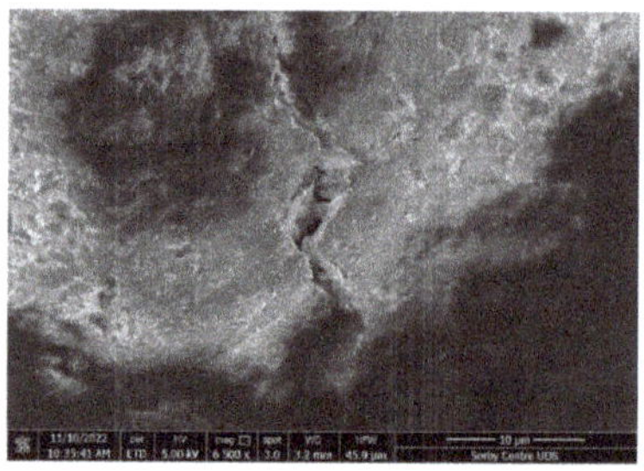

(**b**)

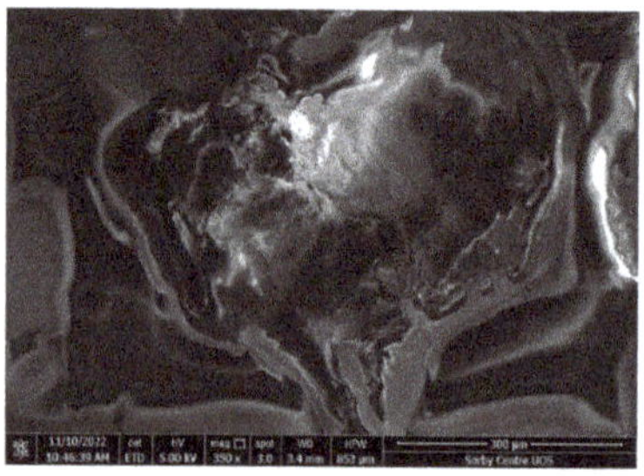

(**c**)

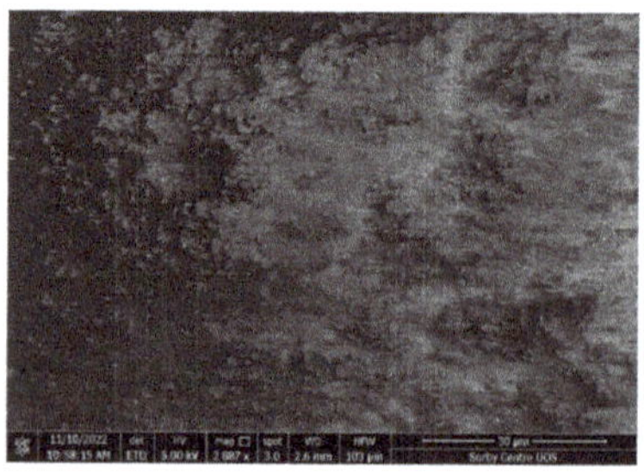

(**d**)

Figure 17. SEM images of solidified clusters of GB and AT sand after conducting high load tests. (**a**) GB sand: large cluster fragment surrounded by glue; (**b**) GB sand: zoom in; (**c**) AT sand: large cluster fragment surrounded by glue; (**d**) AT sand: zoom in.

4. Discussion

This study gives new insights in the crushing behaviour of sand grains with the application of sanded wheel–rail contacts. Sand grains and their fragments are repeatedly crushed during roll-over. Obviously, adhesion can be improved only by those fragments which stay in the contact zone, in contrast to those which are expelled during fracture. The amount of these fragments in the contact zone depends on the type of sand, which differ in grain size, and the contact conditions (dry vs. wet conditions). Even in the first breakage tests clear differences were seen in the spreading behaviour of fragments of GB and AT sand under dry and wet conditions, see Figure 8.

Under realistic service conditions, wheel–rail rolling contact is a rolling-sliding state involving creepage. In this study, creepage is not considered. In order to separate the different effects the focus is on pure vertical loading of the sand grains. It is assumed that there is little influence on the initial breakage by creepage occurring in parallel to the vertical loading.

When a sand grain is entrained in the wheel–rail contact, it will be crushed, with its fragments spreading out. In the conducted initial breakage tests, differences in the spreading behaviour for both types of sand and dry or wet contact conditions were seen. This breakage-spreading process happens repeatedly for those fragments which stay in the running band. Under the extremely high loads in the wheel–rail contact, the sand fragments in the contact zone form clusters, as it was seen in the conducted high load crushing tests. In these tests, the amount of fragments/cluster size depend again on the type of sand, which differ in grain size, and the contact condition, see Figures 13–16. Under the applied high load, the clusters reached a typical thickness in the order of 100 to a 220 μm. Under such extreme loading conditions these sand clusters seem to behave like a solid with certain elasto-plastic material behaviour. For example, Figure 15 shows a cluster with cracks growing form the border towards the centre, which would be a result expected from crushing a solid elasto-plastic material. The clusters' solid-like behaviour is also confirmed by the SEM images shown in Figure 17 revealing that the clusters are obviously not a loose conglomerate of smaller fragments.

It is expected that these solid-like sand clusters are encapsulated between wheel and rail surfaces. The huge local stresses in their vicinity are likely to cause plastic deformations and thus change the surface roughness of wheel and rail. This assumption was indirectly confirmed by the high load crushing tests. For example, in the Alicona scans in Figures 15 and 16 there are indentations around the clusters with a depth in the order of 20–40 μm. These indentations stem from previous tests, as the same upper and lower plate made of hardened steel were used for all tests.

To summarise, this work supports the assumption that the positive effect of sand on the friction in the wheel–rail interface under low adhesion conditions comes about as follows. Dependent on the type of sand and the contact condition, in sanded wheel–rail contacts sand clusters of different size will form. These clusters will be encapsulated between wheel and rail surface, they become plastically deformed and will also plastically indent into the wheel and rail surfaces causing some kind of form-closure effects. Under slip conditions, the relative motion between wheel and rail could occur (i) at the border between the sand cluster and the wheel and rail surfaces, (ii) in shear bands within the sand cluster or (iii) a combination of both.

The described behaviour and the results from the breakage tests can be related to some extent to the sand's influence on the adhesion coefficient in HPT tests, see Figure 2. Under dry conditions, GB sand leaves the AC mostly unchanged, while a clear reduction in the AC can be seen for AT sand. Under wet conditions, both types of sand increase the measured AC, while this effect is more pronounced for GB than for AT sand. For AT sand the ACs obtained under dry and wet conditions are similar—with the AC under dry conditions being slightly higher than that obtained under wet conditions.

One hypothesis to explain these measured ACs in the HPT tests is: If there is enough sand in the contact, then the AC will be determined by the shearing behaviour of the formed sand clusters.

For GB sand under dry conditions, it seems likely that large part of the initial mass of GB sand is expelled from the HPT test. This is supported by the initial crushing tests, see Figure 8a, and also by most results of the high loading tests, see Figure 13. If only little amounts of sand fragments are present during the HPT tests, this will explain why the measured AC for GB sand is close to the unsanded case.

For GB sand under wet conditions, the conducted high loading tests showed large clusters of sand formed, Figure 15, which will increase the AC in the way described above.

For AT sand, dry and wet contact conditions gave similar results: the initial crushing tests showed that nearly all fragments stayed in the contact zone, see Figure 8b, and in the high load tests large clusters formed, see Figures 14 and 16. It seems likely that the shearing behaviour of these clusters determines the measured AC values and thus give similar results both under dry and wet conditions. Compared to the unsanded case, under dry conditions the AT sand can be thought as dry lubrication, reducing the AC. In contrast, under wet conditions it increases the AC compared to the unsanded case.

It is an open question, why under wet conditions GB sand increases the AC more than AT sand does. On the one hand, it could be related to GB sand's larger initial grain size which leads to larger clusters in the high loading tests. On the other hand, also other properties of the sand types can be expected to influence the measured AC.

Based on the results of this study future tests are planned: small scale shear-box experiments should give information about the shearing behaviour of solidified clusters. This is important for the physics-based models to be developed.

Author Contributions: Conceptualization, B.S., W.A.S., R.L. and K.S.; methodology, B.S., W.A.S., R.L. and K.S.; formal analysis, B.S.; investigation, W.A.S.; resources, R.L.; data curation, B.S. and W.A.S.; writing—original draft preparation, B.S.; writing—review and editing, B.S., W.A.S, R.L. and K.S.; visualization, B.S. and W.A.S.; supervision, K.S. and R.L.; project administration, B.S.; funding acquisition, B.S. All authors have read and agreed to the published version of the manuscript.

Funding: This research was funded in whole, or in part, by the Austrian Science Fund (FWF) project P 34273: DEM modelling of adhesion in sanded wheel–rail contacts. For the purpose of open access, the author has applied a CC BY public copyright licence to any Author Accepted Manuscript version arising from this submission. The publication was written at Virtual Vehicle Research GmbH in Graz and partially funded by the COMET K2—Competence Centers for Excellent Technologies Programme of the Federal Ministry for Climate Action (bmk), the Federal Ministry for Digital and Economic Affairs (bmdw), the Austrian Research Promotion Agency (FFG), the Province of Styria, and the Styrian Business Promotion Agency (SFG).

Data Availability Statement: The data set generated and analysed during the current study is openly available in the zenodo.org repository: Suhr, Bettina, Six, Klaus, Skipper, William, & Lewis, Roger. (2023). Single grain crushing tests on two sand types used for wheel-rail sanding [Data set]. Zenodo. DOI 10.5281/zenodo.7547518.

Conflicts of Interest: The authors declare no conflict of interest.

Abbreviations

The following abbreviations are used in this manuscript:

3BL	third body layer
AC	adhesion coefficient
HPT	high pressure torsion
SEM	Scanning Electron Microscope
UMT	universal mechanical tester

References

1. Six, K.; Meierhofer, A.; Trummer, G.; Bernsteiner, C.; Marte, C.; Müller, G.; Luber, B.; Dietmaier, P.; Rosenberger, M. Plasticity in wheel-rail contact and its implications on vehicle-track interaction. *Proc. Inst. Mech. Eng. Part F J. Rail Rapid Transit* **2017**, *231*, 558–569. [CrossRef]
2. Eden, H.C.; Garnham, J.E.; Davis, C.L. Influential microstructural changes on rolling contact fatigue crack initiation in pearlitic rail steels. *Mater. Sci. Technol.* **2005**, *21*, 623–629. [CrossRef]
3. Alwahdi, F.; Kapoor, A.; Franklin, F. Subsurface microstructural analysis and mechanical properties of pearlitic rail steels in service. *Wear* **2013**, *302*, 1453–1460. [CrossRef]
4. Garnham, J.; Davis, C. Very early stage rolling contact fatigue crack growth in pearlitic rail steels. *Wear* **2011**, *271*, 100–112. [CrossRef]
5. Trummer, G.; Marte, C.; Scheriau, S.; Dietmaier, P.; Sommitsch, C.; Six, K. Modeling wear and rolling contact fatigue: Parametric study and experimental results. *Wear* **2016**, *366-367*, 71–77. [CrossRef]
6. Trummer, G.; Six, K.; Marte, C.; Dietmaier, P.; Sommitsch, C. An approximate model to predict near-surface ratcheting of rails under high traction coefficients. *Wear* **2014**, *314*, 28–35. [CrossRef]
7. Trummer, G.; Six, K.; Marte, C.; Meierhofer, A.; Sommitsch, C. Automated measurement of near-surface plastic shear strain. In Proceedings of the 2nd International Conference on Railway Technology: Research, Development and Maintance, Ajaccio, Corsica, France, 8 April–11 April 2014; p. 104.
8. Polach, O. Creep forces in simulations of traction vehicles running on adhesion limit. *Wear* **2005**, *258*, 992–1000. [CrossRef]
9. Six, K.; Meierhofer, A.; Müller, G.; Dietmaier, P. Physical processes in wheel-rail contact and its implications on vehicle-track interaction. *Veh. Syst. Dyn. Int. J. Veh. Mech. Mobil.* **2015**, *53*, 635–650. [CrossRef]
10. Meierhofer, A.; Hardwick, C.; Lewis, R.; Six, K.; Dietmaier, P. Third body layer–experimental results and a model describing its influence on the traction coefficient. *Wear* **2014**, *314*, 148–154. [CrossRef]
11. Lewis, R.; Dwyer-Joyce, R.; Lewis, S.; Hardwick, C.; Gallardo-Hernandez, E. Tribology of the Wheel-Rail Contact: The Effect of Third Body Materials. *Int. J. Railw. Technol.* **2012**, *1*, 167–194. [CrossRef]
12. Descartes, S.; Desrayaud, C.; Niccolini, E.; Berthier, Y. Presence and role of the third body in a wheel-rail contact. *Wear* **2005**, *258*, 1081–1090. [CrossRef]
13. Niccolini, E.; Berthier, Y. Wheel-rail adhesion: Laboratory study of "natural" third body role on locomotives wheels and rails. *Wear* **2005**, *258*, 1172–1178. [CrossRef]
14. Buckley-Johnstone, L.; Trummer, G.; Voltr, P.; Meierhofer, A.; Six, K.; Fletcher, D.; Lewis, R. Assessing the impact of small amounts of water and iron oxides on adhesion in the wheel/rail interface using High Pressure Torsion testing. *Tribol. Int.* **2019**, *135*, 55–64. [CrossRef]
15. White, B.; Nilsson, R.; Olofsson, U.; Arnall, A.; Evans, M.; Armitage, T.; Fisk, J.; Fletcher, D.; Lewis, R. Effect of the presence of moisture at the wheel-rail interface during dew and damp conditions. *Proc. Inst. Mech. Eng. Part F J. Rail Rapid Transit* **2018**, *232*, 979–989. [CrossRef]
16. White, B.; Lewis, R. Simulation and understanding the wet-rail phenomenon using twin disc testing. *Tribol. Int.* **2019**, *136*, 475–486. [CrossRef]
17. Trummer, G.; Buckley-Johnstone, L.; Voltr, P.; Meierhofer, A.; Lewis, R.; Six, K. Wheel-rail creep force model for predicting water induced low adhesion phenomena. *Tribol. Int.* **2017**, *109*, 409–415. [CrossRef]
18. Lewis, R.; Dwyer-Joyce, R.; Lewis, J. Disc machine study of contact isolation during railway track sanding. *Proc. Inst. Mech. Eng. Part F J. Rail Rapid Transit* **2003**, *217*, 11–24. [CrossRef]
19. Lewis, R.; Dwyer-Joyce, R. Wear at the wheel/rail interface when sanding is used to increase adhesion. *Proc. Inst. Mech. Eng. Part F J. Rail Rapid Transit* **2006**, *220*, 29–41. [CrossRef]
20. Skipper, W.A.; Chalisey, A.; Lewis, R. A review of railway sanding system research: Adhesion restoration and leaf layer removal. *Tribol.-Mater. Surf. Interfaces* **2018**, *12*, 237–251. [CrossRef]
21. Skipper, W.; Chalisey, A.; R., L. A Review of Railway Sanding System Research: Wheel/Rail Isolation, Damage, and Particle Application. *Proc. Inst. Mech. Eng. Part F J. Rail Rapid Transit* **2019**, *234*, 567–583. [CrossRef]
22. Arias-Cuevas, O.; Li, Z.; Lewis, R. Investigating the Lubricity and Electrical Insulation Caused by Sanding in Dry Wheel-Rail Contacts. *Tibol. Lett.* **2010**, *37*, 623–635. [CrossRef]
23. Wang, W.; Zhang, H.; Wang, H.; Liu, Q.; Zhu, M. Study on the adhesion behavior of wheel/rail under oil, water and sanding conditions. *Wear* **2011**, *271*, 2693–2698. [CrossRef]
24. Wang, W.; Liu, T.; Wang, H.; Liu, Q.; Zhu, M.; Jin, X. Influence of friction modifiers on improving adhesion and surface damage of wheel/rail under low adhesion conditions. *Tribol. Int.* **2014**, *75*, 16–23. [CrossRef]
25. Huang, W.; Cao, X.; Wen, Z.; Wang, W.; Liu, Q.; Zhu, M.; Jin, X. A Subscale Experimental Investigation on the Influence of Sanding on Adhesion and Rolling Contact Fatigue of Wheel/Rail Under Water Condition. *J. Tribol.* **2017**, *139*, 011401. [CrossRef]
26. Lewis, S.R.; Riley, S.; Fletcher, D.I.; Lewis, R. Optimisation of a Railway Sanding System, Part 2: Adhesion Tests. In Proceedings of the 10th International Conference on Contact Mechanics (CM2015), Colorado Springs, CO, USA, 30 August–3 September 2015.
27. Skipper, W.; Nadimi, S.; Chalisey, A.; Lewis, R. Particle characterisation of rail sands for understanding tribological behaviour. *Wear* **2019**, *432–433*, 202960. [CrossRef]

28. Skipper, W. Sand Particle Entrainment and its Effects on the Wheel/Rail Interface. Ph.D. Thesis, University of Sheffield, Sheffield, UK, 2021.
29. Vollebregt, E.; Six, K.; Polach, O. Challenges and progress in the understanding and modelling of the wheel–rail creep forces. *Veh. Syst. Dyn.* **2021**, *59*, 1026–1068. [CrossRef]
30. Carter, F.W. On the Action of a Locomotive Driving Wheel. *Proc. R. Soc. Lond. Ser. A Contain. Pap. Math. Phys. Character* **1926**, *112*, 151–157. [CrossRef]
31. Fromm, H. Berechnung des Schlupfes beim Rollen deformierbarer Scheiben. *Z. Angew. Math. Mech.* **1927**, *7*, 27–58. [CrossRef]
32. Johnson, K. The effect of a tangential contact force upon the rolling motion of an elastic sphere upon a plane. *J. Appl. Mech.* **1958**, *25*, 339–346. [CrossRef]
33. Johnson, K. The effect of spin upon the rolling motion of an elastic sphere upon a plane. *J. Appl. Mech.* **1958**, *25*, 332–338. [CrossRef]
34. Kalker, J. The computation of three-dimensional rolling contact with dry friction. *Int. J. Numer. Methods Eng.* **1979**, *14*, 1293–1307. [CrossRef]
35. Kalker, J.J. On the Rolling Contact of Two Elastic Bodies in the Prescence of Dry Friction. Ph.D. Thesis, Delft University of Technology, Delft, The Netherlands, 1967.
36. Kalker, J.J. Simplified Theory of Rolling Contact. *Delft Progr. Rep. Ser. C Mech. Aeronaut. Eng. Shipbuild.* **1973**, *1*, 1–10.
37. Kalker, J.J. A Fast Algorithm for the Simplified Theory of Rolling Contact. *Veh. Syst. Dyn.* **1982**, *11*, 1–13. [CrossRef]
38. Kalker, J.J. *Three-Dimensional Elastic Bodies in Rolling Contact*; Kluwer Academic Press: Dordrecht, The Netherlands, 1990; ISBN 978-94-015-7889-9.
39. Knothe, K. History of wheel/rail contact mechanics: From Redtenbacher to Kalker. *Veh. Syst. Dyn.* **2008**, *46*, 9–26. [CrossRef]
40. Poritsky, H. Stresses and deflections of cylindrical bodies in contact with application to contact of gears and of locomotive wheels. *J. Appl. Mech.-Trans. ASME* **1950**, *17*, 191–201. [CrossRef]
41. Meymand, S.Z.; Keylin, A.; Ahmadian, M. A survey of wheel-rail contact models for rail vehicles. *Veh. Syst. Dyn.* **2016**, *54*, 386–428. [CrossRef]
42. Vollebregt, E.A.H.; van der Wekken, C.D. *Advanced Modeling of Wheel–Rail Friction Phenomena*; Technical Report TR19-11; VORtech: Irwindale, CA, USA, 2019.
43. Van der Wekken, C.; Vollebregt, E. Local plasticity modelling and its influence on wheel-rail friction. In Proceedings of the Proceedings of the 11th International Conference on Contact Mechanics and Wear of Rail/Wheel Systems, Delft, The Netherlands, 24–27 September 2018; pp. 1013–1018.
44. Allen, T. *Particle Size Measurement*, 3rd ed.; Chapman & Hall: London, UK, 1981.
45. *BS 1377-2:1990*; Methods of Test for Soils for Civil Engineering Purposes: Part 2: Classification Tests. British Standards Institute: London, UK, 1990.
46. Schindelin, J.; Arganda-Carreras, I.; Frise, E.; Kaynig, V.; Longair, M.; Pietzsch, T.; Preibisch, S.; Rueden, C.; Saalfeld, S.; Schmid, B.; et al. Fiji: An open-source platform for biological-image analysis. *Nat. Methods* **2012**, *9*, 676–682. [CrossRef]
47. The GIMP Development Team. GIMP Version 2.10.22. Available online: https://www.gimp.org (accessed on 15 December 2022).
48. Bernsteiner, C.; Müller, G.; Meierhofer, A.; Six, K.; Künstner, D.; Dietmaier, P. Development of white etching layers on rails: simulations and experiments. *Wear* **2016**, *366–367*, 116–122. [CrossRef]
49. McDowell, G.R.; Bolton, M.D. On the micromechanics of crushable aggregates. *Géotechnique* **1998**, *48*, 667–679. [CrossRef]
50. Lim, W.; McDowell, G.; Collop, A. The application of Weibull statistics to the strength of railway ballast. *Granul. Matter* **2004**, *6*, 229–237. [CrossRef]
51. Laufer, I. Grain crushing and high-pressure oedometer tests simulated with the discrete element method. *Granul. Matter* **2015**, *17*, 389–412. [CrossRef]
52. Weibull, W. A statistical distribution of wide applicability. *J. Appl. Mech.* **1951**, *18*, 293–297. [CrossRef]

lubricants

Article

Mechanical and Tribological Behaviors of U75VG Rail Flash–Butt Welded Joint

Bin Rong [1], Shaopeng Liu [1], Qiuping Li [1], Jinfang Peng [2] and Mingxue Shen [1,2,*]

1 State Key Laboratory of Performance Monitoring and Protecting of Rail Transit Infrastructure, East China Jiaotong University, Nanchang 330013, China
2 Traction Power State Key Laboratory, Southwest Jiaotong University, Chengdu 610031, China
* Correspondence: shenmingxue@126.com

Abstract: Flash–butt welded rail is widely used in railway transportation; however, the welded joint is vulnerable after a long time of service, and its damage mechanism is controversial. Here, tensile and reciprocating friction tests were carried out to analyze the mechanical and tribological behaviors between the welded joint and the base metal of a U75VG rail. The results show that flash–butt welding promotes the pearlite to transform into ferrite, leading to a relatively low hardness value but high plasticity. In addition, the yielding and strength of the all–weld–metal specimen are 385 MPa and 1090 MPa, respectively, which are about 24.51% and 7.63% lower than that of the base metal specimen. It is worth noting that the elongation of the all–weld–metal specimen is 57.1% higher than that of the base metal specimen, and more dimples and tearing ridges can be detected on the fracture morphology of the all–weld–metal specimen, while the fracture morphology of the base metal specimen is filled with shallow dimples and cleavage planes. Moreover, the weld metal has a relatively higher COF (coefficient of friction), and its fluctuation amplitude is 1.25 times higher than that of the base metal, which is due to the rougher worn surface. Furthermore, the introduction of flash–butt welding changes the wear mechanism of the U75VG rail from adhesive wear and oxidation to fatigue wear and slight oxidation, and ultimately leads to more serious damage.

Keywords: U75VG rail; flash–butt welding; mechanical property; tribological behavior; damage mechanism

Citation: Rong, B.; Liu, S.; Li, Q.; Peng, J.; Shen, M. Mechanical and Tribological Behaviors of U75VG Rail Flash–Butt Welded Joint. *Lubricants* **2023**, *11*, 41. https://doi.org/10.3390/lubricants11020041

Received: 20 December 2022
Revised: 23 January 2023
Accepted: 25 January 2023
Published: 27 January 2023

1. Introduction

The key role of the railway transportation site is the rail, which, in normal service, guarantees train guidance, traction, and braking. A continuous welded rail makes the high–speed and heavy axle load of the train develop rapidly, leading to the need for higher welding quality. Currently, most rail manufacturers use flash–butt welding [1], and flash–butt welded rails are widely used in railway transportation. However, the welded joint becomes vulnerable with long–time service, and its damage mechanism is controversial [2]. Moreover, the microstructure and characteristics of welded rail joints are greatly different from those of the base metal, resulting in crushing, fracture, and bending deformation of the welded joints [3–6]. In other words, the failure of the welded rail joints will lead to abnormal contact between the wheel and rail. Subsequently, the vertical acceleration of the train suddenly increases, affecting passenger comfort and endangering traffic safety [7].

At present, much of the research has focused on the damage mechanism of a rail's flash–butt welded joints. On the one hand, some scholars have tested the performance of treated rail and analyzed the microstructure evolution, mechanical properties, and fatigue fracture mechanisms of laser shock peening and corrosive environments [8,9]. On the other hand, there are other scholars who established finite element models of welded rail joints based on different working conditions, to simulate damage behavior. A significant body of research results show that factors, such as vehicle speed, axle load, joint height, residual stress, and rail roughness, have great influence on the plastic deformation, crack behavior, and stress–strain behaviors of welded rail joints [10–13]. Zhao et al. [14] studied the significant stress concentration in welded rail joints due to the impact effect. Their research showed that the distribution of an equivalent plastic strain in weld metal, heat–affected zones, and base metal is uneven during wheel–rail rolling contact. Xu et al. [10] pointed out that the maximum load on welded rail joint during wheel–rail rolling contact is 4–12 times that under quasi–static conditions, and the maximum contact pressure, maximum equivalent stress and equivalent plastic strain increase with the train axle load and the running speed. They obtained the mechanical parameters of welded rail joint by static tests, modified the rail dynamic constitutive finite element model, and obtained the simulation results of welded rail joint in wheel–rail rolling contact. Prakash et al. [15,16] also stated that the friction between wheel and rail increases with the adhesion coefficient, the increase of stress intensity factor leads to an increase of crack enlargement and the fatigue life decrease with weld length. This conclusion is obtained by simulating the crack behavior of rail under different working conditions. Besides, welded rail joint is the easy failure part due to defects such as structural changes, inclusions, and pores [17]. Thus, welded rail joint is more prone to collapse, fracture, serious wear, and other damage problems resulting in continuous welded rail appearing irregularly.

And besides, two–point contact of the wheel–rail is likely to occur when the train passes through the curved section. Sliding friction occurs between the wheel rim and rail head side, resulting in the frictional shear stress on the rail will be significantly higher than that in rolling contact [18]. Based on current research, few reports are on the reciprocating friction test of rail weld joints, and carrying out tribological tests of welded rail joint is of great significance.

In this paper, the reciprocating friction and wear test of U75VG rail weld metal and base metal was carried out. Focusing on the microstructure, mechanical properties, and surface damage behavior of weld and base metal reveals the damage mechanism.

2. Experimental Details

2.1. Experimental Materials

The U75VG is usually used for high–speed trains in China. In this paper, the U75VG welded rail joint was conducted by a flash–butt welding machine. In order to rail weld joint with good quality, the welding process parameters are selected, as shown in Table 1. The specimens of tensile and tribological tests were extracted from the welded rail joint and the base metal (head of U75VG rail, shown in Figure 1).

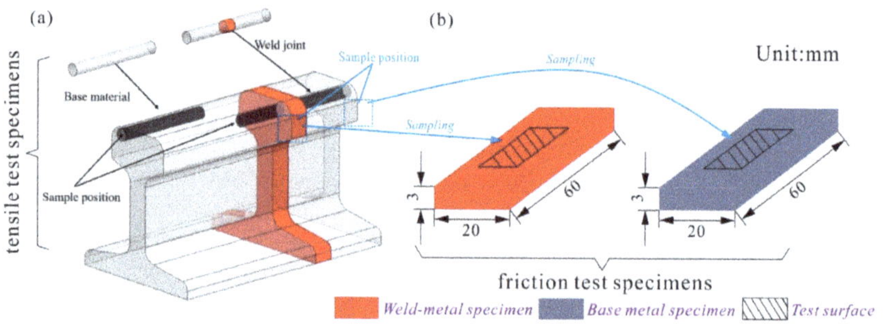

Figure 1. Sample position with (**a**) tensile specimens and (**b**) tribological specimens.

GCr15 steel balls (GB/T18254–2002) with a diameter of 6 mm and surface roughness (Ra) of 0.04 mm were used as the counter–bodies, and the chemical element content is shown in Tables 2 and 3, respectively. All the tests were repeated three times for the same test parameter to reduce inevitable errors and ensure the reliability of the test results.

Table 1. Welding process parameters.

upset force/KN	35
upset length/mm	10.5
welding time/s	85–95
burning speed (mm/s)	13.5–15.5
clamping length/mm	130–150
weld width/mm	20–25
input heat/MJ	8.6

Table 2. Chemical composition of U75VG rail (wt.%).

C	Si	Mn	S	P	V
0.71~0.8	0.5~0.8	0.7~1.05	≤0.03	≤0.03	0.04~0.12

Table 3. Chemical composition of GCr15 steel ball (wt.%).

C	Si	Mn	Cr	Mo	S/P	Ni + Cu
0.95~1.05	0.15~0.35	0.25~0.45	1.45~1.65	≤0.1	≤0.025	≤0.5

2.2. Experimental Method

The tribological test was conducted on a self−made, multi−functional friction and wear testing machine (UNT−3, shown in Figure 2a), and the load, frequency, and duration time of the friction test were 10 N, 2 Hz, and 3 h, respectively. The experiment was conducted in an atmospheric environment (T = 25 °C, RH = 60%). The specimens were cleaned using an ultrasonic cleaner before and after the test. The tensile test by the tension−torsion multi−axis electric servo fatigue testing machine (Walter + BaiLFV−100−HH, 100 KN) and the size of tensile test specimens are shown in Figure 2b.

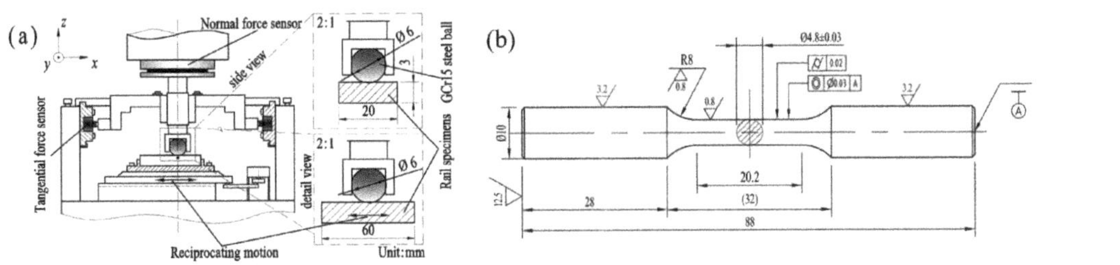

Figure 2. (**a**) Structure diagram of reciprocating friction and wear test bench, dimensions, and contact configuration of rail and GCr15 steel ball specimens, (**b**) size of tensile test specimens.

The specimens were prepared by polishing and etching in a 4% nitric acid and alcohol solution. Then, the microstructure on the cross−section was observed using an optical microscope. The nano−indentation tests on the cross−section were analyzed on a nano indenter using a Berkovich diamond indenter with a constant loading–unloading rate of 30 mN/min. The surface morphologies were observed with a scanning electron microscope, and components of the worn surface were detected through energy−dispersive spectroscopy. The 3D topographies and surface roughness were observed with a white light interferometer.

3. Results

3.1. Microstructure and Mechanical Characteristic

Figure 3 displays the microstructure of weld and base metals of a U75VG rail. There are significant differences between the weld and base metals. The length of the weld metal is about 200 μm and has a large amount of ferrite (Figure 3a), and the Figure 3b,c are the enlarged graph of the Figure 3a, respectively. According to field test and finite element simulations [2,17,19,20], the temperature value differs in different regions in the rail welding process, and the temperature of the weld metal will exceed 1000 °C, resulting in the pearlite transforming into austenite. Moreover, for weld metals, the austenite decomposes into pearlite and ferrite during cooling. Thus, the weld metal has a large amount of ferrite; however, the temperature of the heat−affected zone and base metal does not exceed Ac3, resulting in a significant reduction in the amount of ferrite in the heat−affected zone (Figure 3a), with the main microstructure in the base metal being lamellar pearlite (Figure 3c). Simultaneously, the temperature, grain growth rate, and nucleation rate in different regions of the weld metal pool are not very different, leading to the ferrite in the weld center being more uniform [21]. In any case, to confirm that the microstructure of the base metal of about 500 μm is not affected too much by welding, it is found that the base metal of about 500 μm is the same morphology as 20 cm from the weld zone (Figure 3c,d).

Figure 3. Microstructure of U75VG rail of (**a**) welded joint, and (**b**) weld metal, (**c**,**d**) base metal.

Figure 4 shows the nano−scale hardness, elastic modulus, and load−displacement curves of the weld and base metals. The nano−scale hardness value of the weld metal is significantly lower than that of the base metal, and the elastic modulus of the weld metal is slightly lower (Figure 4a), while the indentation deformation of the weld metal is slightly higher (Figure 4b). This is mainly the recrystallization during the welding process leading to a significant change in the proportion of deformed grains, sub−structure grains, and recrystallized grains [22,23]. It is worth noting that the hardness of ferrite is lower than that of pearlite, whereas the toughness is better. Simultaneously, the proportion of ferrite in weld metal is significantly higher (Figure 3b), while the base metal is mostly composed of lamellar pearlite (Figure 3c,d). For the record, the high proportion of ferrite in weld metal, resulting in the deformation displacement of weld metal in the holding and unloading stage, is higher than that of the base metal (Figure 4b).

Figure 5 shows the stress–strain curves of the welded rail joint specimen and base metal specimen. The yield and tensile strength values of the all−weld−metal specimen are 385 MPa and 1090 Mpa, respectively, which is about 24.51% and 7.63% lower than that of the base metal specimen. In addition, the elongation of the all−weld−metal specimen is 57.1% higher than that of the base metal specimen. This is not surprising since the all−weld−metal specimen has a strong undermatching in the weld region, as clearly indicated by hardness measurements. Similar behavior has also been reported for the welded joints with a strong undermatching in the weld region, such as friction stir−, laser−, or electron−beam−welded Al−alloys [24–32]. Undoubtedly, the variation trend of stress–strain curves is closely related to the microstructure of the specimens. The ferrite number at welded rail joints increases significantly (Figure 3a), resulting in lower strength but better plasticity (Figure 4). In other words, the yield and tensile strength of the all−weld−metal specimen is low, but the plasticity is relatively better.

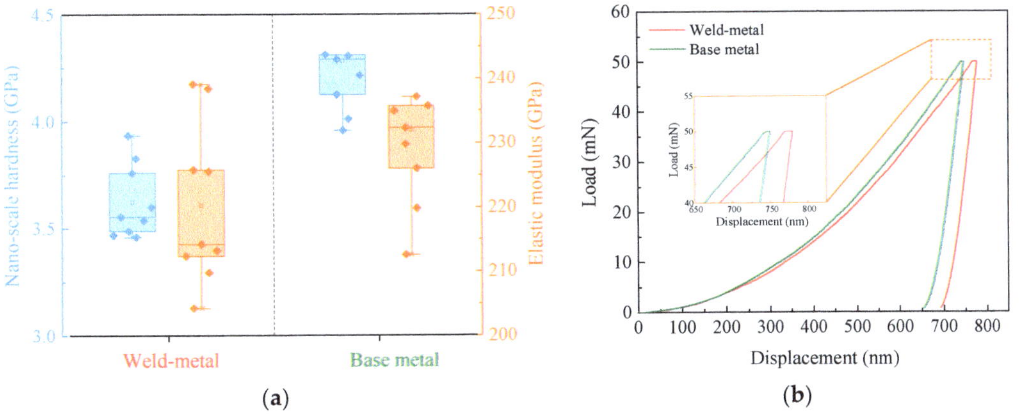

Figure 4. Mechanical characteristic of U75VG rail of (**a**) nano−scale hardness and elastic modulus, and (**b**) load−displacement curves.

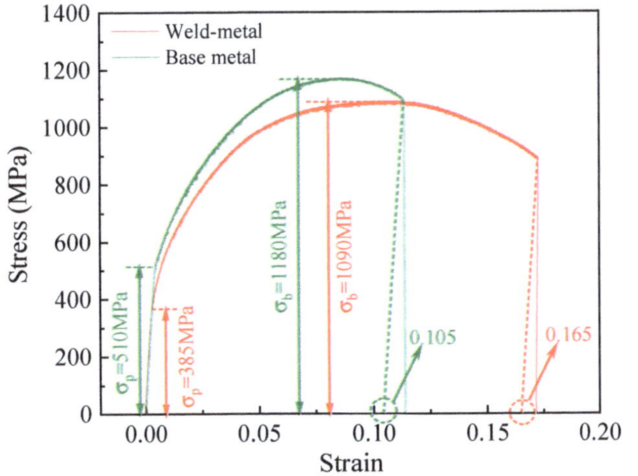

Figure 5. The stress–strain response curves with all−weld−metal and base metal specimens.

3.2. Fracture Morphology

Figure 6 shows the tensile fracture morphology of all−weld−metal and base metal specimens. On the whole, the two specimens are divided into smooth and rough zones (Figure 6a,d), and the Figure 6c,f are the enlarged graph of the Figure 6b,c, respectively. Simultaneously, more dimples and tearing ridges can be detected on the fracture morphology of the all−weld−metal specimen (Figure 6b,c), while the fracture morphology of the base metal specimen is filled with shallow dimples and cleavage planes (Figure 6e,f). Obviously, the fracture mode of the all−weld−metal specimen is mainly plastic fractures, while the base metal specimen is a typical brittle fracture. This phenomenon is due to the low strength and good plasticity of the all−weld−metal specimen (Figure 4b), and the higher plasticity makes the material break later [33].

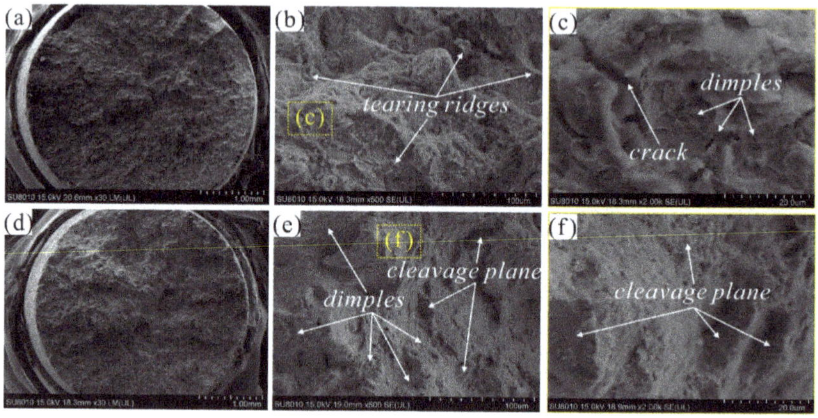

Figure 6. Fracture morphology of the tensile specimens of (**a–c**) all−weld−metal, and (**d–f**) base metal specimens.

3.3. *Coefficient of Friction*

Figure 7 shows the COF (coefficient of friction) time−varying curves of the weld and base metal specimens during a reciprocating tribological test. In general, the COF shows a slow upward trend as the number of cycles increase, following which the coefficient of friction tends to be stable (the number of cycles increased to 15,000). In the stable stage, the average COF of the weld and base metal specimens are 0.5 and 0.45, respectively. Moreover, the weld joint has a relatively higher COF, and its fluctuation amplitude is 1.25 times higher than the base metal. It is mainly the hardness value of the base metal that is relatively high, and the wear mechanism does not change significantly during the cycle, resulting in a relatively more stable COF.

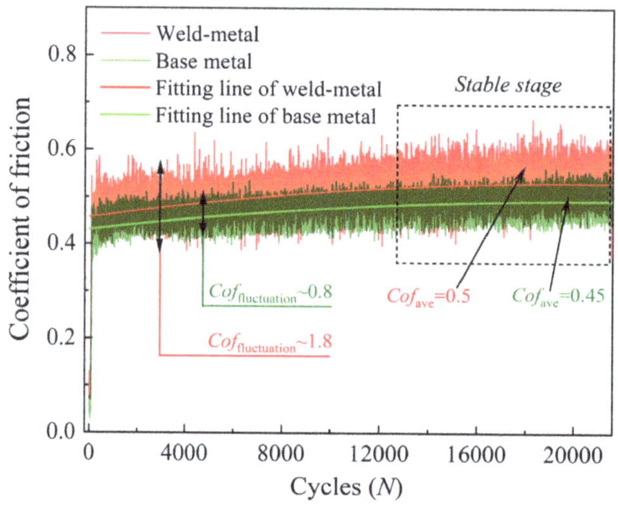

Figure 7. Time−varying curves of coefficient of friction with weld and base metal specimens.

3.4. *Surface Damage Morphology*

Figure 8 shows three−dimensional surface topographies of a weld metal, base metal, and GCr15 steel ball (tribological pair specimens). The obvious ploughing and peeling pit can be observed on the worn surface, and the ploughing is deeper than that of the base metal (Figure 8a,b,e). A two−dimensional, cross−sectional profile comparison shows that the weld metal's maximum wear depth is 9.1 µm, while the base metal's is 6 µm (Figure 8e). Moreover, ploughing can be observed on the worn surface of the GCr15 steel ball (Figure 8c,d), and the cross−section width of the weld metal pair is relatively small (Figure 8f). The hardness value of the weld metal is relatively lower, resulting in a decrease in the material removal effect of the tribological pair steel ball, and the wear width of the tribological pair steel ball is relatively narrower. For the record, the wear debris, as the third body medium at the friction interface, will aggravate the damage of the worn surface, resulting in ploughing appearing on the worn surface of the rail and the GCr15 steel ball [34,35].

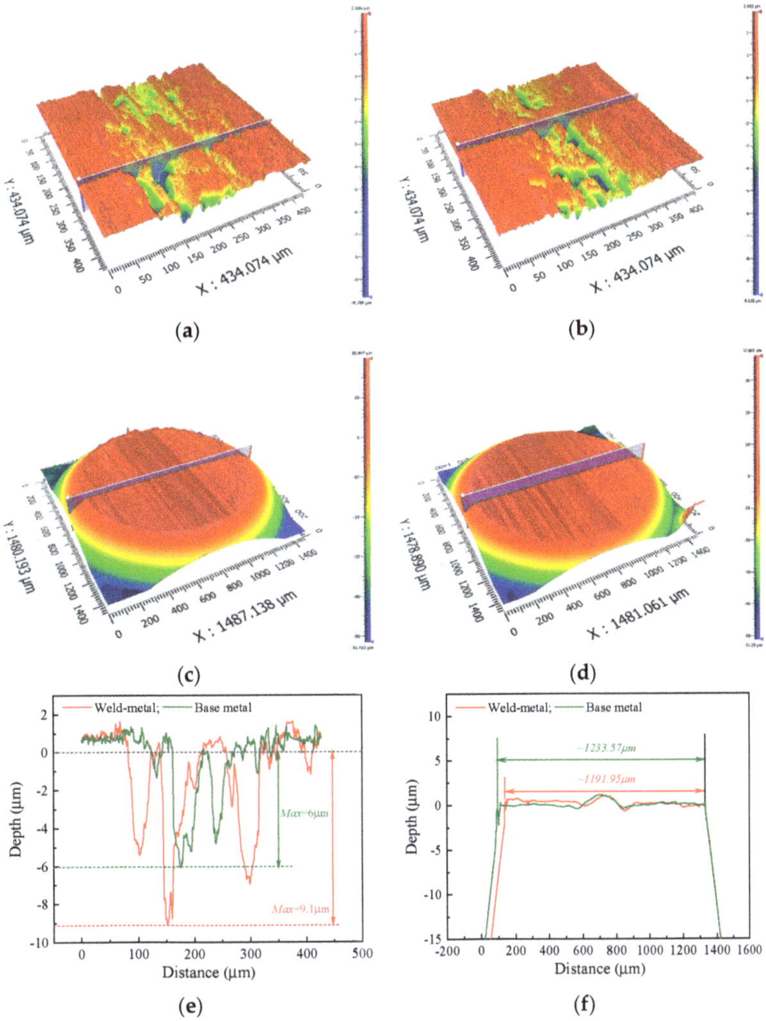

Figure 8. Three−dimensional topographies of the worn surface of (**a**,**c**) weld metal specimen and GCr15 steel ball, (**b**,**d**) base metal and GCr15 steel ball, cross−section profiles of worn surface of (**e**) rail and (**f**) GCr15 steel ball.

In order to reveal the damage mechanism of the weld metal and base metal, the surface damage morphology is shown in Figure 9, and the Figure 9b,h are the enlarged graph of the

Figure 6a,g, respectively. In Figure 9a,b, the worn surface of the weld metal is uneven; there are obvious cracks and serious delamination characteristics, and the ploughing is narrow and deep. The worn surface of the base metal is relatively flat, and no obvious peeling pits are observed, but there are obvious friction films and many adhesions (Figure 9g,h). Moreover, there is no obvious difference in the content of iron and carbon between the weld and base metals (Figure 9d,e,j,k), while the oxygen content in the base metal is higher than that in the weld metal (Figure 9f,i and Figure 10). In addition, the damage to the weld metal is mainly fatigue wear, while the base metal is adhesive wear. Due to the hardness value of the weld metal being relatively lower, the external force causes plastic deformation of the material, and microcracks form after plastic depletion. Thus, the continuous penetration of microcracks leads to delamination characteristics of the worn surface [36]. On the contrary, the hardness value of the base metal is relatively higher and the friction film phenomenon occurs at the initial friction stage due to the increase in temperature, which results in the friction film of the base metal protecting the worn surface and the relatively flat surface is more conducive to the adhesion of the oxide film. However, as the friction continues, the friction film is broken and mixed with the wear debris to form the third medium in the friction interface, resulting in wide and shallow ploughing characteristics that will appear in the later stage of friction [34,35]. Subsequently, the surface damage of the weld metal is more serious than that of the base metal, and the wear resistance is worse.

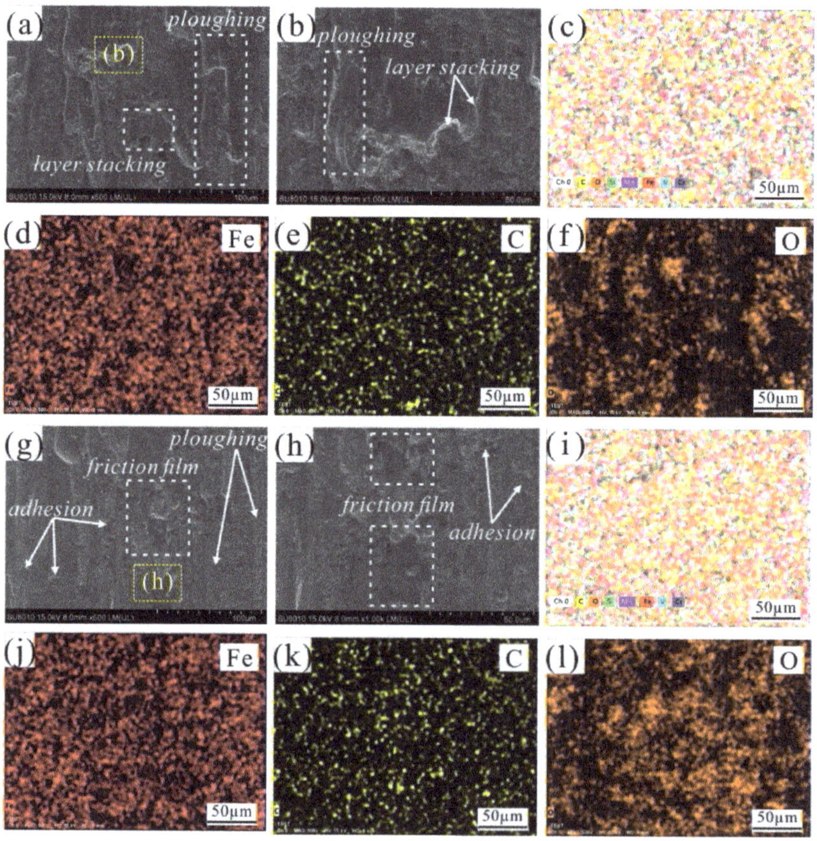

Figure 9. SEM and EDS map of the worn surface of (**a–f**) weld metal, (**g–l**) base metal specimens.

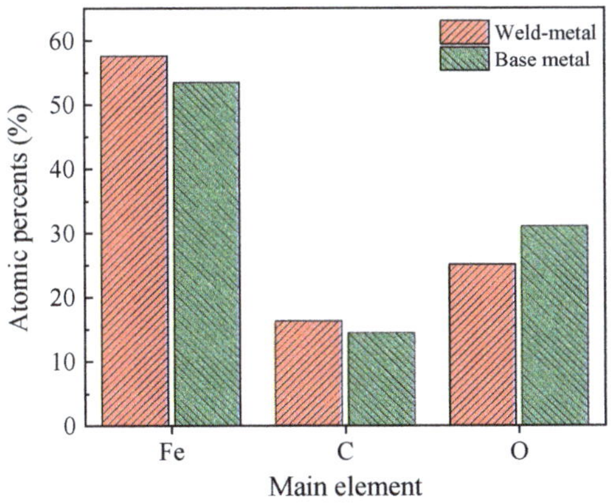

Figure 10. Element content on worn surfaces of weld and base metal specimens.

4. Discussion

Generally, the damage of the weld metal and base metal is related to friction and contact states. Figure 11 displays the initial and stable friction stage. It is well known that the two specimens each contact other specimens which will produce deformation, and the contact spot area is mainly determined by the hardness of the rail material as the friction pair material tested is the GCr15 steel ball. The welding process causes the microstructure of materials to change (Figure 3), resulting in the hardness value of the weld metal being relatively lower (Figure 4a). Therefore, the deformation of the softer weld metal (Figure 4a) is relatively larger, but in the GCr15 steel ball it is relatively smaller, resulting in a smaller contact spot area and deeper wear (Figure 8e,f). At the initial friction stage, the weld metal specimen with better plasticity (Figures 4b and 6) produces greater deformation, and the debris on the worn surface does not easily to fall off, resulting in the worn surface to be rougher (Figure 9a,b). Moreover, the lamellar debris of the weld metal induced by delamination wear is directly involved in friction, resulting in a higher average COF and friction force [37] (Figures 7 and 11a,b). However, the yield and tensile strength values of the all−weld−metal specimen are 385 MPa and 1090 MPa, respectively, and the elongation of the all−weld−metal specimen is 57.1% higher than that of the base metal specimen (Figure 5). Thus, the base metal has poor plasticity and a higher hardness value and yield strength (Figures 4–6), resulting in a relatively flat worn surface (Figure 9g,h).

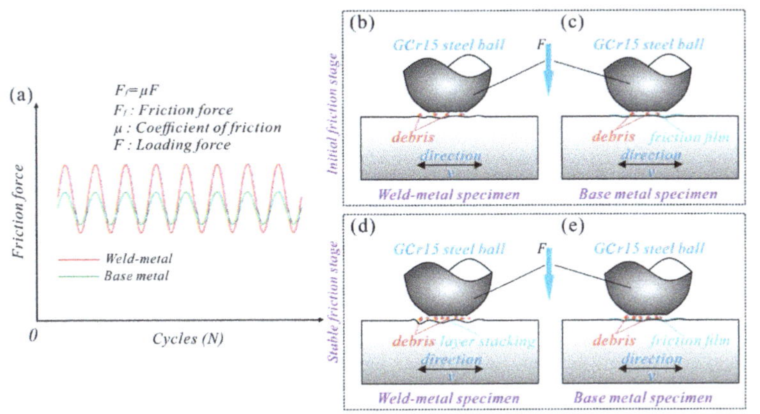

Figure 11. Schematic diagram of (**a**) friction force and damage mechanism of (**b**), (**c**) initial friction stage and (**d**,**e**) stable friction stage.

As the same time, the friction film appears on the worn surface of the base metal due to the increase in temperature, and the friction film plays a protective role (Figure 11c), resulting in the COF and its fluctuation amplitude being relatively lower [38] (Figure 7). As the friction progresses, the friction film mixes with wear debris to form a third body medium which participates in the friction (Figure 11e). As shown in Figure 9, there is no obvious difference in the content of iron and carbon between the weld and base metals (Figure 9d,e,j,k), while the oxygen element of the obviously base metal increases (Figure 9f,l). This phenomenon shows that the weld and base metals will produce debris peeled from the material during the wear process, while the worn surface of the base metal occurs due to the obvious oxide film. It has been found that the debris at the friction interface is embedded in the dual surface and they interlock with each other, resulting in ploughing on the contact surface [39]. Thus, there are obvious ploughing characteristics on the worn surface of the GCr15 steel ball (Figure 8d). In the stable friction stage, the oxide film on the worn surface of the base metal is continuously generated and removed to achieve a dynamic balance, and the oxide film on the surface of the base metal reduces damage (Figure 9g,h and Figure 11e), leading to the COF and its fluctuation amplitude being relatively lower [38] (Figure 7). Obviously, the weld metal has better plasticity, but the hardness and yield strength values are lower than that of the base metal (as shown in the Figures 4 and 5), and the friction force and fluctuation amplitude are relatively lager, resulting in more serious damage of the weld metal than the base metal (Figures 8 and 9). In the tensile test, the more dimples and tearing ridges that can be detected on the fracture morphology of the all−weld−metal specimen (Figure 6b,c), the relatively larger deformation production of the weld metal during friction. Therefore, during the friction and wear process, the material deformation of the weld metal is relatively larger and obviously removed, and the worn surface appears with deeper ploughing and thicker accumulation layers (Figure 9a,b and Figure 11d). In short, the damage mechanism of the weld metal is mainly fatigue wear, and the base metal is mainly adhesive wear.

5. Conclusions

The tensile and tribological tests were conducted on the tension–torsion multi−axis electric servo fatigue testing machine and self−made multi−functional friction and wear testing machine, respectively. The fracture morphology and surface wear behavior of the weld metal and base metal rails were studied by their microstructure, hardness, yield and tensile strength, elongation, coefficient of friction, and element content on their worn surfaces. The following conclusions can be drawn:

(1) The length of the weld metal is about 200μm and has a large amount of ferrite, while the base metal is lamellar pearlite and no obvious ferrite was observed, leading to a significantly higher nano−scale hardness value. Simultaneously, the high proportion of ferrite in the weld metal results in higher plasticity than that of the base metal.
(2) The yielding and strength of the welded specimen are 385 MPa and 1090 MPa, respectively, which are about 24.51% and 7.63% lower than that of the base metal specimen. More dimples and tearing ridges can be detected on the fracture morphology of the all−weld−metal specimen, while the fracture morphology of the base metal specimen is filled with shallow dimples and cleavage planes.
(3) The all−weld−metal specimen has a relatively higher COF, and its fluctuation amplitude is 1.25 times higher than that of the base metal specimen, which was due to the rougher worn surface. In the stable stage, the average COF of the all−weld−metal and base metal specimens are 0.5 and 0.45, respectively.
(4) The introduction of flash−butt welding will change the wear mechanism of the U75VG rail from adhesive wear and oxidation to fatigue wear and slight oxidation, leading to slighter wear damage. Therefore, the worn surface of the weld metal is uneven; there are obvious cracks and serious delamination characteristics, and the ploughing is narrow and deep, while the worn surface of the base metal is relatively

flat and no obvious peeling pits are observed, and there are obvious friction films and many adhesions.

Author Contributions: Conceptualization, M.S.; methodology, S.L. and J.P.; software, B.R.; validation, M.S. and S.L.; formal analysis, M.S. and Q.L.; investigation, M.S. and B.R.; resources, M.S.; data curation, B.R.; writing—original draft preparation, B.R.; writing—review and editing, M.S.; visualization, S.L.; supervision, M.S. and Q.L.; project administration, J.P.; funding acquisition, M.S. All authors have read and agreed to the published version of the manuscript.

Funding: This research was funded by the National Natural Science Foundation of Jiangxi Province (nos. 20212ACB214003 and 20224ACB204012), Technology Research and Development Project from the CHINA RAILWAY (no. N2021T012), National Natural Science Foundation of China (no. 52061012), and the Open Fund of Traction Power State Key Laboratory (TPL1906).

Data Availability Statement: No new data were created or analyzed in this study. Data sharing is not applicable to this article.

Acknowledgments: The authors thank Fengjun Gong from Jiangxi Ruichang Railway Construction Co. LTD for providing the technical support and samples of welded rail with welded joints.

Conflicts of Interest: The authors declare no conflict of interest. The funders had no role in the design of the study; in the collection, analyses, or interpretation of data; in the writing of the manuscript; or in the decision to publish the results.

References

1. Farhangi, H.; Mousavizadeh, S.M. Horizontal split−web fractures of flash butt welded rails. In *Proceedings of the 8th International Fracture Conference*; Yildiz Technical University: Istanbul, Turkey, 2007.
2. Porcaro, R.R.; Faria, G.L.; Godefroid, L.B.; Apolonio, G.R.; Candido, L.C.; Pinto, E.S. Microstructure and mechanical properties of a flash butt welded pearlitic rail. *J. Mater. Process. Technol.* **2019**, *270*, 20–27. [CrossRef]
3. Shi, S.C.; Wang, W.C.; Ko, D.K. Influence of Inclusions on Mechanical Properties in Flash Butt Welding Joint of High−Strength Low−Alloy Steel. *Metals* **2022**, *12*, 242. [CrossRef]
4. Skyttebol, A.; Josefson, B.L.; Ringsberg, J.W. Fatigue crack growth in a welded rail under the influence of residual stresses. *Eng. Fract. Mech.* **2005**, *72*, 271–285. [CrossRef]
5. Stone, D.H.; Lwand, H.C.; Kristan, J.; Lehnhoff, G.R. Flash Butt Rail Weld Vertical Fractures. *J. Fail. Anal. Prev.* **2015**, *15*, 33–38. [CrossRef]
6. Bauri, L.F.; Alves, L.H.D.; Pereira, H.B.; Tschiptschin, A.P.; Goldenstein, H. The role of welding parameters on the control of the microstructure and mechanical properties of rails welded using FBW. *J. Mater. Res. Technol.* **2020**, *9*, 8058–8073. [CrossRef]
7. Kabo, E.; Ekberg, A.; Maglio, M. Rolling contact fatigue assessment of repair rail welds. *Wear* **2019**, *436*, 203030. [CrossRef]
8. Zhao, X.H.; Fan, Y.J.; Liu, Y.; Wang, H.Y.; Dong, P. Evaluation of fatigue fracture mechanism in a flash butt welding joint of a U75V type steel for railroad applications. *Eng. Fail. Anal.* **2015**, *55*, 26–38. [CrossRef]
9. Li, X.Y.; Ma, R.; Liu, X.; Lv, Q.B.; Wang, X.; Tian, Z. Effect of laser shock peening on fatigue properties of U75VG rail flash−butt welding joints. *Opt. Laser Technol.* **2022**, *149*, 107889. [CrossRef]
10. Xu, J.M.; Wang, P.; Gao, Y.; Chen, J.Y.; Chen, R. Geometry evolution of rail weld irregularity and the effect on wheel−rail dynamic interaction in heavy haul railways. *Eng. Fail. Anal.* **2017**, *81*, 31–44. [CrossRef]
11. Li, W.; Xiao, G.W.; Wen, Z.F.; Xiao, X.B.; Jin, X.S. Plastic deformation of curved rail at rail weld caused by train–track dynamic interaction. *Wear* **2011**, *271*, 311–318. [CrossRef]
12. Fang, X.Y.; Zhang, H.N.; Ma, D.W.; Wu, Z.J.; Huang, W. Influence of welding residual stress on subsurface fatigue crack propagation of rail. *Eng. Fract. Mech.* **2022**, *271*, 108642. [CrossRef]
13. Cai, W.; Wen, Z.F.; Jin, X.S.; Zhai, W.M. Dynamic stress analysis of rail joint with height difference defect using finite element method. *Eng. Fail. Anal.* **2007**, *14*, 1488–1499. [CrossRef]
14. Zhao, J.Z.; Peng, X.; Fu, P.L.; Wang, Y.; Kang, G.Z.; Wang, P.; Kan, Q.H. Dynamic constitutive model of U75VG rail flash−butt welded joint and its application in wheel−rail transient rolling contact simulation. *Eng. Fail. Anal.* **2022**, *134*, 106078. [CrossRef]
15. Sen, P.K.; Bhiwapurkar, M.; Harsha, S.P. A 3−D numerical simulation of fatigue crack growth in an alumino thermite welded UIC60 rail joint under different loading conditions. *Mater. Today Proc.* **2022**, *59*, 405–412. [CrossRef]
16. Sen, P.K.; Bhiwapurkar, M.; Harsha, S.P. UIC60 rail alumino thermite weld's semi elliptical head crack and stress intensity factor using ANSYS. *Mater. Today Proc.* **2022**, *56*, 3058–3064. [CrossRef]
17. Pang, Y.; Grilli, N.; Su, H.; Liu, W.C.; Ma, J.; Yu, S.F. Experimental investigation on microstructures and mechanical properties of PG4 flash−butt rail welds. *Eng. Fail. Anal.* **2022**, *141*, 106650. [CrossRef]
18. Zhang, H.; Zhang, S.Y.; Zhong, H.; Wang, W.J.; Meli, E.; Cui, X.L. Damage mechanism of a long−wavelength corrugated rail associated with rolling contact fatigue. *Eng. Fail. Anal.* **2022**, *136*, 106173. [CrossRef]

19. Tawfik, D.; Mutton, P.J.; Chiu, W.K. Experimental and numerical investigations: Alleviating tensile residual stresses in flash−butt welds by localised rapid post−weld heat treatment. *J. Mater. Process. Technol.* **2008**, *196*, 279–291. [CrossRef]
20. Ma, N.S.; Cai, Z.P.; Huang, H.; Deng, D.A.; Murakawa, H.; Pan, J.L. Investigation of welding residual stress in flash−butt joint of U71Mn rail steel by numerical simulation and experiment. *Mater. Des.* **2015**, *88*, 1296–1309. [CrossRef]
21. Mansouri, H.; Monshi, A. Microstructure and residual stress variations in weld zone of flash−butt welded railroads. *Sci. Technol. Weld. Join.* **2004**, *9*, 237–245. [CrossRef]
22. Su, Y.; Li, W.Y.; Wang, X.Y.; Ma, T.J.; Ma, L.; Dou, X.M. The sensitivity analysis of microstructure and mechanical properties to welding parameters for linear friction welded rail steel joints. *Mater. Sci. Eng. A* **2019**, *764*, 138251. [CrossRef]
23. Su, H.; Li, J.; Lai, Q.; Pun, C.L.; Mutton, P.; Kan, Q.H.; Kang, G.Z.; Yan, W.Y. Ratcheting behaviour of flash butt welds in heat−treated hypereutectoid steel rails under uniaxial and biaxial cyclic loadings. *Int. J. Mech. Sci.* **2020**, *176*, 105539. [CrossRef]
24. Çam, G.; Javaheri, V.; Heidarzadeh, A. Advances in FSW and FSSW of dissimilar Al−alloy plates. *J. Adhes. Sci. Technol.* **2022**, *37*, 162–194. [CrossRef]
25. Çam, G. Prospects of producing aluminum parts by wire arc additive manufacturing (WAAM). *Mater. Proc.* **2022**, *62*, 77–85. [CrossRef]
26. Luo, G.Y.; Cheng, M.P.; Liu, C.H.; Li, S.M.; Wang, X.G.; Song, L.J. Improving mechanical properties of quasi−continuous−wave laser beam welded 7075 aluminum alloy through microstructural refinement and homogenization of the fusion zone. *Opt. Laser Technol.* **2022**, *153*, 108221. [CrossRef]
27. Liu, F.C.; Zhou, B.S.; Mao, Y.Q.; Huang, C.P.; Chen, Y.H.; Wang, Z.T. Microstructure and mechanical properties of laser welded joints between 2198/2060 Al–Li alloys. *Mater. Sci. Technol.* **2018**, *34*, 111–122. [CrossRef]
28. Zhou, X.H.; Zhao, H.Y.; Liu, F.Y.; Yang, B.A.; Xu, B.X.; Chen, B.; Tan, C.W. Effects of beam oscillation modes on microstructure and mechanical properties of laser welded 2060 Al−Li alloy joints. *Opt. Laser Technol.* **2021**, *144*, 107389. [CrossRef]
29. Ipekoğlu, G.; Çam, G. Formation of weld defects in cold metal transfer arc welded 7075−T6 plates and its effect on joint performance. *IOP Conf. Ser. Mater. Sci. Eng.* **2019**, *629*, 012007. [CrossRef]
30. Çam, G.; Ventzke, V.; Dos Santos, J.F.; Koçak, M.; Jennequin, G.; Gonthier−Maurin, P. Characterisation of electron beam welded aluminium alloys. *Mater. Sci. Technol.* **1999**, *4*, 317–323. [CrossRef]
31. Ancona, A.; Lugara, P.M.; Sorgente, D.; Tricarico, L. Mechanical characterization of CO2 laser beam butt welds of AA5083. *J. Mater. Process. Technol.* **2007**, *191*, 381–384. [CrossRef]
32. EI−Batahgy, A.; Kutsuna, M. Laser Beam Welding of AA5052, AA5083, and AA6061 Aluminum Alloys. *Adv. Mater. Sci. Eng.* **2009**, *2009*, 974182. [CrossRef]
33. Yang, Z.J.; Li, J.X.; Hou, S.S.; Cao, J.H.; Wang, G.L.; Lang, S.T.; Ding, P. Microstructural characteristics and mechanical properties of Ti–6Al–2Nb–2Zr–0.4B alloy welded joint using tungsten inert gas welding. *J. Mater. Res. Technol.* **2022**, *21*, 3129–3139. [CrossRef]
34. Lewis, R.; Dwyer−Joyce, R.; Lewis, S.R.; Hardwick, C. Tribology of the Wheel−Rail Contact: The Effect of Third Body Materials. *Int. J. Railw. Technol.* **2012**, *1*, 167–194. [CrossRef]
35. Chen, J.F.; Chu, J.Y.; Jiang, W.C.; Yao, B.; Zhou, F.; Wang, Z.B.; Zhao, P.C. Experimental and Numerical Simulation to Study the Reduction of Welding Residual Stress by Ultrasonic Impact Treatment. *Materials* **2020**, *13*, 837. [CrossRef] [PubMed]
36. Hardwick, C.; Lewis, R.; Stock, R. The effects of friction management materials on rail with pre existing rcf surface damage. *Wear* **2017**, *384*, 50–60. [CrossRef]
37. Li, X.; Dong, M.; Jiang, D.Y.; Li, S.F.; Shang, Y. The effect of surface roughness on normal restitution coefficient, adhesion force and friction coefficient of the particle−wall collision. *Powder Technol.* **2020**, *362*, 17–25. [CrossRef]
38. Xiao, Y.L.; Cheng, Y.; Zhou, H.B.; Liang, W.H.; Shen, M.X.; Yao, P.P.; Zhao, H.P.; Xiong, G.Y. Evolution of contact surface characteristics and tribological properties of a copper−based sintered material during high−energy braking. *Wear* **2021**, *488–489*, 204163. [CrossRef]
39. McColl, I.R.; Ding, J.; Leen, S.B. Finite element simulation and experimental validation of fretting wear. *Wear* **2004**, *256*, 1114–1127. [CrossRef]

MDPI
St. Alban-Anlage 66
4052 Basel
Switzerland
www.mdpi.com

Lubricants Editorial Office
E-mail: lubricants@mdpi.com
www.mdpi.com/journal/lubricants

www.ingramcontent.com/pod-product-compliance
Lightning Source LLC
Chambersburg PA
CBHW061520210926
51850CB00008B/1982